AF843762

# FROM BARROW TO BOOTHIA

The Arctic Journal of Chief Factor Peter Warren Dease, 1836–1839

In 1835 the map of the arctic coast of North America was still far from complete, with unmapped gaps of 280 kilometres from Return Reef to Point Barrow in Alaska and 550 kilometres from Point Turnagain to Boothia Peninsula in the Central Canadian Arctic. The Hudson's Bay Company developed a plan to fill the gaps and two of the Company's officers were chosen to carry it out: the veteran Chief Factor Peter Dease – efficient, competent, steady, and with an excellent rapport with Indians and the "servants," mostly Métis – and Thomas Simpson, young, energetic, ambitious, arrogant, and cousin and secretary to George Simpson, the Company's governor in North America.

Over a three-year period from 1837 to 1839, operating from a base-camp at Fort Confidence on Great Bear Lake, the expedition achieved its goal. Despite serious problems with sea ice, Dease and Simpson, in some of the longest small-boat voyages in the history of the Arctic, mapped the remaining gaps in a model operation of efficient, economical, and safe exploration. Thomas Simpson's narrative, the standard source on the expedition, claimed the expedition's success for himself, stating "Dease is a worthy, indolent, illiterate soul, and moves just as I give the impulse." In *From Barrow to Boothia* William Barr shows that Dease's contribution was absolutely crucial to the expedition's success and makes Dease's sober, sensible, and modest account of the expedition available.

Dease's journal, reproduced in full, is supplemented by a brief introduction to each section and detailed annotations that clarify and elaborate the text. By including relevant correspondence to and from expedition members, Barr captures the original words of the participants, offering insights into the character of both Dease and Simpson and making clear what really happened on this successful expedition.

William Barr is professor emeritus of geography, University of Saskatchewan, and a research associate for the Arctic Institute of North America, University of Calgary.

*Dr John Rae's sketch of Fort Confidence, as refurbished by John Bell for Rae's and Dr Richardson's expedition in search of Franklin, 1848. Courtesy Hudson's Bay Company Archives.* HBCA P-194 (N 14526).

# From Barrow to Boothia

...

## The Arctic Journal of Chief Factor Peter Warren Dease, 1836–1839

Edited and Annotated by
William Barr

McGill-Queen's University Press
Montreal & Kingston • London • Ithaca

© McGill-Queen's University Press 2002
ISBN 0-7735-2253-0

Legal deposit first quarter 2002
Bibliothèque nationale du Québec

Printed in Canada on acid-free paper

This book has been published with the help of a grant from the Humanities and Social Sciences Federation of Canada, using funds provided by the Social Sciences and Humanities Research Council of Canada.

McGill-Queen's University Press acknowledges the financial support of the Government of Canada through the Book Publishing Industry Development Program (BPIDP) for its activities. It also acknowledges the support of the Canada Council for the Arts for its publishing program.

The Warren Dease Journal, 1836-1839, M2714, is published by permission of the McCord Museum of Canadian History, Montreal. The McCord has recently changed its accession numbers; in the present volume the book is referred to under its previous number, M214.

---

**National Library of Canada Cataloguing in Publication Data**

Dease, Peter Warren
From Barrow to Boothia : the Arctic journal of Chief Factor Peter Warren Dease, 1836–1839

(Rupert's Land Record Society series ; 7)
Includes bibliographical references and index.
ISBN 0-7735-2253-0

1. Dease, Peter Warren—Diaries. 2. Simpson, Thomas, 1808–1840. 3. Arctic Coast (Canada)—Discovery and exploration—English. 4. Arctic Coast (Alaska)—Discovery and exploration—English. 5. Arctic Regions (Canada)—Discovery and exploration—English. 6. Nunavut—Discovery and exploration—English. 7. Hudson's Bay Company. I. Barr, William II. Title. III. Series.

FC3961.D42 2002 917.19'5 C2001-900777-9
F1060.8.D42 2002

# Contents

# Abbreviations

BCA British Columbia Archives, Victoria, BC

FNB Field Notebook (refers to one of Peter Dease's field notebooks owned by Mr Warren Baker of Montreal)

HBCA Hudson's Bay Company Archives (Provincial Archives of Manitoba), Winnipeg, Manitoba

MCM McCord Museum, McGill University, Montreal, Quebec

SPRI Scott Polar Research Institute, Cambridge, England

# Preface

The standard source on the Hudson's Bay Company's expedition to explore the arctic coast of North America in 1837–39 is the posthumously published journal of the second-in-command, Thomas Simpson.[1] The journal of the expedition leader, Chief Factor Peter Warren Dease, has, till now, remained unpublished, despite the fact that it adds substantially to the record of the expedition's activities. In February 1843, over three years after the expedition successfully completed the survey of most of the arctic coast, and some two and a half years after Thomas Simpson's death, Sir Edward Sabine, who was editing Simpson's journal at the request of Sir John Henry Pelly, governor of the Hudson's Bay Company, wrote to the veteran arctic explorer Sir John Richardson, asking him to read the proofs of the book. On so doing Richardson was clearly irritated by Simpson's arrogant treatment of his colleague, Peter Dease, noting that Simpson

> by no means puts his companion prominent: indeed he speaks of himself at all times so manifestly as the director of all proceedings, as to have struck me, tho' wholly unacquainted with the matter previously, as somewhat objectionable. Of this, you can form a better judgment than I can, as the two individuals are quite unknown to me, even by character. I see no reason why occasional passages from Mr. Deases journal may not be introduced in notes, adding to the store of information, and doing justice to all parties, but I should prefer to be guided by your opinion in this respect.[2]

In fact, no attempt was made to include any information from Dease's journal, in notes or otherwise, in the published version of Simpson's journal.

Richardson's assessment was patently correct; insertion of some passages from Dease's journal would undoubtedly have added useful information to Simpson's text, and would have partially restored the balance, in terms of the relative contributions of the two men, as has been very effectively stated by Dr Ian MacLaren in his assessment of the significance of Dease's field notebooks.[3]

The present work represents an attempt to improve even on Richardson's suggestion, by publishing Dease's journal in its entirety, with the addition of explanatory notes and relevant correspondence. Unfortunately, very few of Dease's own letters have survived, and in their absence I have been obliged to include coauthored

---

1 Simpson, *Narrative of the Discoveries on the North Coast of America.*

2 Letter from John Richardson to Sir Edward Sabine, 9 February 1843, SPRI MS 1503/26/2.

3 MacLaren, "The HBC's Arctic Expedition 1836–1839."

correspondence and relevant letters from Thomas Simpson.

Publication of Dease's journal is all the more necessary in light of the fact that while Richardson's complaint about the manner in which Simpson blows his own trumpet is well founded, vastly more objectionable are his egregious and often totally unfounded criticisms of Dease in what he no doubt assumed would be private correspondence with his brother, Alexander (who promptly published much of it),[4] his cousin (and superior) George Simpson, the Company's governor in North America, and his close friend Chief Factor Donald Ross at Norway House. These criticisms will be discussed in some detail in chapter 12.

In several respects, Dease's journal complements Simpson's published narrative very usefully. First there are periods when Simpson and Dease were operating independently, such as their respective journeys to Fort Chipewyan at the start of the expedition, and Dease's sojourn at Fort Chipewyan prior to Simpson's arrival. Then there are the two occasions when Simpson proceeded on foot, leaving Dease to guard the boats when they were blocked by ice, namely the last lap to Point Barrow in the summer of 1837 and the stretch of coast beyond Point Turnagain in the summer of 1838. In the latter case, Simpson did Dease a grave injustice, as was his wont. It had been agreed that if the ice permitted, Dease would take one boat and catch up with the hiking party with a view to pushing on eastwards by water. Dease makes it clear that the ice prevented him from any such attempt. On the day that Simpson returned (29 August 1838), Dease noted "much heavy ice remains about us yet" (see p. 190). Next day the ice still prevented them from making any move in the boats, and even on the following day (31 August) they had to "push through a considerable quantity of floating ice with a heavy swell" before they could start on their homeward trip. According to Simpson, however, the ice "only obstructed the shores leaving everywhere a clear offing" (see p. 204) and they could easily have explored farther eastwards by boat.

Even when both men were together, Dease's journal commonly adds a great deal of useful information to Simpson's published narrative. For example, Dease's journal provides us with vastly more detail of the daily activities during the two winterings at Fort Confidence (especially the first winter). Dease kept a careful tally of all the meat and fish brought in either by the expedition's hunters and fishermen, or by the local Indians. For the first wintering (26 September 1837 until 5 June 1838), this tally of "country food" totalled some 25,954 pounds of caribou meat, i.e. almost 13 tonnes (at roughly 100 pounds of meat per animal, this represents some 260 animals), plus 379 caribou tongues, 3636 pounds of muskox meat, 5153 fish (of varying sizes), 74 geese, and 3 swans. This information is missing from Simpson's account.

During the three coastal trips, too, Dease kept a careful record of hunting activities and success, whereas Simpson specifically writes that "I shall pass them over

4 Alexander Simpson, *The Life and Travels of Thomas Simpson, the Arctic Discoverer.*

entirely, as they were now become mere matters of course, while our whole thoughts were bent upon subjects of far higher interest."[5] Dease also provided a much more detailed description of all the Inuit camps and caches encountered on the coast than did Simpson.

One hopes that these reasons are sufficient to justify the publication of Dease's journal. It is only unfortunate that over 150 years had to elapse before Dease's own version of events was made public, to counter the egotistical narrative of Simpson and to give the lie to the latter's disparaging remarks about Dease.

My debt of gratitude to Dr Ian MacLaren, in connection with the preparation of Dease's journal for publication, is enormous. It was he who located Dease's field notebooks, in the possession of Mr Warren Baker of Montreal. He then produced a careful transcription and wrote and published an analysis of their contents.[6] Having next located Dease's journal, in the possession of McCord Museum, McGill University, Dr MacLaren arranged for it to be transcribed; this transcription was executed in exemplary fashion by Mr Kenneth Reynolds, and checked by Ms Mary Beth Wolicky. I am very grateful to both. After thus completing some of the most crucial and tedious steps in preparing Dease's materials for publication, Dr MacLaren, over-extended with other commitments and duties, invited me to finish the task. I cannot properly express my gratitude for this magnanimous invitation. I also wish to thank Ms Pam Little and other staff members at the McCord Museum for providing access to Dease's journal and for giving permission to publish it. And I am extemely grateful to Mr Warren Baker of Montreal for permission to publish extracts from Dease's field notebooks and the photos of Dease's commissions, as well as for his own hospitality.

I should also like to thank the staff of the Hudson's Bay Company Archives, Provincial Archives of Manitoba, in Winnipeg, especially the Keeper of the Archives, Ms Judith Hudson Beattie, and the head of Research and Information Services, Ms Ann Morton, for their unfailing assistance. To the staff of the British Columbia Archives, Victoria, I am particularly indebted for permitting me access to the Donald Ross Papers during a week when the archives were officially closed for inventory taking. And I am also indebted to Mr Bob Headland, Archivist at the Scott Polar Research Institute, Cambridge, for access to relevant documents in his care. In like manner, I wish to thank Dr Richard Davis, Department of English, University of Calgary, for providing me with proof copies of John Franklin's correspondence with Peter Dease, prior to publication of Franklin's journal and correspondence from his expedition of 1825–27.

Finally I wish to express my very sincere thanks to Mr Keith Bigelow of the Department of Geography, University of Saskatchewan, for drafting the maps which illustrate this book. As always, his workmanship is of the highest quality.

5 Simpson, *Narrative*, 358.
6 MacLaren, "The HBC's Arctic Expedition."

# Illustrations and Maps

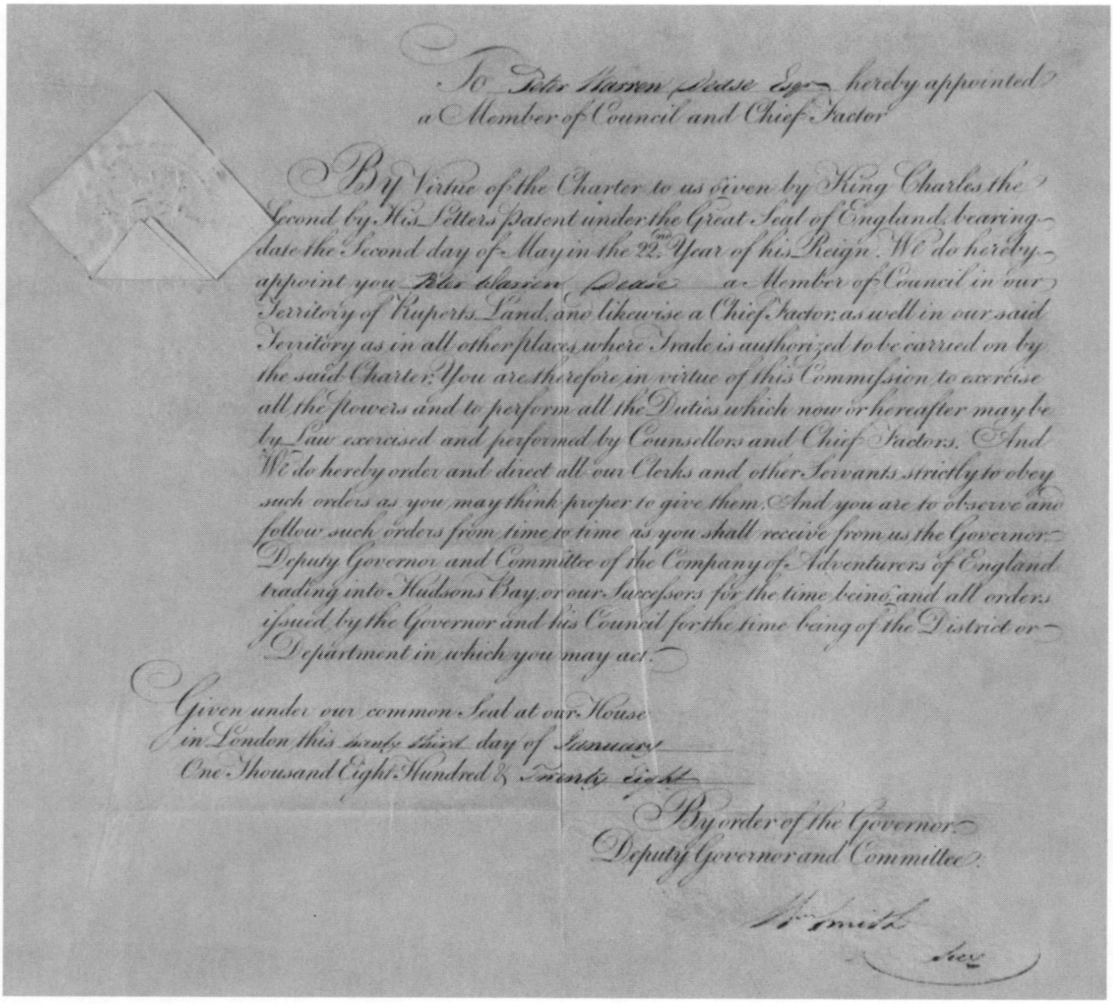
To Peter Warren Dease Esqr hereby appointed
a Member of Council and Chief Factor

By Virtue of the Charter to us given by King Charles the Second by His Letters Patent under the Great Seal of England bearing date the Second day of May in the 22nd Year of his Reign. We do hereby appoint you Peter Warren Dease a Member of Council in our Territory of Ruperts Land, and likewise a Chief Factor, as well in our said Territory as in all other places where Trade is authorized to be carried on by the said Charter; You are therefore in virtue of this Commission to exercise all the Powers and to perform all the Duties which now or hereafter may be by Law exercised and performed by Counsellors and Chief Factors. And We do hereby order and direct all our Clerks and other Servants strictly to obey such orders as you may think proper to give them. And you are to observe and follow such orders from time to time as you shall receive from us the Governor, Deputy Governor and Committee of the Company of Adventurers of England trading into Hudsons Bay, or our Successors for the time being, and all orders issued by the Governor and his Council for the time being of the District or Department in which you may act.

Given under our common Seal at our House
in London this twenty third day of January
One Thousand Eight Hundred & Twenty eight

By order of the Governor,
Deputy Governor and Committee.

W. Smith
Secy

1 *Peter Dease's commission as Chief Trader, Hudson's Bay Company. Courtesy Mr Warren Baker, Montreal.*

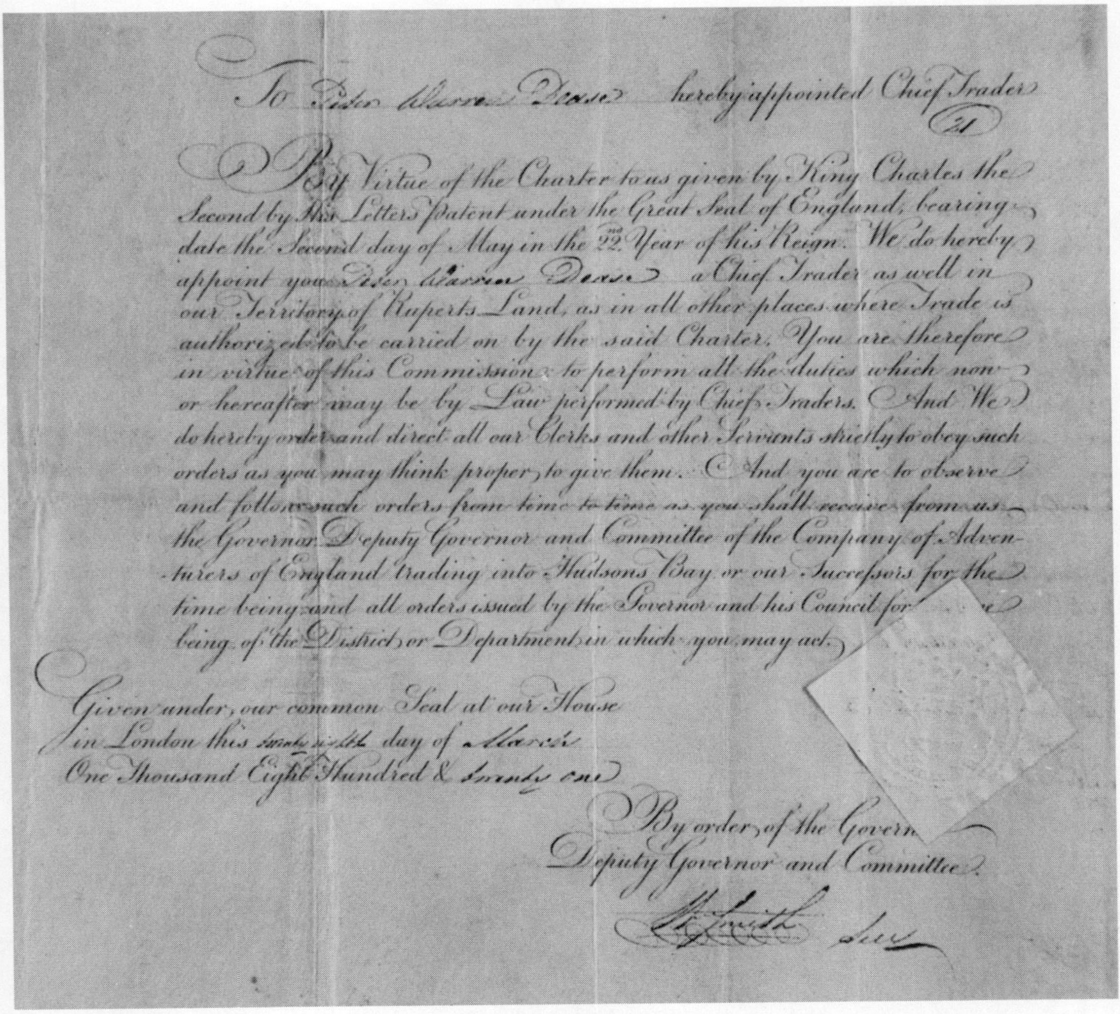

To Peter Warren Dease hereby appointed Chief Trader (21)

By Virtue of the Charter to us given by King Charles the Second by His Letters Patent under the Great Seal of England, bearing date the Second day of May in the 22nd Year of his Reign. We do hereby appoint you Peter Warren Dease a Chief Trader as well in our Territory of Ruperts Land, as in all other places where Trade is authorized to be carried on by the said Charter. You are therefore in virtue of this Commission to perform all the duties which now or hereafter may be by Law performed by Chief Traders. And We do hereby order and direct all our Clerks and other Servants strictly to obey such orders as you may think proper to give them. And you are to observe and follow such orders from time to time as you shall receive from us the Governor Deputy Governor and Committee of the Company of Adventurers of England trading into Hudsons Bay or our Successors for the time being and all orders issued by the Governor and his Council for the time being of the District or Department in which you may act.

Given under our common Seal at our House in London this [illegible] day of March One Thousand Eight Hundred & twenty one

By order of the Governor Deputy Governor and Committee

W. Smith Secy

2 *Peter Dease's commission as Chief Factor and Member of Council, Hudson's Bay Company. Courtesy Mr Warren Baker, Montreal.*

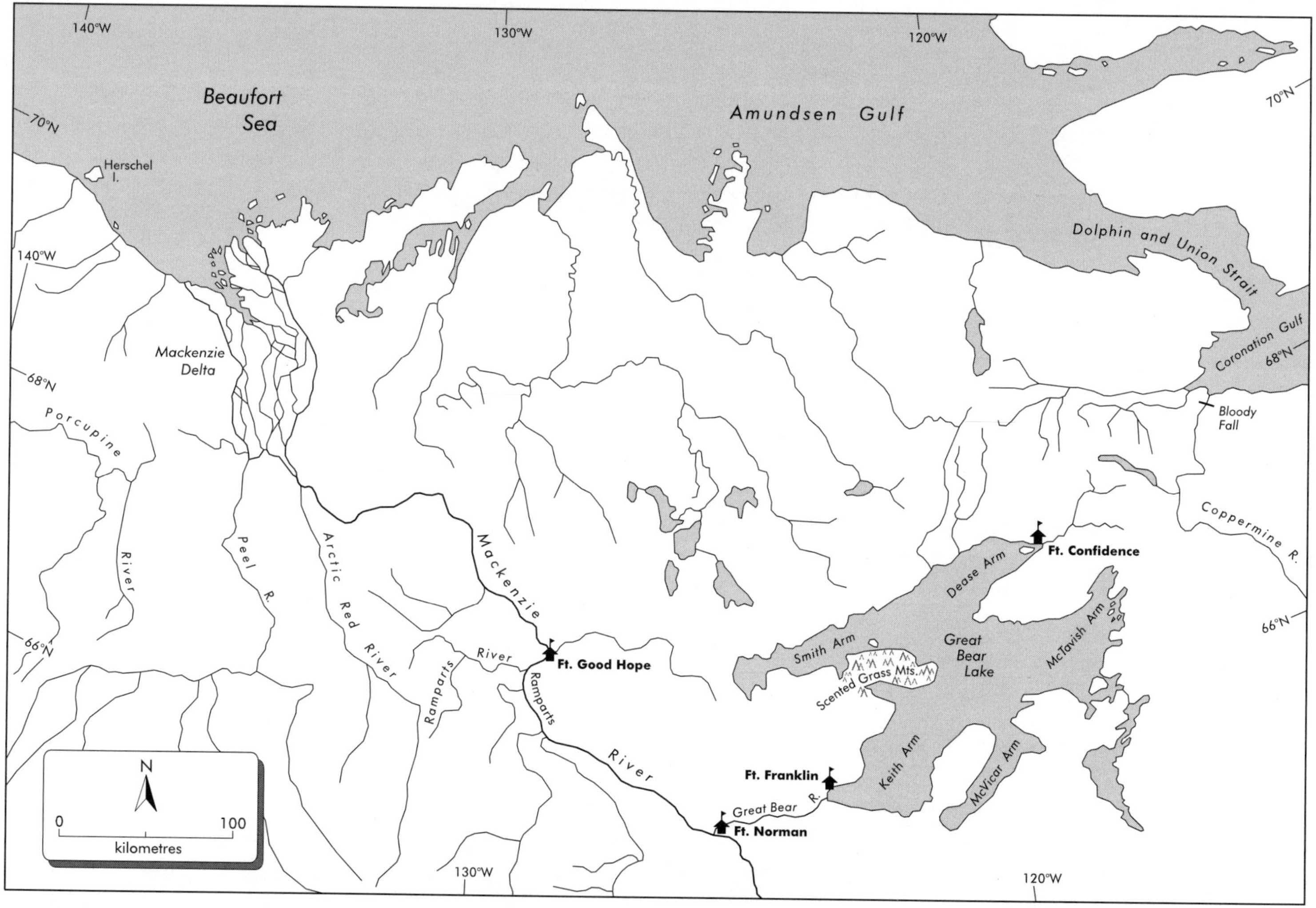


3 *Map of Lower Mackenzie River and Great Bear Lake.*

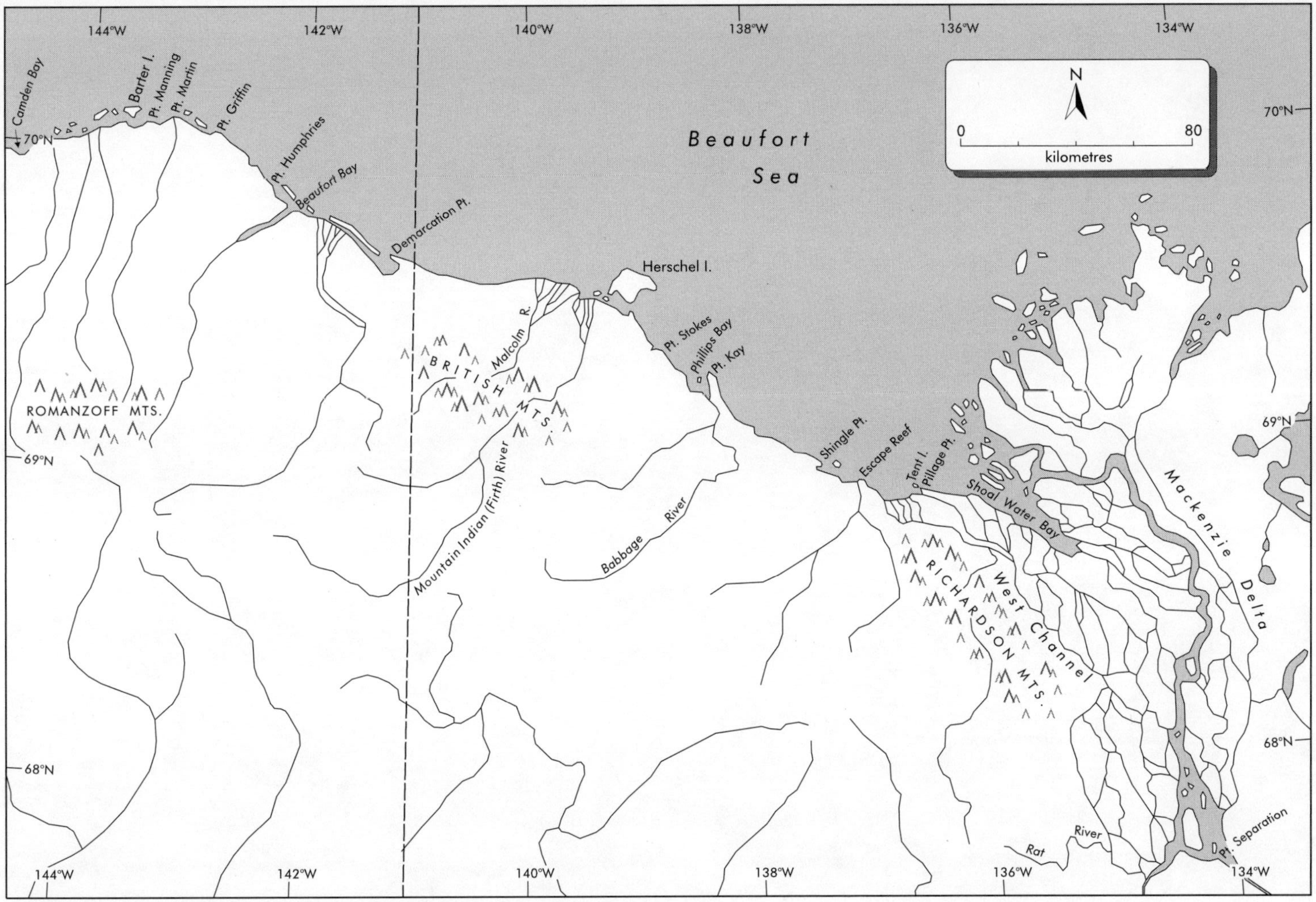


4 *Expedition's area of operations, Mackenzie Delta to Barter Island, 1837.*

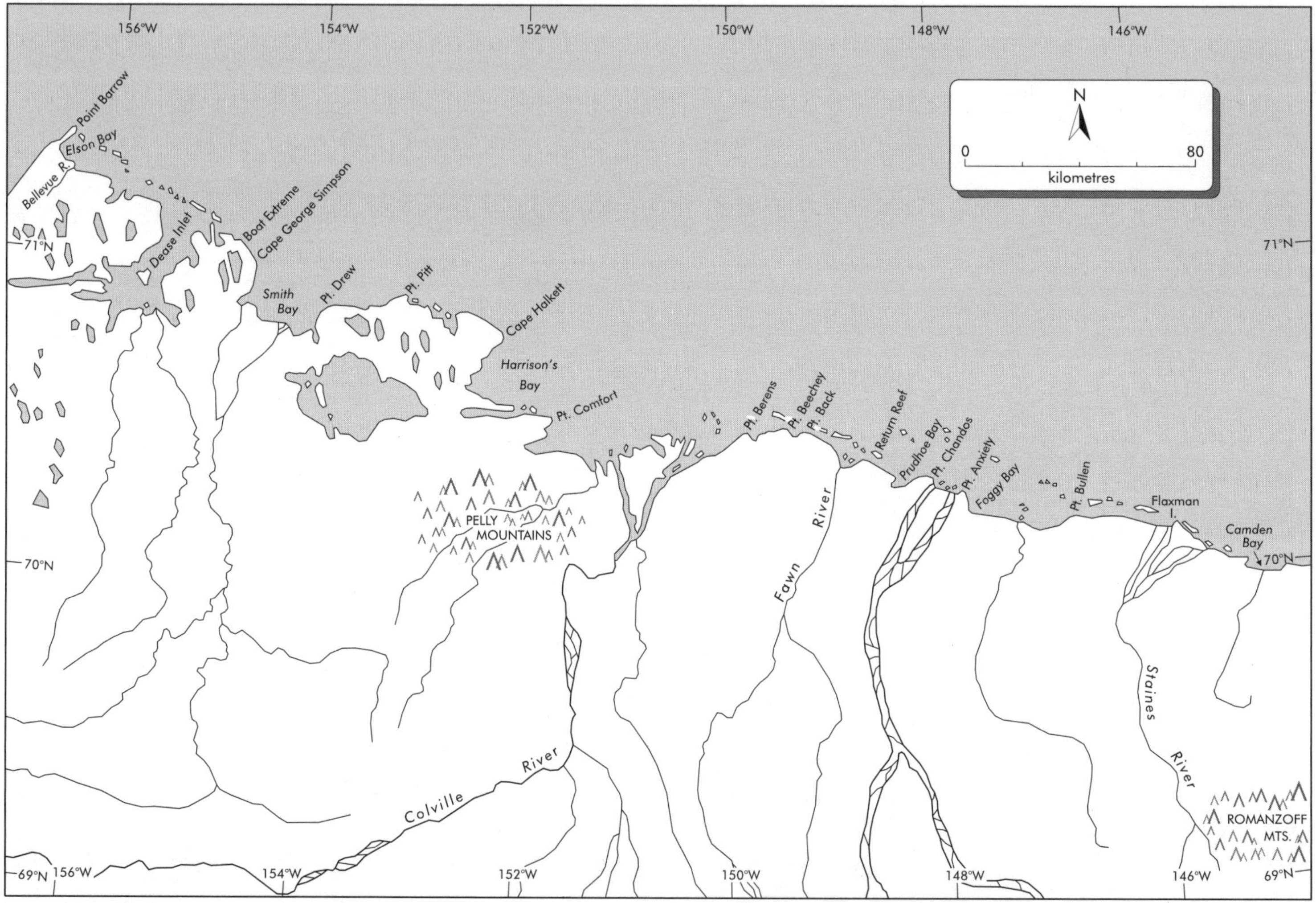


5 *Expedition's area of operations, Camden Bay to Point Barrow.*

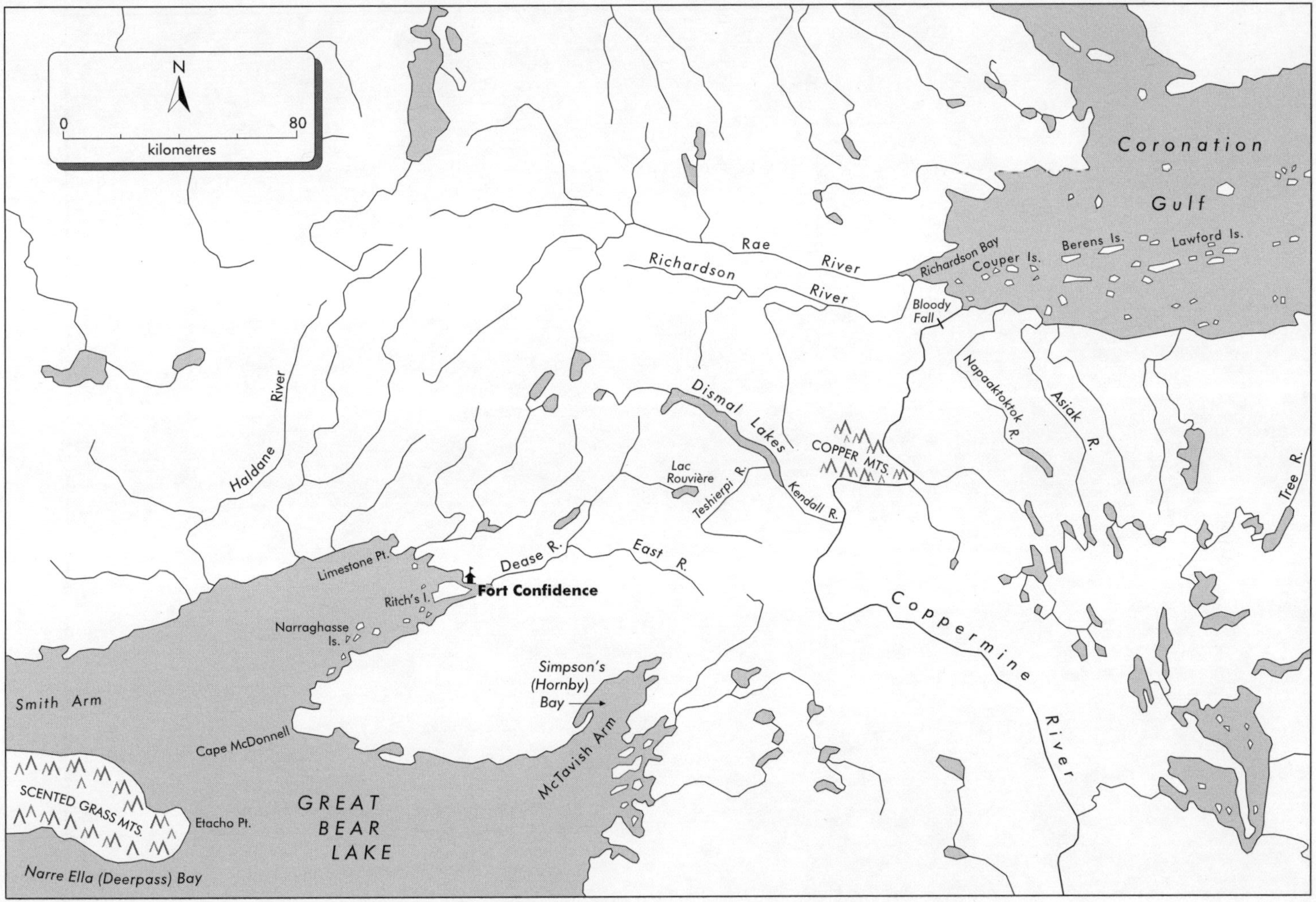


6 *Expedition's area of operations, Great Bear Lake to Coronation Gulf.*

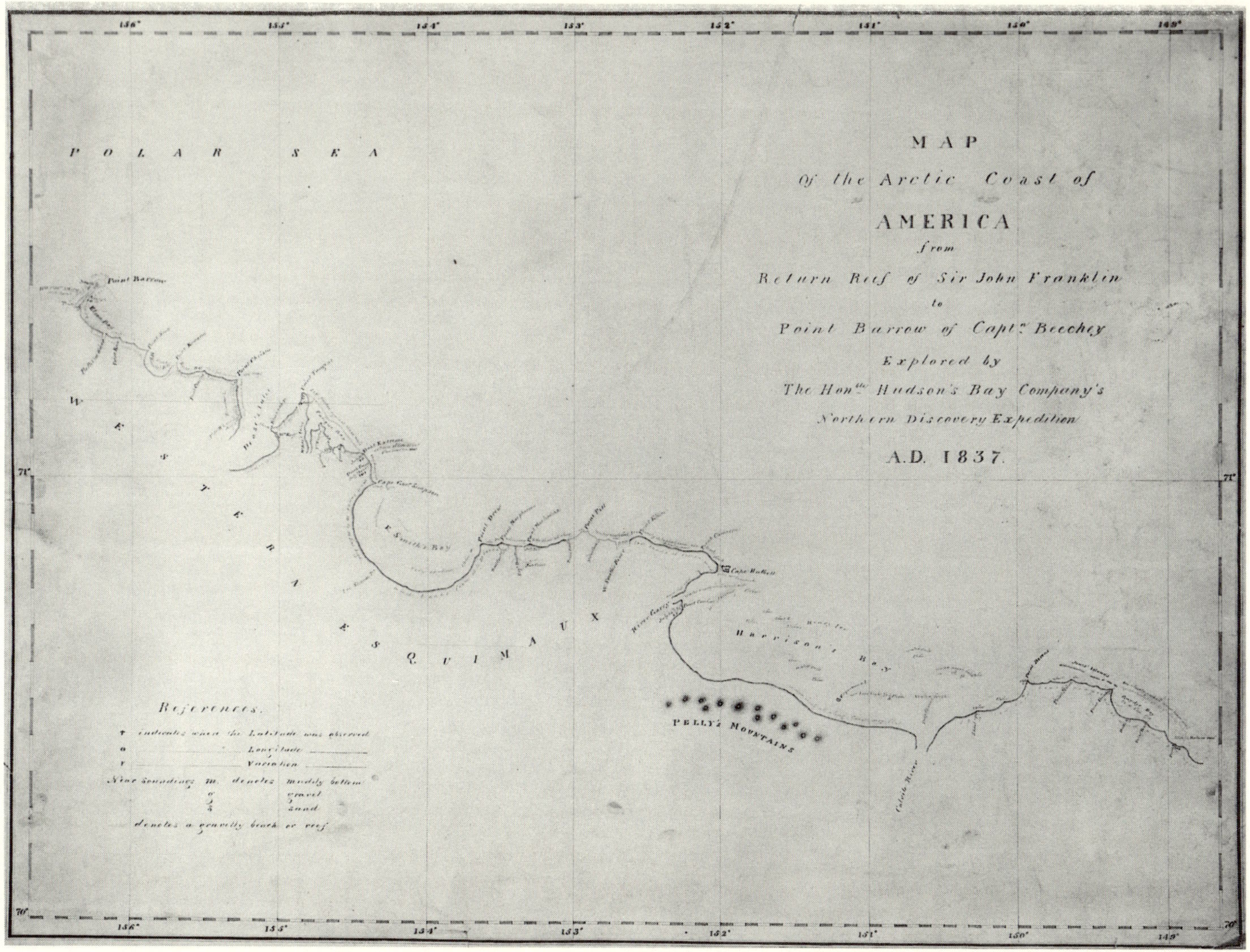


7 *Thomas Simpson's map of the arctic coast, Return Reef to Point Barrow. Courtesy Hudson's Bay Company Archives.* G.1/180 (N 7482).

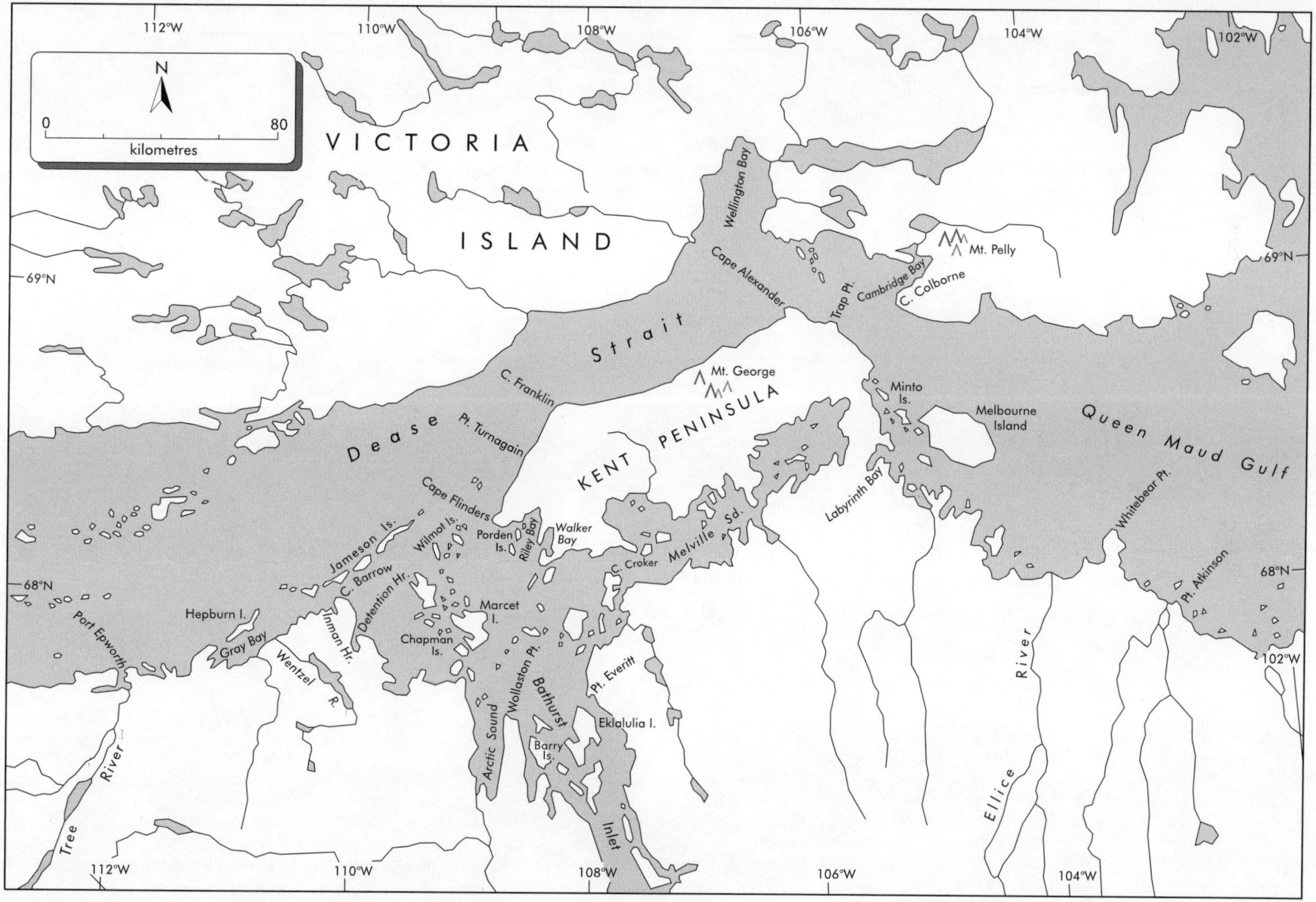


8 *Expedition's area of operations, Coronation Gulf to Queen Maud Gulf, 1838 and 1839.*

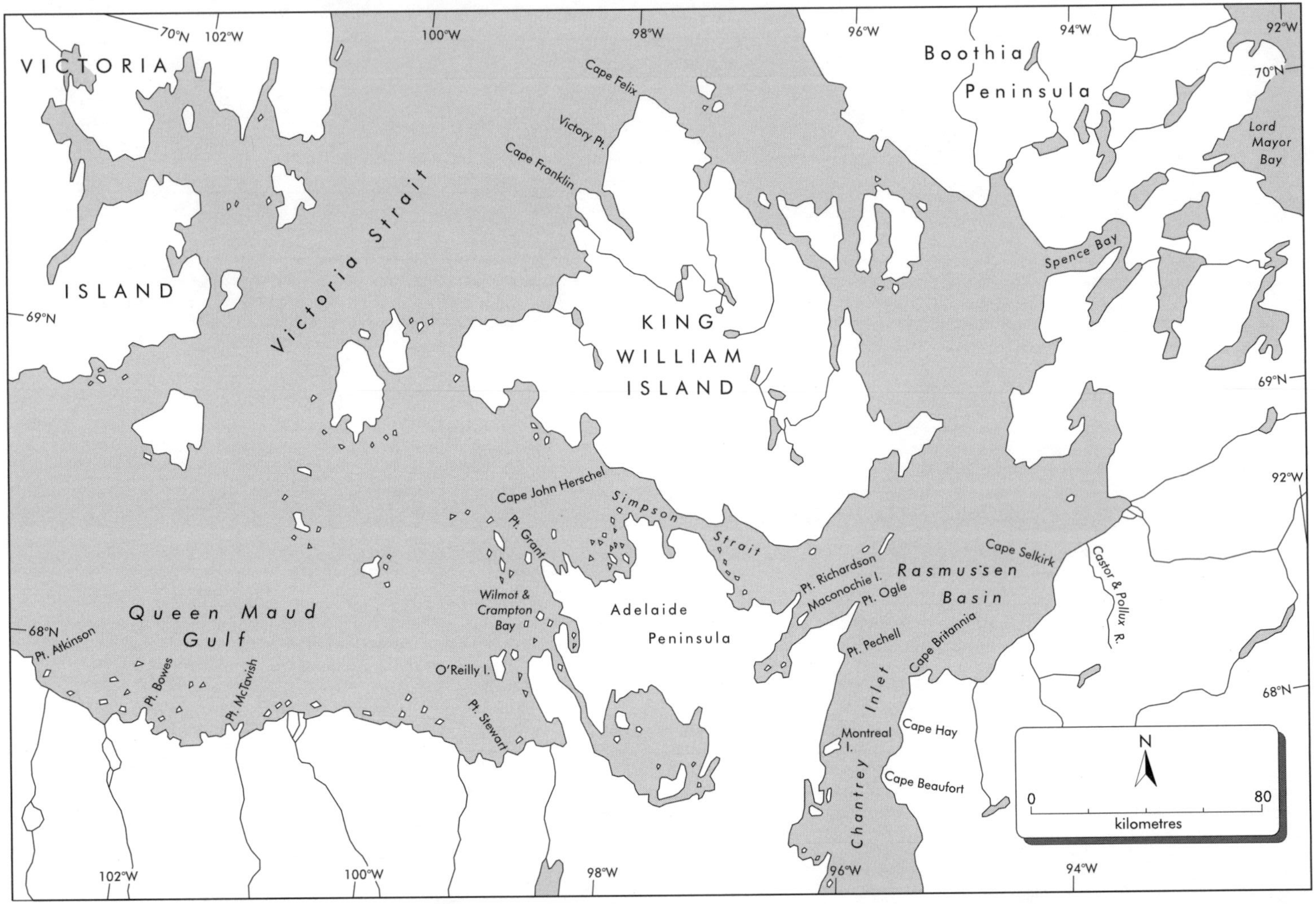


9 *Expedition's area of operations, Queen Maud Gulf to Boothia Peninsula, 1839.*

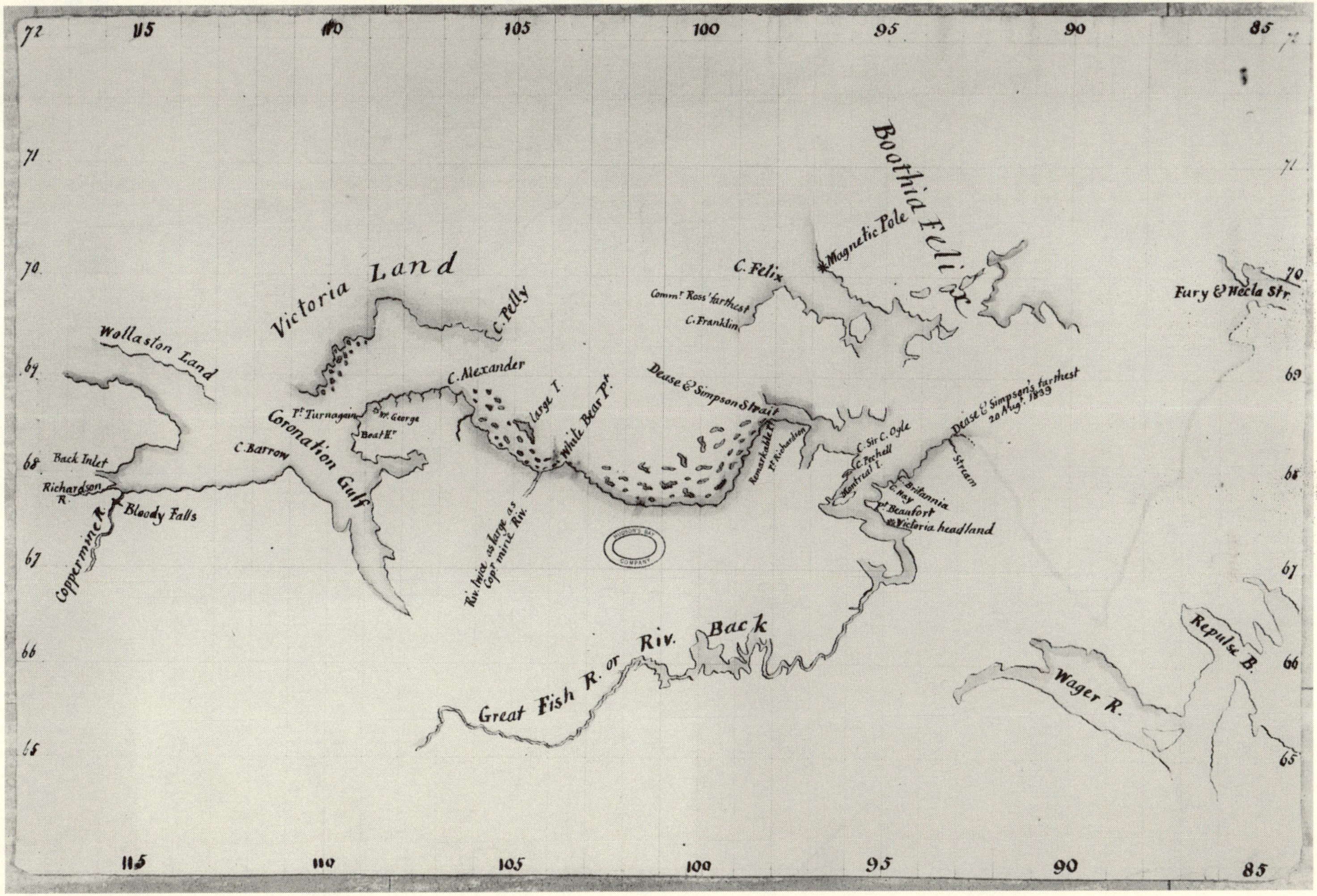


10 *Contemporary map of area explored by Dease and Simpson, Coppermine River to Boothia Peninsula. Courtesy Hudson's Bay Company.* HBCA G.1/5 (N 3640).

# FROM BARROW TO BOOTHIA

# Chapter One
# The Men and Their Task

## INTRODUCTION

By the mid-1830s, substantial parts of the arctic mainland coast of North America (and of the islands to the north) had been explored and mapped by a whole series of expeditions, some private, others dispatched by the British Government and especially by the Admiralty. In 1742 most of the west coast of Hudson Bay (as far north as Repulse Bay) had been effectively explored and charted by Captain Christopher Middleton.[1] Five years later, with minor additions (for example, the lower part of Chesterfield Inlet), these surveys were confirmed by Captains William Moor and Francis Smith.[2]

Later in the century the position of the arctic coast was fixed at two widely separated points farther west. In 1771 Samuel Hearne, working for the Hudson's Bay Company, reached the mouth of the Coppermine River in Coronation Gulf,[3] and in 1789 Alexander Mackenzie, of the North West Company, reached tidewater in the Mackenzie Delta.[4] In the interim the west end of the arctic coast had been fixed by Captain James Cook, who reached Icy Cape on the northwest coast of Alaska in 1778.[5]

The next addition to the map of the arctic coast was made by Lieutenant John

1 Barr and Williams, eds., *The Voyage of Christopher Middleton 1741–1742*, vol. 1.

2 Barr and Williams, eds., *The Voyage of William Moor and Francis Smith, 1746–1747*, vol. 2.

3 Hearne, *A Journey from Prince of Wales's Fort in Hudson's Bay, to the Northern Ocean undertaken by Order of the Hudson's Bay Company, for the Discovery of Copper Mines, a North West Passage etc. in the Years 1769, 1770, 1771 and 1772.*

4 Mackenzie, *Voyages from Montreal, on the River St. Lawrence, through the Continent of North America, to the Frozen and Pacific Oceans; in the years 1789 and 1793.*

5 Beaglehole, ed., *The Journals of Captain James Cook on His Voyages of Discovery*, vol. 3, *The Voyage of the Resolution and Discovery 1776-1780* (2 vols.).

Franklin during his first overland expedition in 1821; he and his party, travelling by canoe, explored the coast from the mouth of the Coppermine River to Port Turnagain on Kent Peninsula, including the intricacies of Bathurst Inlet, before embarking on their disastrous overland trek back to their winter quarters at Fort Enterprise.[6]

In 1819–20 Captain William Edward Parry's remarkable penetration west through Parry Channel, almost reaching the Beaufort Sea – almost through the entire Arctic Archipelago – was blocked by ice in M'Clure Strait;[7] but on that voyage he added nothing to the knowledge of the arctic mainland coast. On his next expedition (1821–23) he did make a contribution to that knowledge. From winter quarters at Winter Island, then at Iglulik, he explored and mapped the south and east coasts of Melville Peninsula, from Repulse Bay to Fury and Hecla Strait. Although he did not cross the peninsula to Committee Bay, he did acquire some knowledge (and maps) of it from the Inuit.[8]

A few years later, in 1826, several more large pieces of the puzzle fell into place. In the western Arctic, Captain Frederick Beechey (HMS *Blossom*) pushed north through Bering Strait and coasted north to Icy Cape. From there the mate, Thomas Elson, continued northwards by boat, using shore leads, and reached Point Barrow.[9] It had been hoped that Beechey's expedition might achieve a rendezvous with another expedition coming from the east. In 1825 an expedition led by Captain John Franklin and Dr John Richardson had descended the Mackenzie River and wintered at Fort Franklin on Great Bear Lake. From there, travelling by boat in the summer of 1826, Franklin had led a party westwards from the Mackenzie Delta, but was forced to turn back by ice at Return Reef, only some 280 kilometres from Elson's farthest at Point Barrow. In the meantime a second party, also travelling by boat, and led by Dr Richardson, had turned eastwards from the mouth of the Mackenzie and had explored the coast east to the mouth of the Coppermine before returning overland to its winter quarters.[10]

The next addition to the map of the arctic coast was made in 1829–33 by a small, private expedition, led by Captain John Ross on board *Victory*. Pushing south down Prince Regent Inlet, Ross wintered three times in the vicinity of Lord Mayor Bay. During a number of sledge trips he and his party explored the east coast of Boothia

6 Franklin, *Narrative of a Journey to the Shores of the Polar Sea in the Years 1819, 20, 21 and 22*; Davis, ed., *Sir John Franklin's Journals and Correspondence: The First Arctic Land Expedition 1819–1822.*

7 Parry, *Journal of a Voyage for the Discovery of a North-West Passage from the Atlantic to the Pacific; performed in the Years 1819–20, in His Majesty's Ships Hecla and Griper.*

8 Parry, *Journal of a Second Voyage for the Discovery of a North-West Passage from the Atlantic to the Pacific, performed in the Years 1821–22–23, in His Majesty's Ships Fury and Hecla, under the Orders of Captain William Edward Parry.*

9 Beechey, *Narrative of a Voyage to the Pacific and Beering's Strait, to co-operate with the Polar Expeditions: performed in His Majesty's Ship Blossom, under the Command of Captain F. W. Beechey … in the Years 1825, 26, 27, 28.*

10 Franklin, *Narrative of a Second Expedition to the Shores of the Polar Sea in the Years 1825, 1826, and 1827 … Including an Account of the Progress of a Detachment to the Eastward by John Richardson*; Davis, ed., *Sir John Franklin's Journals and Correspondence: The Second Arctic Land Expedition 1825–1827.*

Peninsula (and of Somerset Island) from the southern limits of Lord Mayor Bay north to Port Leopold, and in a number of sledging expeditions crossed the Isthmus of Boothia and explored parts of the west coast of the peninsula. In 1830 Ross's nephew and second-in-command, James Clark Ross, explored the north coast of King William Island as far as Victory Point, but did not detect what is now called Rae Strait, and left the impression that King William Island was in fact a peninsula (or even the arctic mainland coast itself), extending west from Boothia Peninsula.[11]

In the light of growing public concern at the protracted absence of the Ross expedition, in 1833 Captain George Back, with Dr Richard King as his second-in-command, mounted a search expedition by boat down the Back River. They reached the mouth of the river in Chantrey Inlet and explored much of that inlet as far as Point Ogle, including Montreal Island, in the summer of 1834.[12]

Thus, in the autumn of 1835 there were still three substantial gaps in the map of the arctic coast of North America. In the west there was a gap of some 280 kilometres between Franklin's Return Reef and Point Barrow, reached by Elson in 1826. In the central Arctic there was a gap of some 480 kilometres between Franklin's Point Turnagain, on Kent Peninsula, and the Spence Bay area, explored by John and James Clark Ross on the west side of the Isthmus of Boothia, with the fairly limited exception of the Chantrey Inlet area, explored by Back in 1834. In this "blank" on the map, the picture was somewhat confused by James Ross's failure to detect Rae Strait, which left the impression that King William Island was joined to Boothia Peninsula to the east. The third remaining gap extended from Ross's farthest south, in Lord Mayor Bay on the east side of Boothia Peninsula, to the west end of Fury and Hecla Strait, including the extensive embayments of Pelly and Committee bays.

Almost immediately after Captain Back's return to England, in September 1835,[13] several proposals surfaced in Britain for expeditions aimed at completing the exploration of the arctic coast of North America.[14] The Royal Geographical Society formed a committee to evaluate the various proposals and published the plans of such arctic experts as Sir John Barrow, Captain Sir John Franklin, Dr John Richardson, Captain John Ross, and Sir Francis Beaufort, hydrographer to the Navy.[15] Undoubtedly the most persistent advocate of a further expedition was Dr Richard King, who had accompanied Back on his expedition. On Friday 15 January 1836 he addressed a public meeting held at the Literary Institution, Aldersgate Street, on his proposed expedition,[16] whereby he planned to explore the coast from the mouth of the Back River west to Point Turnagain, and also to solve the mystery of the geography of the Isthmus of Boothia, all in three

11 Ross, *Narrative of a Second Voyage in Search of a North-West Passage, and of a Residence in the Arctic Regions during the Years 1829, 1830, 1831, 1832, 1833.*

12 Back, *Narrative of the Arctic Land Expedition to the Mouth of the Great Fish River, and along the Shores of the Arctic Ocean, in the Years 1833, 1834 and 1835*; King, *Narrative of a Journey to the Shores of the Arctic Ocean in 1833, 1834 and 1835, under the Command of Capt. Back, R.N.*

13 Back, *Narrative of the Arctic Land Expedition.*

14 Wallace, *The Navy, the Company and Richard King.*

15 Royal Geographical Society, "Communications on a North-West Passage."

16 "North-West Passage," *The Times*, 16 January 1836, p. 6; King, *Narrative of a Journey.*

seasons. He proposed leading a party of six men, which would travel by the usual fur-trade route to Lake Athabasca; from there he proposed heading north via a series of rivers, lakes, and portages, of which he had heard from the Chipewyans, to a river running parallel to the Back River (almost certainly the Thelon). According to his information, the valley of this river was well wooded and supported a substantial population of moose. Here he would winter and in the following spring he would travel north to the Back River by a convenient system of rivers and portages, descend it to the sea, and explore the coast west to Point Turnagain and east to Boothia. His estimate of total cost was £1000. Dr Thomas Hodgkin proposed a motion for opening a public subscription towards this cost; the motion was carried unanimously. Subsequently, however, Dr King managed to antagonize so many people at the Colonial Office, the Admiralty, and the Hudson's Bay Company that a year later (January 1837) his fund had risen to only £400, and, for the moment at least, King abandoned the project.[17]

The motivation of the Hudson's Bay Company for mounting an exploring expedition has been analysed in considerable detail by Coates.

The Company's commitment to Arctic exploration, and specifically its sponsorship of the Simpson and Dease expedition was, ultimately, a public relations undertaking of a most unusual sort. The key lies in the timing. In 1836 the London Committee of the Hudson's Bay Company decided to seek an extension of their Exclusive Licence to Trade, not due to expire until 1841. Chartered enterprise, once the engine of the Empire, no longer found favour with the British public, and in the eyes of many politicians and government officials, the Hudson's Bay Company was becoming increasingly anachronistic. The firm anticipated the criticism, and knew of the government's likely reservations about extending the licence. While they could, and did, identify significant improvements in Rupertsland over the pre-merger era,[18] the Company's officers also recognized that more dramatic measures were necessary.[19]

The attitude of the Company towards further exploration was undoubtedly coloured by the fact that exploration was specified as one of its obligations in its original charter.[20] In 1836 the Committee informed Lord Glenelg, the colonial secretary, that they were planning an expedition to complete the map of the arctic coast. As Coates further explains:

The officers of the Hudson's Bay Company believed that much could be gained by sponsoring a major Arctic survey. The British public's fascination with Arctic travel remained unabated, and any northern venture, particularly one destined to complete the map of the North American coastline, would undoubtedly generate much favourable coverage.

17 Wallace, *The Navy, the Company and Richard King.*

18 The allusion here is to the merger of the North West Company and the Hudson's Bay Company in 1821.

19 Coates, *The Commerce of Discovery: The Hudson's Bay Company and the Simpson and Dease Expeditions*, 8–9.

20 Rich, *The History of the Hudson's Bay Company 1670–1870*, vol. 2: *1763-1870*, 650.

For years the Hudson's Bay Company had provided occasional assistance to other explorers; now the time was right to claim a greater share of the glory, and as much of the credit as possible. The Board of Trade and the Colonial Office, it was hoped, could not help but take note of the Company's unselfish commitment to scientific adventure.[21]

The governor of the Company, Sir Henry Pelly, wrote to George Simpson, the governor in North America, on 9 March 1836, advising him to be prepared to assist a government expedition, if one were mounted, or otherwise to make arrangements for the Company itself to mount an exploring expedition (see p. 20).

Governor George Simpson chose two men to lead the expedition, Chief Factor Peter Warren Dease, and his own young cousin, Thomas Simpson. He evidently solicited plans for an expedition from both men (as reported in the minutes of the Northern Council); two different plans by Thomas Simpson have survived,[22] but no plan by Dease.

## *Peter Warren Dease*

The two leaders selected by George Simpson could scarcely have been more different. Peter Warren Dease was born on 1 January 1788 at Michimackinac (now Mackinac Island, Michigan), the fourth son of Dr John B. Dease, an Irishman and the controversial deputy superintendent general of the Western Indians,[23] and Jane French, possibly a Mohawk from Caughnawaga.[24] Three of Peter's brothers, Charles, John, and Francis, also served with the Hudson's Bay Company. Peter Warren Dease was named after a relative of his father's, Admiral Sir Peter Warren, who captured Louisbourg from the French in 1745.

Having spent his childhood at Mackinac and in Montreal, on 11 April 1801, at the age of thirteen, Peter Dease joined the XY Company for a six-year term and was posted to Great Slave Lake.[25] With the amalgamation of the XY and North West companies in 1804 he was posted as a clerk first to Athabasca and then, after 1817, to the Mackenzie District, where he took charge of Fort Good Hope and was later stationed at Fort Chipewyan and various Mackenzie posts.

With the merger of the North West and Hudson's Bay companies in 1821, both Peter Warren and John Dease were appointed chief traders in the restructured company; Peter's commission (fig. 1) is dated 28 March 1821. He attended the council meeting of the Northern Department at Norway House in August 1821[26] and was appointed to the Athabasca District. Thus in October 1823, writing to William McKintosh at Dunvegan, James Keith, chief factor in charge of Fort Chipewyan,

21 Coates, *The Commerce of Discovery*, 9–10.

22 Alexander Simpson, *Life and Travels*, 178–83.

23 Armour, "Dease, John."

24 Sampson, "Dease, Peter Warren"; Fleming, ed., *Minutes of Council, Northern Department of Rupert Land, 1821–31*, 434–6; MacLaren, "The HBC's Arctic Expedition," 465–79.

25 Sampson, "Dease, Peter Warren"; Keith, *North of Athabasca*, 368.

26 Fleming, *Minutes of Council*.

reported that "The winter appointments, in this quarter as contemplated by Council have been strictly conformed to vizt. ... Mr. P. Dease, 3 Junior Clerks, myself & 51 men at this place."[27]

Earlier that year, on 9 February, George Simpson had written to Dease with instructions to mount an expedition to explore the basin of the Finlay River, i.e. present-day north-central British Columbia, with a view to developing its fur resources:

The expense will be inconsiderable, not exceeding the amount of 8 or 10 pieces of Goods, and the services of 4 Men in addition to yourself and a Clerk, with 4 Guides & Hunters. I am therefore anxious that the Country on the West of the Mountains, North of Babine river or Lake and running parallel or nearly so with the former, should be explored in the course of next Summer, as far as the Season admits, which may probably be until you get to about the latitude of Riviere aux Liards and then recross it, so as to fall on the head Waters of that Stream, or any other leading into McKenzie's River. If the Country is found to be rich in Fur bearing Animals, it will be necessary to establish a friendly intercourse with the Natives for the purpose of Trade, and to ascertain the best means of settling it to advantage, and it affords me great satisfaction, that you undertake the Mission with so much Zeal, as I do not know any Gentleman in the Country better qualified for an enterprise of such importance.[28]

Simpson's closing remarks here are especially significant, indicating that Dease had already made a very favourable impression on him; and George Simpson was not easily impressed.

To be able to take full advantage of the summer of 1823 for exploration, Simpson recommended that Dease travel west up the Peace to Fort St John on the ice, prior to breakup. But Simpson's letter must have been delayed on its way to Fort Chipewyan, and Dease was forced to postpone his start on the expedition: "the advanced state of the Season at which your letter came to hand, with the few days necessary to take the People's Arrangements ... have rendered it impracticable for us to reach Dunvegan on the Ice."[29] While Simpson accepted this (and other reasons mentioned in later letters) as being valid, the delay meant that the task of mounting the expedition went to somebody else; writing on 14 July 1823, Simpson noted:

The reasons you assign for not proceeding on this mission at such an advanced period of the season, are satisfactory and the postponement of it under all circumstances, I consider highly judicious.

The Council have reconsidered this subject and resolved on putting the charge of the expedition under Mr. Black's management, your services being considered necessary in Athabasca, to which District you are appointed.[30]

27 Letter from James Keith to William McKintosh, Dunvegan, 2 October 1823, HBCA B.39/a/22, f. 7.
28 Letter from George Simpson to Peter Dease, Fort Chipewyan, 9 February 1822, HBCA D.4/2, f. 50.
29 Letter from Peter Dease to George Simpson, 22 March 1823, HBCA D.4/117, ff. 53d–54.
30 Letter from George Simpson to Peter Dease, 14 July 1823, HBCA, D.4/2, f. 57.

Hence it was Samuel Black, rather than Peter Dease, who led the expedition which explored the Upper Finlay and the headwaters of the Stikine in the summer of 1824.[31]

But soon afterwards Dease was to play a subordinate yet extremely important role in another exploring expedition, which would become much better known to the world at large. Almost exactly a year after the above-quoted letter, on 26 July 1824, Simpson wrote to Dease:

> ...you are appointed to an Expedition fitting out by Government under the command of Capt. Franklin for the purpose of descending McKenzie's River and exploring the coast to the westward. I have therefore to request that you will consider yourself attached and belonging to that Expedition after the rect. of this letter, and take charge of the men brought up from Canada by Mr. McVicar.[32]

Franklin had met Dease at Fort Chipewyan during Franklin's first expedition (1819–22) and had been greatly impressed by his knowledge of the Indians and the geography of the country through which the expedition would be travelling;[33] on this basis he specifically requested Dease's services to establish and run the expedition's base-camp at Fort Franklin on Great Bear Lake.[34]

Writing to Franklin on 8 August 1824, Simpson approved his choice:

> In selecting Mr. Dease as a Fellow-Traveller you have made a happy choice, as I do not know a Gentleman in the country better qualified for the duty he has to perform. He is now in charge of Fort Chipewyan but we have passed a Resolution in Council attaching him to the Expedition which I enclosed to him, and he will consider himself as belonging thereto from the rect. of that instruction.[35]

Franklin also wrote to Dease directly, requesting:

> that you will do me the favour and kindness of uniting your services to the Expedition. I have obtained the permission of all the Directors for you to do so ... At the same time they have assured me that your being attached to the Expedition will be considered as a service rendered to the Company, and which of course entitle you to a continuance of your share in the profits of the Concern while so employed as much as if you were stationed at any of their forts, and more, they have intimated to me "entre nous" that your going with us will give you the strongest claim on their favourable recommendation for advancement in the Company.[36]

---

31 Rich, ed., *A Journal of a Voyage from Rocky Mountain Portage in Peace River to the Sources of Finlays Branch and North West Ward in Summer 1824.*

32 Letter from George Simpson to Peter Dease, 26 July 1824, HBCA D.4/3, f. 163; D.4/7, ff. 51–2; D.4/8, ff. 83–4.

33 Sampson, "Dease, Peter Warren."

34 Franklin, *Narrative of a Second Expedition*, xxii.

35 Letter from George Simpson to John Franklin, 8 August 1824, HBCA D.4/4, f. 15.

36 Letter from John Franklin to Peter Dease, 4 March 1824, SPRI MS 248/281/1, pp. 105–20; Davis, *Franklin's Second Arctic Land Expedition*, 299–304.

A year later, on 10 July 1825, Simpson wrote to Franklin:

My object in now addressing you in this private and confidential manner is to entreat that you do not part with Mr. Dease under any circumstances, but that he be allowed to form one of the party immediately under your command. This Gentleman is one of our best voyageurs, of a strong, robust habit of body, possessing much firmness of mind joined to a great suavity of manners, and who from his experience in the country, and being inured to the hardships and privations incident thereto, would be a most valuable acquisition to the party in the event of its being unfortunately placed in trying or distressing circumstances; his presence would, moreover, give a confidence to the people which that of strangers to the country cannot inspire ...[37]

On his return from Franklin's second expedition in the spring of 1827, Dease was appointed to take charge of Fort Good Hope, almost on the Arctic Circle, one of the most remote and by far the most northerly of the Company's posts. While there, on 3 January 1828, he was made chief factor and member of Council (fig. 2).

Part of his mandate was to expand the Company's territory, and on his appointment he was directed by the Council of the Northern Department "to occupy the ensuing Summer in acquiring information about the Country and Natives of a River falling into the Mouth of McKenzie's River discovered by Captain Franklin, and that the necessary steps be taken to establish a Post thereon if considered expedient."[38]

In July 1829 Dease was able to report that, in keeping with these instructions,

I have made particular enquiries regarding the Natives and countries about Peels river (the one I am informed alluded to) and by all we can collect from the lower Squint Eyes [Loucheux, now Gwich'in] no other Tribe but themselves frequent that Stream, they generally remain to the Westward between it and where the Mountains dip into the Ocean which is their Hunting grounds, they tell us that towards the Sources of that River, Beaver is to be found pretty numerous, but the distance great for them to hunt and bring their hunts to this establishment; that was a post established for them they would be able to make better hunts, but at the same time the Whites would require to be strong, as they would be subject to the visits from the Esquimaux who they represent as a very treacherous and hostile People; they come up the River annually in barges as far as [Arctic] Red River to traffic with the Squinteyes, and by last accounts from the latter Tribe, they had every reason, they said, to think the Esquimaux were hostilely inclined towards them – some shots with arrows were exchanged in the Spring between the two Tribes, and in the Fall they say two Esquimaux in a Canoe came up to their Camp apparently as Spies and without going on shore or speaking to them turned back – this they look upon as a certain mark of hostile intentions.[39]

---

37 Letter from George Simpson to John Franklin, 10 July 1825, HBCA D.4/5, f. 48d.

38 Fleming, *Minutes of Council*, 209.

39 Letter from Peter Dease to the governor, chief factors, and chief traders, 31 January 1829, HBCA D.4/122, ff. 42–3.

Later that same year, by which time Dease had handed over Fort Good Hope to John Bell and was stationed at Fort Simpson, Dease and Edward Smith reported:

The Esquimaux last Summer treacherously murdered one of the Loucheux Indians, an insult which the latter will not forgive. Any hostilities between those Tribes will affect the Trade of that Post [Fort Good Hope] and should the former prove Victors will endanger the safety of the Establishment, as in the latter case there is no barrier between us and the Esquimaux who are now acquainted with our residence near their lands.[40]

It would be another decade before a post (Peel's Post, later Fort McPherson) was established on the Peel River, near the head of the Mackenzie Delta right at the southern margin of Inuit territory, and for several decades thereafter this post operated under constant fear of Inuit aggression.

But despite these fears and tensions, under Dease's management Fort Good Hope showed a respectable profit. Thus for the 1828 Outfit Fort Good Hope returned a profit of £2000 in a year when the entire Mackenzie District (comprising Fort Simpson, Fort Norman, Fort Good Hope, and Fort des Liards) showed a profit of just under £12,000.[41]

Soon after Smith and Dease submitted their report, Dease left Fort Simpson on 3 December 1829 to spend the bulk of the winter in an advisory capacity at the new post at Fort Halkett, established by John Hutchinson the previous summer.[42] Dease started back for Fort Simpson in February 1830 to report favourably on the trade with the Sekanee Indians thus far, and on the prospects for the future.

In July 1830 Dease attended the Council meeting at York Factory[43] and discovered that he had been appointed to Fraser's Lake in New Caledonia District,[44] where he would serve under Chief Factor William Connolly for a year before taking charge of the district, with headquarters at Fort St James, in 1831.[45] This district, encompassing a large area of present-day north-central British Columbia, included (apart from Fort St James) the posts of Babine Lake, Fraser Lake, McLeod Lake, Connolly's Post, Fort George, and Fort Alexandria. It was considered to be remote, and manning and supply of provisions and trade goods represented ongoing problems. Rather than pemmican or caribou meat, salmon was the staple, and when the fisheries occasionally failed, it caused serious distress. Even when salmon were abundant, it was found that a steady diet of dried salmon was injurious to health, and altogether New Caledonia had an extremely bad reputation among both servants and gentlemen.[46]

---

40 Letter from E. Smith and P.W. Dease to the governor, chief factors, and chief traders, 29 November 1829, HBCA D.4/123, f. 45.

41 Letter from E. Smith and P.W. Dease to the governor, chief factors, and chief traders, 29 November 1829, HBCA D.4/123, f. 45.

42 Fort Simpson journal, 3 December 1829, HBCA B.200/a/11.

43 Fleming, *Minutes of Council*, 271.

44 Fleming, *Minutes of Council*, 252.

45 Fort St James journal, 7 May 1831, HBCA B.188/1/17.

46 For further details see Gibson, *The Lifeline of the Oregon Country*, 43–58.

Servants in New Caledonia were paid higher wages than those in the Columbia District, but even with this inducement, it was difficult to recruit men.[47]

Dease had a hard act to follow: in his final year Connolly had achieved returns of 150 packs of fur, which, as Simpson pointed out in a letter to Dease:

> leaves but little room for his successor in the way of improvement, but from your habits of business & ability as an Indian trader we look with confidence to its being kept in the high order in which it has fallen into your hands, and if zeal & perseverance can be productive of still further improvement we are satisfied it will in due time shew itself.[48]

While the personnel and correspondence travelled to and from New Caledonia overland, from York Factory via Norway House, Portage la Loche, Fort Chipewyan, and the Peace River, the outfits and the returns travelled from and to Fort Vancouver on the lower Columbia, for transport by sea to England; the round trip distance from Fort St James to Fort Vancouver by the route followed was at least 2730 kilometres, and Dease made this trip annually from 1831 until 1834.[49]

His journal for one of these trips, in 1831, has survived.[50] In that year he set off down the Stuart River from Fort St James on Stuart's Lake on 7 May, reaching Fort George (now Prince George) on the afternoon of the 9th; the final assembling of the brigade occurred here, and it set off down the Fraser River on the 10th: one bateau with five men and five canoes with four men in each. Next day, at the Big Rapid, the man in charge of the bateau (Louis Paul) was thrown overboard by his steering oar and drowned. Nonetheless the brigade reached Fort Alexandria that afternoon.

Here the brigade left the Fraser River; the returns were loaded onto pack horses for the next leg, across country to Kamloops and Fort Okanagan Post (at the confluence of the Okanagan and the Columbia). The caravan consisted of ten strings of nine pack horses and two of six horses; two men were assigned to each string. The cavalcade set off on the morning of 15 May, reaching Kamloops on the 25th and Fort Okanagan on 3 June. Here they transferred to nine boats, and on 5 June they started down the Columbia River, reaching Walla Walla on the 6th and Fort Vancouver on the 11th.

After a respite of a couple of weeks, Dease and his party started back north on 28 June in their nine boats, each with seven men and holding forty-six or forty-eight pieces of the new outfit for the New Caledonia District. Unfortunately, as the boats were being lined up the Cascades, one boat was swept away and capsized, drowning the two men on board. The brigade reached Fort Okanagan on 21 July and here the outfit was transferred to the pack horses. The cavalcade consisted of 110 pack horses, each string of ten being in the charge of two men. They reached

47 Gibson, *Lifeline*, 48.

48 Letter from George Simpson to Peter Dease, 2 July 1831, HBCA D.4/19, f. 1.

49 The brigade system whereby New Caledonia was supplied, including precise details of the route and schedule, has been exhaustively analysed by Gibson (*Lifeline*).

50 Fort St James journal, 19 April to 13 September 1831, HBCA B.188/1/17.

Kamloops on 5 August and Fort Alexandria on the 18th. Once the canoes were repaired, they were loaded and the flotilla set off up the Fraser on 24 August, reaching Fort George on 3 September and Fort St James on the 13th.

Thus this extremely strenuous trip (of which only the major events have been noted) occupied over four months of almost constant travelling. Dease made the gruelling trip each summer up until and including 1834. He also devoted the same amount of energy and, undoubtedly, a great deal of tact to the affairs of the district throughout the year.

George Simpson was greatly impressed by Dease's management of this challenging district; on 3 July 1834, writing from the annual Council meeting at York Factory, he recorded "our unqualified approbation of your excellent management of the affairs of the Caledonia, which appears to be in a very prosperous and promising state."[51] And on the eve of Dease's departure from the district, in 1835, the governor noted that "The information your letter ... conveys on the affairs of the District is very satisfactory, and now, on your departure therefrom, we consider it a duty both owing to you and ourselves to record, in this public manner, our entire and unqualified approbation of your whole management, while it has been under your direction."[52]

It says much for Dease that this level of efficiency was not achieved at the expense of his relations with his subordinates. In April 1832, Charles Ross, who had previously been at the extremely remote Connolly's Post, wrote to James Hargrave at York Factory:

To my great delight I quitted that dreary solitude in October last, and came down to Stuart's Lake, the Emporium of these parts. Here I passed the greater portion of the Winter in the enjoyment of more cheerfulness, and contentment, than I had known for many years. Our new B-s- [Bourgeois] Mr. Dease, I found a most amiable, warm-hearted, sociable man – quite free from that haughtiness & reserve, which often characterises those who have little else to recommend them – and if Fortune always favors the Good, he should enjoy an unusual share of her smiles.[53]

As reported by another of the gentlemen of the New Caledonia District, John McLean:

Mr. Dease, superintendent of the district, played remarkably well on the violin and flute, some of us "wee bodies" could also do something in that way, and our musical soirées, if not in melody, could at least compete in noise, numbers taken into account, with any association of the kind in the British dominions.[54]

Dease also gave dinner parties at which bear, beaver, and marmot were served, and participated in games of backgammon, chess, and whist.[55]

---

51 Letter from George Simpson to Peter Dease, 3 July 1834, HBCA D.4/20, 19–21.

52 Letter from George Simpson to Peter Dease, 1 June 1835, HBCA D.4/21, pp. 110-11.

53 Glazebrook, *The Hargrave Correspondence, 1821-1843*, 92.

54 Wallace, ed., *John McLean's Notes of a Twenty-Five Years' Service in the Hudson's Bay Territory*, 147.

55 Sampson, "Dease, Peter Warren."

The high regard in which George Simpson held Dease is clearly reflected in his entry concerning Dease in his Character Book, compiled in 1832; few individuals earned a more laudatory assessment in that remarkable document:

About 45 years of Age. Very steady in business, an excellent Indian trader, speaks several of the Languages well and is a man of very correct conduct and character. Strong, vigorous and capable of going through a great deal of Severe Service but rather indolent, wanting in ambition to distinguish himself in any measure out of the usual course, inactive until aroused to exertion and over easy and indulgent to his people which frequently occasions a laxity of discipline, but when his temper gets ruffled he becomes furiously violent. His judgement is sound, his manners are more pleasing and easy than those of many of his colleagues, and altho' not calculated to make a shining figure, may be considered a very respectable member of the concern.[56]

In March 1834 Dease wrote to George Simpson to request a furlough in 1835 because of persistent poor health. This was probably deteriorating eyesight, a problem which was to plague him for the remainder of his life.[57] Simpson granted the request, stipulating that Dease travel overland to Red River in the spring of 1835 to attend a Council meeting there prior to starting his leave.[58] Circumstances did not allow Dease to attend the Council meeting but he did reach Norway House on 9 July and left again for Red River on 1 August 1835.[59] From there he proceeded east to Montreal. On receipt of George Simpson's instructions, appointing him to command of the new expedition to explore the arctic coast, he travelled west to Norway House in the spring of 1836, and started his preparations.

## *Thomas Simpson*

Thomas Simpson was born on 2 July 1808, son of Alexander Simpson, the schoolmaster at Dingwall in Ross-shire,[60] and Mary Simpson. Since his father died in 1822 (when Thomas was only thirteen), his teenage years were spent in straitened circumstances. As a child his health was delicate, even sickly,[61] and he was of a mild, docile temperament, but he clearly had a sharp mind. In the autumn of 1824 he was enrolled at King's College, Aberdeen, at the early age of sixteen. Here he studied Greek, Latin, Mathematics, Natural History, Chemistry, Natural Philosophy, Moral Philosophy, and Logic.[62] He attained the degree of Master of Arts in 1828 and also achieved the Huttonian Prize for Mental Philosophy.

---

56 HBCA A.34/2; Williams, ed., *Hudson's Bay Miscellany, 1670–1870*, 184.

57 Letter from Peter Dease to John Stuart, 3 July 1838, Warren Baker collection; see p. 213.

58 Letter from George Simpson to Peter Dease, 18 January 1838, HBCA D.4/20, p. 15.

59 Norway House journal, 1835–36, HBCA B.154/1/26.

60 Alexander Simpson, *Life and Travels*, 5; Harper, *An Unsolved Mystery*, 1; Stone, "Thomas Simpson, Arctic Explorer," 2.

61 Simpson, *Narrative*, viii.

62 Stone, "Thomas Simpson," 2.

Student life had not been restricted to studying, however. Thomas Simpson was a keen member of the Debating Society and in the summer vacations he and his brother Alexander would go on extensive hiking trips, and in the autumn would visit an uncle in Lochbroom (probably George Simpson's father) to indulge in shooting and fishing.[63]

Well before Thomas's birth, his parents had adopted and brought up George Simpson, the illegitimate son of Mary's brother; George had gone south to London around 1800, and by March 1821 had risen to become governor of the Northern Department of the Hudson's Bay Company. While on furlough in 1825 he visited his aunt (and foster-mother) in Dingwall and was so impressed with young Thomas (back from his first year at university) that he tried to persuade him to abandon his studies and to go to North America with him as his private secretary. At this point Thomas declined, but when the invitation was renewed in 1828 (after his graduation) he accepted.

Thomas Simpson reached Lachine (near Montreal), the Company's North American headquarters, on 25 April 1829, and on the 30th started west by the normal canoe route. On 18 June he joined his cousin at Norway House and assumed his duties as George Simpson's secretary at a salary of £40 per year. Thereafter he accompanied the governor on a tour of the Southern Department of Rupert's Land, ending back at Lachine. In April 1830 he was given the task of conducting a party of some one hundred "voyageurs and servants" bound for various posts in the west, as far as Fort William; this he achieved with a much lower rate of desertion than usual. From Fort William he accompanied Chief Factor Christie, the governor and the latter's new bride to Lake Winnipeg and York Factory, arriving there on 27 June 1830.

He remained at York Factory for the first half of the winter but then, on 10 February 1831, on orders from George Simpson, started south for Red River, travelling on snowshoes and accompanied by two dog sledges. He accomplished the whole trip (some 1100 kilometres) in twenty-two days of travelling, with stops totalling six days at Oxford House, Norway House, and Berens River. Simpson reached Red River on 10 March. This trip clearly demonstrates that Simpson, who had no previous experience of extensive snowshoe travel, was fit, active, and energetic.

Thomas Simpson spent the following five years at Red River, with brief trips to either York Factory or Norway House each summer to attend the annual meeting of the Northern Council, where he acted as the governor's secretary. At Red River he was kept busy with the governor's correspondence and with the minutiae of the Company's business, but from his letters to his brother[64] and to close friends such as Donald Ross at Norway House and James Hargrave at York Factory, it is clear that he greatly enjoyed the social life of the Red River settlement.

A brief comparison of the expedition's two co-leaders is quite revealing. In the summer of 1836 Peter Dease was forty-eight years of age; Thomas Simpson was twenty-eight. Dease had probably acquired a basic education in Montreal, but had

---

63 Harper, *Unsolved Mystery*, 20.

64 Alexander Simpson, *Life and Travels*.

started work (with the XY Company) at the age of thirteen; Thomas Simpson had graduated with an MA degree with first-class honours from the University of Aberdeen. Peter Dease had worked in the fur trade since 1801 and had been working and travelling throughout the *pays d'en haut* since 1804, for thirty-two years. Simpson had worked in the fur trade, in the rather limited capacity of Governor Simpson's secretary, since 1829, only seven years.

Dease's travels (or those which can be documented, which represent only a fraction) extended over most of the map of present western and northwestern Canada. While in charge of Fort Good Hope, he made regular annual trips to Fort Simpson, and even to Portage la Loche. For the winters of 1825–27 he wintered with Franklin's second expedition at Fort Franklin on Great Bear Lake. And while in charge of New Caledonia (1830–35), based at Fort St James, he made the strenuous annual trip by canoe, horseback, and bateau south to Fort Vancouver and back.

By contrast, Thomas Simpson's travels had been limited: his initial trip west in 1829; the tour of the Southern Department with Governor Simpson in the fall of that year; the return trip from Lachine to York Factory; and regular annual trips with the governor to the Council meetings at Norway House or York Factory. In short, compared to Dease, Thomas Simpson was a novice at wilderness travel.

Dease had vast experience at running remote fur-trading posts, and of managing an expeditionary winter quarters. Most important of all, he was very experienced with Indians and Métis, and was known to be a sociable and companionable "bourgeois."

On the other hand, as events will show, Thomas Simpson was extremely ambitious and unboundedly egotistical, and, worst of all, given the tasks which he had undertaken, he had a very low opinion of Indians and especially Métis. Indeed his feelings towards the latter group (which encompassed most of the voyageurs employed by the expedition) were akin to sheer hatred.

Some samples of his unguarded remarks in private correspondence will demonstrate this. Writing to Donald Ross from Fort Norman, 8 September 1837: "The D—l take the natives, for my part I owe them hatred and not pity."[65]

Also to Donald Ross, 23 April 1837, from Fort Confidence: "The smallpox that has been raging in the plains has perhaps thinned the Upper Settlement [the Métis area of Red River], which would be no great loss to humanity."[66]

In the following spring, also to Donald Ross, in response to news of fears of severe flooding on the Red River: "That infernal Colony, too, seems to be going to Old Harry with rapid strides. If those in whom we are interested were removed, I should sorrow very little to hear that the expected deluge whelmed all the rest this spring."[67] And the following January, also to Donald Ross: "but frankly confess that I have not the least sympathy with the depraved and worthless half breed population. Unless the Company maintains a tight little garrison at the Forts (say a score of determined

65 Letter from Thomas Simpson to Donald Ross, Norway House, 8 September 1837, BCA ADD.MSS 635.
66 Letter from Thomas Simpson to Donald Ross, Norway House, 23 April 1837, BCA ADD.MSS 635.
67 Letter from Thomas Simpson to Donald Ross, 13 May 1838, BCA ADD.MSS 635.

fellow-Europeans, for what is the use of stone walls without people to defend them) – there will, I much fear, be *another* bloody reckoning ere long."[68]

Simpson's intense dislike of the Métis was undoubtedly exacerbated by an incident that happened around Christmas 1834. A Métis tripman, Antoine Larocque, who had engaged to work for the Company the following season, went to Upper Fort Garry to receive a second advance instalment on his wages for the coming summer, a fairly standard arrangement.[69] Simpson apparently refused the request and used insulting language to the tripman, who responded in kind. Simpson then hit him on the head with a poker and drew blood. A mob of Métis soon gathered, asking that Simpson be handed over to them, or that he be publicly flogged, and threatening that otherwise they would destroy the fort. The reaction of the Company officers was to close the gates of the fort. In an attempt to defuse the situation, Governor Alexander Christie asked Father Belcourt, who was popular with the Métis, to intercede. After lengthy negotiations Belcourt persuaded the Métis to disperse on condition that Larocque be paid without making the trip the following summer, and that ten gallons of rum and tobacco be distributed to the incensed Métis.

Subsequently, at the prompting of George Simpson, the clergy and the élite of the Red River Colony who made up the Council of Assiniboia formed a volunteer militia. It was commanded by Alexander Ross, who had under his command a sergeant-major, four sergeants, and fifty-four privates.[70] The formation of this paramilitary organization cannot have helped to defuse the tensions in the settlement, or to endear Thomas Simpson to the Métis. It seems very probable that the ill feeling harboured by the Métis towards Simpson contributed to his death soon after his return from the Arctic.

## *Peter Dease's Journals*

During most of the expedition, Peter Dease kept a daily journal in a series of leather-bound notebooks, three of which have survived and are in the possession of Mr Warren Baker of Montreal; these three notebooks fit into a moose-hide pouch. The entries in the first notebook start on 20 July 1836 at Norway House and (covering his trip to Fort Chipewyan, which he reached on 28 September) continue on a daily basis until 4 October 1836. Most of the remainder of the notebook is devoted to an English/Inuktitut vocabulary, which he had copied from that included in Parry's narrative of his second expedition, to Winter Island and Igloolik,[71] and to which Dease had made some minor additions, in pencil.

The entries in the second surviving notebook begin on 14 July 1837 at Point Kay (west of the Mackenzie Delta); hence there is a gap of nine and a half months, and

68 Letter from Thomas Simpson to Donald Ross, Norway House, 30 January 1839, BCA ADD.MSS 635.

69 For details of this incident see Ens, *Homeland to Hinterland*, 54; Morton, *A History of the Canadian West to 1870–71*, 683.

70 Pannekoek, *A Snug Little Flock*, 91.

71 Parry, *Journal of a Second Voyage*, 559–71.

we have no information from this source on the wintering at Fort Chipewyan, or of the voyage down the Slave and Mackenzie rivers in the spring of 1837. Conceivably this lacuna was covered in a notebook which has not survived. After 14 July the daily entries continue until 8 March 1838 and thus cover the entire trip west to Point Barrow and back, and the bulk of the first wintering at Fort Confidence.

There then follows a further substantial gap (possibly covered in another notebook which has not survived) to the first entry in the third notebook, which is for 15 June 1839. Thus there is no coverage (from this source) for the remainder of the first wintering at Fort Confidence, for the whole of the first trip eastwards to the east end of Kent Peninsula and back to Great Bear Lake in the summer of 1838, or for the second wintering at Fort Confidence. The third notebook starts with the expedition's departure from Fort Confidence on the second trip eastwards, and the daily entries continue until 14 October 1839 when it reached Fort Simpson, homeward bound. It covers the whole of the second eastward trip to Boothia Peninsula in the summer of 1839, the return trip to Fort Confidence, and the trip up the Mackenzie in the fall of 1839.

On the basis of these field notebooks, at some date after his return from the expedition, Dease wrote out a fair copy of his journal, which is now in the McCord Museum, McGill University, Montreal, having been acquired by David Ross McCord in 1910. The journal was inherited by Peter Dease's daughter Amelia, who married Thomas Lee.[72] The latter lent it to Dr Sir William Hingston, but promised it to David McCord. After the doctor's death, David McCord contacted his widow, Lady Hingston, who gladly relinquished the journal to him. This is the route whereby it reached the McCord Museum.

In the fair copy of the journal the entries also start on 20 July 1836 at Norway House and continue on a daily basis until 3 October 1836 at Fort Chipewyan (where Dease and his party arrived on 28 September). From then until 1 June (i.e. for the period of the wintering at Fort Chipewyan) the entries are quite sporadic, usually totalling only three or four per month. But from 1 June, when the party set off down the Slave River, bound for the arctic coast, the entries are again daily. Thus these daily entries cover the entire progress down the Mackenzie, a progress which is not covered in the surviving field notebooks, which would suggest that a notebook is missing. The daily entries continue for the entire trip west to Point Barrow and back east to Fort Confidence, the first wintering at Fort Confidence (1837–8), the trip east to Kent Peninsula in the summer of 1838, and the start of the second wintering at Fort Confidence (where the party returned on 14 September 1838). Thus the first trip eastwards (not covered in the surviving field notebooks) is fully covered by daily entries in the fair journal, which would strongly suggest that a second field notebook is missing.

Coverage of the second wintering starts off with daily entries until 20 September, but thereafter the entries again become spasmodic, at only seven or eight per month; they end completely on 31 December 1838. The daily entries resume with

72 Note by David McCord, 2 July 1910, MCM M2714, bound in the journal.

the expedition's departure on their second trip eastwards on 15 June 1839, and continue until they reach Fort Simpson on the homeward journey on 14 October 1839. Thus the coverage of the entire trip east and the trip across Great Bear Lake and back up the Mackenzie River is complete.

## *George Simpson's Instructions*

On July 1836, at the close of the meeting of the Northern Council, George Simpson composed a detailed set of instructions for Dease and Simpson (who were both present).[73] In 1837 they were to proceed down the Mackenzie River to Fort Norman; from there they were to send a small party to build winter quarters on the northeast shores of Great Bear Lake, while they themselves continued north to the sea and explored the coast west to Point Barrow. Returning to Great Bear Lake, they were to winter there. Then in the spring of 1838 they were to portage their boats across to the Coppermine River, descend that river to the sea, and explore the coast eastwards from Franklin's Point Turnagain to the mouth of the Back River, returning to Great Bear Lake. If necessary they should be prepared to take a further season if ice conditions or other circumstances dictated this. Dease was to proceed to Fort Chipewyan or Fort Resolution immediately, to superintend the building of the necessary boats and the accumulation of supplies over the winter; Thomas Simpson was to spend the fall and early part of the winter at Red River, acquiring enough knowledge of navigation and surveying to compile a map of the coasts they were to explore.

What neither the Committee in London nor Governor George Simpson realized initially was that a rival expedition was being planned by the Admiralty. Even as the Council of the Northern Department was gathering at Norway House, on 11 June 1836 John Barrow, second secretary to the Admiralty, was signing the orders for this expedition, addressed to Captain George Back.[74] He was to take HMS *Terror* into Hudson Bay, heading for either Wager or Repulse Bay. From there he was to transport boats overland to the south end of Prince Regent Inlet, i.e. Committee Bay. From there, using the boats, he was first to explore the coast north to Fury and Hecla Strait, and, if possible, on northwards to the west coast of Cockburn Island, i.e. the northwest coast of Baffin Island. Secondly he was to explore the mainland coast west to the mouth of the Back River and on westwards to Franklin's farthest at Point Turnagain. Quite unrealistically, Back was ordered to attempt this very ambitious project in one year: *Terror* was expected to return to Britain in the fall of 1836.

As events transpired, Dease and Simpson had little to worry about in terms of rivals on their "turf." Back attempted to reach Repulse Bay via Frozen Strait, around the north end of Southampton Island, as Parry had done successfully in 1821, but

---

73 Letter from George Simpson to Peter Dease and Thomas Simpson, 2 July 1836, HBCA D.4/22, pp. 78–83; Simpson, *Narrative*, 2–9; Alexander Simpson, *Life and Travels*, 184–91.

74 Back, *Narrative of an expedition in H.M.S. Terror, undertaken with a view to geographical discovery on the Arctic shores, in the years 1836-37.*

*Terror* became beset off the northeast coast of the island. Throughout the winter she drifted southeast through Foxe Channel and into the western end of Hudson Strait where she was released by the ice breakup in July 1837. Her hull was badly strained and she was making water at an alarming rate, but Back managed to nurse his ship back across the Atlantic to Lough Swilly in Ireland.

## DOCUMENTS

### *1. Extract from letter from the Governor and Committee to George Simpson, 9 March 1836*[75]

5 ... We understand, several Gentlemen in this country have recently proposed themselves or been proposed as leaders of expeditions, both by sea and land, for completing the survey of the Northern Shores of America. Nothing, however has yet been determined on, and the season is now so far advanced, that no Expedition by Canada is likely to be attempted in the course of next Summer. This subject continues to excite much interest in the public mind, and it is more than probable the Government may determine on fitting out another Expedition to go inland from York next autumn, or via Canada the following Summer. In the event of Government taking it up, we shall afford the Expedition our usual assistance and support, but otherwise we are desirous that it should be undertaken by the Fur Trade, and from the habits of the Gentlemen employed in the service, their knowledge of the Indian character and their being more or less accustomed to the privations and difficulties usually met with in travelling thro' the Wilds we should consider many of them well qualified for conducting such an expedition. You will therefore take measures to assist, or to cooperate with any Expedition that may be fitted out by Government, or an expedition to be outfitted by the Company, should the Government decline meddling with it, with a view to trace the unexplored Shores of the Polar Sea, either East or West, as may hereafter be decided on.

• • •

### *2. Thomas Simpson's plan for the expedition, version 1*[76]

Engage, at Norway House, ten good men, eight of whom to accompany the expedition to the sea, and two to be employed with a party of Indians, during the summer of 1837, in collecting provisions at the point of rendezvous, for the retreat of the expedition, and for the fall fishing at Fort Reliance.[77] With these ten men, and the requisite outfit (as per list), the expedition should start as soon as possible from Norway House in two canoes, and overtake the *Portage-la-Loche* boats and get them

---

75 HBCA A/6/24, ff. 24–24d; D.54 ff. 160–7.

76 Alexander Simpson, *Life and Travels*, 178–80.

77 The winter quarters occupied by Captain George Back for the winters of 1833–34 and 1834–35, at the extreme eastern end of Great Slave Lake.

to embark five bags of pemican each (say twenty bags) at Ile a la Crosse, to be brought on by the Mackenzie River craft to Fort Resolution, and thence by one of the canoes despatched this fall from Fort Reliance. The canoes to push on to establish Fort Reliance, and to procure the co-operation of a party of Indian hunters there, to be employed in collecting provisions during the fall and winter, to aid in transporting the boats, one canoe, and stores, to the head of Great Fish River in the spring, to accompany the expedition down that river to Lake Beechey, there to make a deposit of provisions, and, if they can be so induced, to cross over and meet the expedition on their return at the mouth of Back's River.[78]

If, for fear of the Esquimaux, the Indians could not be prevailed on to meet the expedition at the coast, they must remain at Lake Beechey, at a point to be fixed upon, till the return of the expedition from the sea, and accompany it back to Fort Reliance.

This part of the journey to be effected by means of the canoe already noticed, which would be left at Beechey on the descent. These arrangements having been made with the Indians, the party to proceed in their boat with the utmost despatch down Great Fish River to Back's farthest point, where the survey must be renewed. If the expedition find Ross' Strait as supposed, and can reach Back's River by boat the same season, they must proceed on foot, provided with hides to make temporary canoes if necessary, to the appointed rendezvous at Lake Beechey. If they find a peninsula where Ross' Strait is supposed to exist, they must either, after examining this peninsula, return as they went; or, being provided with an Esquimaux interpreter (forwarded from Churchill by the winter express), if they fall in with that tribe, prosecute their journey during the winter on foot to Back's River, where the same party of Indians who have wintered at Fort Reliance should be appointed to meet them with the requisite provisions for their return early in the spring of 1838.

By the proposed project, it is hoped that that portion of the coast which is most attainable from the interior, and whose delineation is most interesting to science, may be traced in the shortest period and at the least expense.

• • •

### *3. Thomas Simpson's plan for the expedition, version 2*[79]

Engage at Norway House twelve good men; start from thence with the Athabasca brigade, with one additional boat manned by eight of the above hands, and the remaining four distributed among the four Athabasca boats. Take on the requisite outfit from Norway House, any deficiencies in which to be indented for and brought up from York Factory this season, and forwarded next spring by the *Portage-la-loche* brigade. Embark this season, either at Cumberland House or Ile a la Crosse, thirty bags of pemican, and proceed to winter in the Athabasca district. There make the necessary arrangements with the Dogrib and Copper Indians to secure their co-

78 Here Simpson is referring *not* to the present Back River, but to what is now the Western River, flowing into the head of Bathurst Inlet. Named after Lieutenant George Back, a member of his expedition, by Franklin on 5 August 1821 (Franklin, *Narrative of a Journey*, 1823, 377).

79 Alexander Simpson, *Life and Travels*, 181–3.

operation with an establishment to be formed next summer at Dease's River, Great Bear Lake.

At the opening of the navigation in 1837, proceed to Great Bear Lake River in two boats, or one boat and a canoe, manned by the twelve people belonging to the expedition, and two halfbreeds or Indians.

From thence despatch the smaller boat or canoe, with four of the men and the two half-breeds or Indians, to Dease's River, to erect the establishment and make the fishery. The officers and the remaining eight men to proceed to the coast, and complete the survey westward, from Point Beechey to Point Barrow; thence return to winter quarters at Dease's River. The additional outfit to be forwarded, as already noticed, by the portage brigade in the summer of 1837, having been brought to Fort Norman, or to the entrance to Great Bear Lake River, it would be conveyed from thence to winter quarters by the expedition on its return from the sea; likewise any provisions that could be procured from Athabasca or Mackenzie River.

As early as practicable in the spring of 1838, transport the boat intended for the eastern survey (which should be built at Dease's River during the winter) and the necessary stores across to the Coppermine River, and there embark with eight men as before, leaving the remaining four and the Indians to make provision for the ensuing winter.

The expedition then to proceed direct to point Turnagain, and continue the survey from thence to the eastward, so as to connect the discoveries of Captains Ross and Back, if practicable; this effected, or when it becomes unsafe to remain longer on the coast, the party to return to the Coppermine, and make the best of their way back to Dease's River; to winter there, or at Fort Simpson, if the season will permit; and the whole expedition to come out to Norway House in the fall of 1839.

Thomas Simpson

• • •

### *4. Minutes of Council, Northern Department, Norway House, June 1836, Resolutions 79, 80* [80]

78. The delineation of the unexplored parts of the Arctic Coast westward of Mackenzie's River & Eastward of Point Turnagain being an object that has for a long time excited the most lively interest in the public mind and has called forth the energies of many enterprizing and scientific men whose exertions have nevertheless been unsuccessful it is Resolved (79th) That an expedition be fitted out for that purpose by the Honble Hudson's Bay Company to consist of Chief Factor Peter W. Dease & Mr. Thomas Simpson with a party of 12 men and they be provided with such craft and such Provisions or other supplies as may be required to accomplish that object agreeably to the plan submitted by these gentlemen to the Council.

80th. That the Gentlemen in charge of the Districts and posts of Athabasca and Mackenzie's River comply with any demands whatsoever connected with their

80 HBCA B.239/k/2, pp. 140–1

respective Charges that may be made by Chief Factor Dease, or in his absence, by Mr. Thos. Simpson.

• • •

## *5. George Simpson's instructions to Peter Dease and Thomas Simpson, 2 July 1836* [81]

Norway House, 2nd July, 1836

Messrs. P.W. Dease & Thos. Simpson

Gentlemen,

1. By the 79 & 80th Resolves of Council of this Season, copies of which are annexed, you will observe that we have determined on fitting out an Expedition forthwith, for the purpose of endeavouring to complete the discovery and Survey of the Northern shores of this continent.

2. This object has for a great length of time excited the most lively interest in the public mind, and has baffled the exertions of many enterprising men, among whom the names of Parry, Franklin, Ross & Back have of late years appeared conspicuous, but I trust that the honour of its accomplishment is reserved for the Hudson's Bay Company, through your exertions, and in Selecting you for so important a Mission, we give the best proof of the high opinion we entertain of your abilities and qualifications for such an undertaking.

3. The expedition consisting of twelve men, is now placed under your direction, and you will be pleased to conduct it without delay to the Athabasca country, and to pass the ensuing Winter at Fort Chipewyan, or Great Slave Lake,[82] as you may consider expedient, altho', in my opinion, Great Slave Lake would be the preferable wintering Ground, in many respects, as regards the object of the expedition.

4. At the opening of the navigation in June, you will proceed by boat down McKenzie's River to Fort Norman, and there leave 4 men, with directions that they proceed from thence to the North East end of Great Bear Lake, and there erect buildings, establish fisheries, and collect provisions for the accommodating and maintenance of the party during the winter of 1837/8.

5. You will then go down to the Sea with the remaining 8 men, and endeavour to trace the coast to the westward, to Longitude 156° 21' (N Lat. 71° 23' 39") whence Captn. Beechy's barge returned.[83] Should your progress along the coast be obstructed by Ice or fog, as Sir John Franklin's was, you will either put the boat in a place of Security, and proceed on foot with all your party or leave 4 men with the boat for its protection, while you go along shore carrying a sufficient quantity of provisions with you for the journey. It is desirable to take observations as frequently, and to survey the Coast as accurately as possible, without however losing time on your outward Journey in waiting for the appearance of the Sun, moon or stars, which are

81 HBCA D.4/22, pp. 79–83.

82 Here Simpson means Fort Resolution, at the mouth of the Slave River.

83 See p. 4.

frequently obscured by the dense fogs that prevail so much on that Coast, but devoting as much time to these objects as the season and the state of your provisions will admit on your return.

6. At the most westerly point you may reach, you will erect in a conspicuous situation a pillar or mound, and leave deposited in the Earth at its base, a bottle hermetically Sealed, containing an outline of the leading circumstances connected with the voyage.

7. In suggesting that the Boat should be left, in the event of your progress being obstructed by Ice or fog, I beg it to be understood that that ought not to be done, if there be the least probability that by perseverance you may succeed in getting her along shore, as the preservation of the boat I consider it to be highly essential, both to the accomplishment of the voyage and to the protection of the party, but if there be no possibility of getting on with the boat, I beg to recommend that you provide yourselves with axes and cordage to make rafts for crossing rivers, and some parchment sheeting, and oilcloth, to make a couple of Small Canoes for the conveyance of the party should it be found impossible to cross the rivers on rafts, and in order to Secure your retreat in the event of the loss of the Boat.

8. Should you not be able to accomplish the voyage or Journey during the season of open water and that you fall in with friendly Esquimaux or Indians, as many of the party as can be maintained may remain with them, so as to complete the Survey in the course of the winter, or Spring; in this however you will exercise your own discretion, and be guided by circumstances.

9. It is Exceedingly desirable however, that you should return by open water, so as to pass the winter at the establishment to be formed at the North east end of Great Bear Lake, in order to make the necessary preparations for another voyage of discovery to the eastward, at the opening of the navigation, Summer 1838.

10. The object of that voyage is to trace the Coast from Franklin's "Point Turnagain" eastward to the entrance of Back's Great Fish River. To that end you will haul your boat across from the North eastern Extremity of Great Bear Lake to the Coppermine River before the winter breaks up, and at the opening of the navigation proceed to the Sea, and make as accurate a survey of the coast as possible, touching at "Point Turnagain" and proceeding to Back's Great Fish River, if the Strait or passage exists which that officer represents as separating the mainland from Ross's Boothia Felix; but should it turn out on examination that no such strait exists, and that Captn. Ross is correct in his statement that it is a Peninsula, not an Island, you will in that case leave your boat and cross the isthmus on foot, taking with you materials for building two Small Canoes by which you may follow the Coast to Point Richardson, Point Maconochie, or some other given spot that can be ascertained as having been reached by Captn. Back. And you will be regulated in determining whether you will return by Great Fish River or by the Coast, by the period of the Season, at which you may arrive there; the state of the navigation and other circumstances.

11. In order to guard against privation in the event of your returning by Great Fish River, it will be adviseable to make arrangements at Great Slave Lake that a supply

of provisions with ammunition and fishing Tackle, likewise babiche[84] for Snow Shoe lacing, be deposited at Lake Beachy or some other point of that route.

12. Should you be unable to complete the voyage to the eastward from Coppermine River in one Season, you may, as suggested in reference to the other voyage, take up your quarters with the Esquimaux for the Winter so as to accomplish it the following season.

13. In making your arrangements for both voyages, I have to recommend that a considerable quantity of pemmican and flour not less than 100 pieces[85] be provided for voyaging provisions and that you be well supplied with materials for constructing small canoes, leather for shoes, and Snow Shoe netting, likewise with ammunition, axes, crooked knives, fishing hooks, nett thread, backing and setting lines and with warm clothing for yourselves and the people.

14. The necessary astronomical & surveying instruments are provided to enable you to take observations and to make surveys, in which you will be as accurate as possible, and you will be pleased to prepare a full and particular journal or narrative of the voyage, likewise a chart of the Coast, and to take formal possession of the country on behalf of Great Britain in your own names, acting for the Honble Hudson's Bay Company, at every part of the coast you may touch, giving names to the different headlands, mountains, rivers and other remarkable objects you may discover. It is also desirable that you make a collection of minerals, plants, or any specimens of natural history you may fall in with, that appear to be new, curious, or interesting.

15. You are hereby authorised to avail yourselves for the use of the expedition of any assistance whatsoever you may require, at any of the Honble. Company's Establishments you may touch at or have communication with, either by letter or otherwise, and the Gentlemen in charge of those Establishments are hereby instructed to meet all demands you may make upon them.

16. In the event of any accident occurring to prevent either of you from proceeding on this mission, the other will be pleased to follow up the object of it, and to avail himself of the assistance, as a second in command, of any Clerk of the Company he may find within his reach, and such Clerk will be pleased to act in that capacity accordingly.

With fervent prayers for your safety & success,
I remain,
Gentlemen,
Your most Obedt. Servt.
(Signed) Geo. Simpson

• • •

84 Lacing cut from dressed caribou, moose, or elk hide.

85 The *pièce*, weighing ninety pounds, was the standard unit (bale or package) of the fur trade.

### *6. Extract from letter from George Simpson to James Hargrave, York Factory, 20 June 1836*[86]

Norway House
20th June 1836

Dear Sir,

… The services of Mr. Thomas Simpson being required to accompany a discovery expedition to the Polar Sea, renders it necessary to make a change in the counting house Departments of York and Red River. Mr. C.T. Gladman is therefore appointed to the Office of accountant at York in the room of Mr. Ballenden, and the latter Gentleman succeeds Mr. Thomas Simpson at Red River Settlement …

… With reference to the Arctic Discovery Expedition a requisition will be forwarded to you by Mr. Thomas Simpson for Goods to be sent forward by the Athabasca and McKenzie River brigades of next year to which I request the favor of your particular attention …

George Simpson

• • •

### *7. Extract from letter from George Simpson to Alexander Christie, Red River, 20 June 1836*[87]

Norway House 20th June 1836

Dear Sir,

… Flett joins the Polar Discovery expedition and is succeeded by Cromartie in command of one of the Vessels …

We have engaged McKay, Sinclair, Taylor & Flett for the expedition and I think Ritch will likewise agree. It is intended that the party should pass the Winter either at Athabasca or Great Slave Lake, proceed down McKenzie's River next Season and explore the Coast Westwards to Point Beechy, return and winter at Great Bear Lake, cross over to the Copper mine River in Spring 1838 and proceed to the Eastward. As the party winter in Athabasca Department, the pleasure of both the Gentlemen will not immediately be required here. Mr. Thos. Simpson will therefore return to Red River soon after my departure hence, in order to practise the use of the Instruments with Mr. Taylor, and proceed to Athabasca only in the winter …

Geo. Simpson

• • •

---

86 HBCA D.4/22, pp. 30–2.
87 HBCA D.4/22, pp. 32–3.

*8. Extract from letter from George Simpson to Murdoch McPherson, Fort Simpson, 30 June 1836*[88]

Norway House 30th June 1836

Dear Sir,

... Chief Factor Dease and Mr. Thomas Simpson are appointed to conduct an expedition for the purpose of exploring the Coast from the mouth of McKenzie's River round to Beechy's Point Franklin, during the summer of 1837, and of tracing the coast from the mouth of the Coppermine eastwards to the mouth of Back's Great Fish River during the Summer of 1838 as you will observe by the annexed copy of the 79 Resolution of Council, and we have to beg you will afford them every assistance and support in the object of their missions and meet any demands they may make upon you ...

• • •

*9. Extract from letter from George Simpson to James Hargrave, York Factory, 4 July 1836*[89]

Norway House 4th July 1836

Dear Sir,

... By the 79th Resolution of Council you will observe that we are about to fit out an expedition under the command of Mr. Dease and Mr. Thomas Simpson for the purpose of completing the discovery and Survey of the shores of the Polar Sea. A requisition for the use of that Expedition has been sent to you, which you will no doubt comply with, in as far as your means admit, and should they require your assistance in any other way connected with this object, I beg it may be rendered ...

• • •

88 HBCA D.4/22, p. 74.

89 HBCA D.4/22, pp. 88–92.

# Chapter Two
# Norway House to Fort Chipewyan

## INTRODUCTION

Peter Dease set off from Norway House by York boat, manned by nine men, on the morning of 21 July; he was accompanied by Chief Factor Edward Smith, in charge of Athabasca District, and Postmaster Adam McBeath, and by the boats with the outfits bound for Portage La Loche and beyond. Having crossed the north end of Lake Winnipeg, they had reached Grand Rapids, where the Saskatchewan River debouches into Lake Winnipeg, by the evening of the 26th. They reached The Pas on the morning of 3 August and Cumberland House on the morning of the 6th. Getting under way again that afternoon, they followed the standard fur-trade route northwards up the Sturgeon Weir River; owing to the steady succession of rapids and portages, progress up this stream was usually quite slow, but the party was further delayed when a boat was stove, a man almost drowned, and some considerable time spent on repairs to the boat and on drying the freight. It was thus the 16th before they reached Frog Portage, the very low portage across the divide to the Churchill (English) River. Then followed the long, slow succession of portages and rapids of the Churchill River, but the party made the post at Ile-à-la-Crosse safely on 1 September. Here they stocked up with pemmican, then pushed on again on the afternoon of the 2nd.

They reached the south end of Methye Portage (Portage La Loche), which crosses the divide to the Clearwater River, and thus the Mackenzie Basin, on the evening of 13 September. With the help of some Indians who were camped here, the long, tedious task of carrying the goods across the long portage and down the precipitous drop to the Clearwater began. The York boats from Norway House were left at the south end of the portage, and a new set picked up on the Clearwater. Finally, on the afternoon of the 21st, the party set off down the Clearwater and Athabasca rivers. Travelling with the current, they made excellent progress, reaching Fort Chipewyan on the north shore of the west end of Lake Athabasca on the afternoon of the 28th. The entire trip had thus taken two months and one week.

Thomas Simpson left Norway House, accompanied by two men, Peter Taylor and George Sinclair, on 20 July, travelling with the Winnipeg traders Andrew McDermot and James Sinclair in their two York boats. They parted from Dease and his party on the 23rd and headed south up Lake Winnipeg. By 7 August Simpson was settling in at Red River; here he spent the following four months, learning surveying and navigation from Mr Taylor, and dividing his time between Upper Fort Garry near the Forks of the Red and Assiniboine rivers and the New Fort (Lower Fort Garry). By 1 December he had completed his preparations and was ready to start for Fort Chipewyan to join Dease.

## DOCUMENTS

### *1. Peter Dease's journal*[1]

**20 July, 1836.** All the Settlements being completed, to proceed in Company with the Athabasca Boats[2] under C.F. Smith.[3] It was discovered this morning that Anderson, who has been hired as fisherman for the Expedition (in consequence of Linklater being from lameness unable to go), was missing, having apparently absconded. I am therefore under the necessity of leaving this without a fisherman in hopes of getting one among the retiring Servants from McKenzie's River who we will meet on the way. Three of the 4 Athabasca Boats & mine were sent off, the former manned by 7 men & mine with 8,[4] Exclusive of John Ritch[5] the Boat builder who is Exempted from work or Carrying over the portages. Peter Taylor[6] & Geo[rge] Sinclair[7] remain with Mr. Thos. Simpson, who returns to Red River and will leave it in December to Join [me] at Athabasca in Winter.

Mr. C.F. Smith & Self with Mr. A. McBeath P.M.[8] for MK[9] River will join them in the morning. Messrs McDermot[10] & Sinclair[11] for RR[12] in 2 boats for that Place; Mr Simpson & the 2 men forementioned accompany them.

**21 July 1836.** Left Norway House & Joined the Boats at 9 A.M. Proceeded to Play Green Lake, and were obliged by a Strong Wind ahead to put up on Lower Kettle

1 MCM M214.

2 The boats bound for Fort Chipewyan (headquarters of the Athabasca District) with the annual stock of supplies and trade goods.

3 See biography, p. 308.

4 Seven of these eight were K. McKay, Charles Begg, George Flett, P. Morin, Laurent Curtin, François Flamand, and John McKey (FNB A, p. 1).

5 See biography, p. 305.

6 See biography, p. 308.

7 See biography, p. 307.

8 See biography, p. 303.

9 Mackenzie.

10 See biography, p. 303.

11 See biography, p. 308.

12 Red River.

Island,[13] where we were detained the remainder of the day by the violence of the wind & heavy rain which came on.

**22 July, 1836.** The Same violent weather prevented our departure today, but the boats being Safe from the Surf, sustained no Injury.

**23 July 1836.** Weather having Cleared up pushed off at 4.30 A.M. & put ashore at Norwegian Point. After breakfast Separated from our Friends[14] who proceeded on for RR. The wind being thought too Strong to round the Mossy Point, we passed the day here. The 2 vessels for RR came up.

**24 July, 1836, Sunday.** The wind having moderated left Norwegian Point at 4 A.M.; passed the Mossy Banks, and obliged to put ashore & unload from the wind increasing to a Strong Gale at the commencement of Sandy Point.

**25 July 1836.** Pushed off at 3.30 A.M. and got as far as the Islands off MacKintosh Point where we were again Stopped by the wind for the rest of the day.

**26 July 1836.** At 3.30 left the Islands under Sail with a SE wind. Stopped 1 hour for breakfast, and proceeded with an increasing gale to the River[15] which we reached at 4 P.M. The Boats towed up to the foot of the rapid,[16] where we found a Camp of Indians & freemen, from whom we procured some Sturgeon & put up for the night.

**27 July 1836, Wednesday.** The Boats performed 2 trips with half Cargoes up to the Portage and 2 of the 5 were launched across the Portage; the Baggage and other boats up the hill, ready to carry in the morning, all Safe.

**28 July 1836.** The Boats & baggages were all taken over the portage and prepared to Start in the morning.

**29 July 1836.** Embarked at 4.30. Carried at the Red Rock half Cargoes and proceeded to the Islands near Grand Décharge of Cedar Lake or Cross Lake[17] where a violent Shower of Rain, & Strong wind coming on, put up for the night.

**30 July 1836.** Carried 2 trips at Gd. Décharge. Had a Strong head wind, which veered round & became fair after passing the narrows where we took breakfast. Sailed to the Islands, where we were Stopped 4 hours. Pushed off again at 6 PM & got to Rabbit Point under Sail at 8 P.M. Put up for the night.

**31 July 1836.** A Strong wind from Land rather than ahead to 9 A.M. Plyed the oars & got to the River Au Pas at 3 P.M. Passed Muddy Lake & put up at Fir Point at 7.30 there being no good Encampt. for some distance from it.

**1 August 1836.** Left our Encampment at 3.30; little wind. Laboured at the Oars untill 7.30. Put up. Weather cloudy, with appearance to Rain.

**2 August 1836, Tuesday.** Had a Shower of Rain over night. Embarked at 4 A.M.; wind ahead. Got to the middle of the Channel of Isle du Pas at 8 P.M. & put up.

---

13 Also known as Goose Island (FNB A, p. 1).

14 McDermot and Sinclair; Thomas Simpson was travelling in Sinclair's boat (FNB A, p. 1).

15 The Saskatchewan River, draining Cross Lake via Grand Rapids to Lake Winnipeg.

16 Grand Rapids.

17 Between Cross Lake and Cedar Lake. The term *décharge* implies that cargoes had to be unloaded and portaged, but the boats towed up empty. As a result of the dam at Grand Rapids, this set of rapids has been submerged and Cedar Lake and Cross Lake are now continuous.

**3 August 1836.** Started at 3.30. Got to the Pas at 8.40 and left it at 12; wind ahead. Put up at 8 P.M. above the Leaf Channel.

**4 August 1836.** Left our Encampment at usual hour. Stopped 3 hours by head wind & put up at the Channel of the Barrier.

**5 August 1836.** Passed by the main Channel and lost some time among the Shoals of Little Muddy Lake, or just below it. Put up at 8. Strong head wind.

**6 August 1836, Saturday.** Got to Cumberland House at 10 A.M. Had the boats unloaded as 2 of the Atha. boats [18] are to be left here to take new ones which were left here in Spring by C. F. Rowand[19] & require to be pitched & caulked partially, which they began to do.

**7 August 1836.** The Work going on about the Boats. Strong N E Wind.

**8 August 1836.** The boats were not quite finished yet today. All other wants Supplied[20] & ready as soon as that work is done. Strong wind all day.

**9 August 1836.** This morning C.F. Rowand with Mr. B. McKenzie,[21] latter for Columbia Dep. arrived, one boat, the Saskatchewan brigade[22] being sent on by the River. I am sorry to hear that a considerable loss was sustained by one of his Boats having been driven in the night from its moorings in Lake Ounupique[23] & nearly swamped. 12 Pieces were thrown overboard to lighten her; most of her Cargo wet. 2 men were on board at the time, and the boat eventually Saved, by the Sacrifice of the Pieces thrown out. Mr Rowand left this at 3 P.M. & all being ready, we Embarked at 4 and got to the 2nd narrows where we put up for the night.

**10 August 1836.** Got to the Rat Portage in Sturgeon River[24] at 9 A.M. Carried all the Cargoes and proceeded, the water in a fine State. Got above the Shallows of the Barrier, full load, & Slept in the Still water between it & upper Rapids.

**11 August 1836, Thursday.** Got up the Rapids[25] at 4 P.M. into Beaver Lake[26] at 5.30. Sailed across the Lake and reached Portage La Pente at 8. Proceeded across the Portage; Boats & Cargoes remained at lower End.

**12 August 1836.** The boats loaded at 8.30 at the Grand Rapid La Pente. The Guide's (Cadotte's) boat in getting up a Strong rush of the Rapid, Sheered and driving against a Rock below was Stoven in, about 3 feet of one of the Planks. One of the men of the tow line drawn in & nearly drowned.[27] Had to unload, repair & dry the property partially. The other boats were taken up the Rapid Safe.

**13 August 1836.** This morning at 6 A.M. the boat & Cargo being up the Rapid we

---

18 Athabasca boats.

19 See biography, p. 305.

20 A calf was even killed for Dease and his men (FNB A, p. 4).

21 See biography, p. 304.

22 The Saskatchewan brigade, bound for Fort Edmonton, the headquarters of that district, continued west up the Saskatchewan River.

23 Lake Winnipeg.

24 Now Sturgeon Weir River, flowing south from Amisk Lake.

25 Crooked Rapid (FNB A, p. 4).

26 Now Amisk Lake (*amisk* being Cree for *beaver*).

27 This man, Roy, was also quite badly hurt (FNB A, p. 4).

proceeded to Marten Lake where a proper Spot was fixed upon to dry the property thoroughly which was done by 6 P.M., when we proceeded, and got to the Carp Portage, where we put up for the night. Cargoes Carried, but boats remained for the morning. Mr. McBeath took 4 White fish,[28] with a hook & line.

**14 August 1836.** The boats launched & loaded by 6.15 A.M. Passed the Birch Portage, Island P[ortage], and got to the Fir or Pine Portage. Cargoes Carried, but the boats not taken up, being too late to get up the Rapids.

**15 August 1836.** Started at the usual hour, passed Lake Miron[29] & in Pellican Lake met G. Sandison Guide in Charge of the Portage la Loche Boats, with the return of McKs River.[30] Stopped 1 hour & Engaged J. Norquay one of the Retiring Servants[31] as fisherman for the Expedition for 3 years. Placed him in one of the Atha[basca] boats. Passed the Décharge and got to the 3 Little Portages. Cargoes passed over; the last boats below.

**16 August 1836.** Passed Woody Lake & Décharge and got to the Frog Portage.[32] Boats & Cargoes taken across. Saw 2 Indians today who had nothing to give us fresh. Killed 3 Ducks in the Evening.

**17 August 1836, Wednesday.** Passed the Frog Portage Lake, Décharge, and Barrel Portage[33]. Put up about 4 miles below Island Portage.

**18 August 1836.** Portage Island was passed and got to Rapid River Portage where one of the Boats received some damage[34] towing up the Rapid, which was temporarily arranged. Proceeded & put up at 7h.45. Cloudy & Windy weather; had some rain last night.

**19 August 1836.** Got to the Décharge of Rapid River at 6 A.M. One of the men was hurt towing up the Boats & bled. Strong head Wind; Cloudy threatening weather. Had a little Close hauled Sail Wind, in Lake Queue Depouille (or Cloud Berry). Got to the Mountain Portage 2.30 P.M. Strong head wind. Got to Little Rocky Portage; wind too Strong to proceed, and one of the Boats[35] injured in the Stem & Stern Posts & plates. Put up for the night. Killed 2 Partridges[36] and 1 Wood Pigeon.[37] Saw some Cree Indians [who] have no Provisions.[38] Passed at the Establishment of Rapid River & had the boat Repaired there; one man in charge from whom we got a few White fishes. Put up at 8.30.

---

28 Probably lake whitefish (*Coregonus clupeaformis*).

29 Now Mirond Lake.

30 The annual harvest of furs from the Mackenzie River District, bound for York Factory.

31 One of the Company employees whose contract had expired.

32 The low, and hence convenient, portage across the major divide between the Saskatchewan River system to the south and the Churchill River system to the north.

33 Now Keg Portage.

34 The damage was a broken stern post (FNB A, p. 5).

35 Charbonneau's boat (FNB A, p. 6).

36 One of three possible grouse species: most likely either the spruce grouse (*Canachites canadensis*) or the ruffed grouse (*Bonasa umbella*), but possibly also the sharp-tailed grouse (*Pedioecetes phasianellus*).

37 Almost certainly the passenger pigeon (*Ectopistes migratorius*), now extinct.

38 They had no food for trade.

**20 August 1836.** With a Strong head wind got to the Mountain Portage from there to Little Rock of the Mountain where the wind was too violent to proceed.
**21 August 1836.** Had a heavy Shower of Rain in the night & violent wind. About 6 A.M., weather having moderated, pushed off. Got to the 3 little Rapids where the whole cargoes were carried from the Shallowness of the water.[39] Got to the Devils Portage; whole Cargoes Carried from the lower to upper End. All the Boats were got up & 3 of them loaded. Rained this Evening.
**22 August 1836.** Started at 5.15. Passed Grosse Roche rapid which was towed up. The rapids below the Ecors half Cargoes taken up. Put up at Petit Rocher de la Truite. Boats & Cargoes all across.
**23 August 1836.** Started at 5 A.M. Passed Trout Portage & Lake [and] Hallier Portage & put up for the night in Bear Island Lake. Wind ahead all day.
**24 August 1836, Wednesday.** Left our Encampment at 4.35. Passed the Silent Rapid at 1.30 P.M. Had a Chase after a Bear which was killed after Several Shots fired at it. Passed the Canot tourné portage & put up for the night at upper End of Birch portage; Boats remain below yet.
**25 August 1836.** The Boats got up without any accident. Took breakfast at Pin Portage. Passed the Pin,[40] Mouse & Serpent Lakes & put up for the night at the foot of the Serpent Rapids. Strong wind, & Rain in the Evening.
**26 August 1836.** Got above the Serpent Rapids at 11 A.M., safe. Found an Indian[41] with his family who traded a little meat for amunition & Tobacco. The wind ahead so Strong we could not proceed, which continued all day.
**27 August 1836.** Left our encampment at 3.30; wind yet Strong. Sail was tried, but gave us little help. Passed the Hay River. Saw Indians (Chipewyans) at the Entrance of Rein Deer River, and proceeded to 7.35 when we put up for the night in Knee Lake. Some of the Indians came to us in the Evening, but have nothing to give or to get.
**28 August 1836, Sunday.** Put off at 4.15. Stopped 3 hours to allow the men to make oars to replace several that have been broken, and got up the lower & middle Knee Rapids where we Stopped for the night. Got 6 Ducks & 2 Pigeons today.
**29 August 1836.** Got up the last of the Knee rapids by 8.15. Took breakfast & entered Primeau Lake. Saw some Indians who traded a little fresh meat. Had a Close Sail wind[42] part of the day, but turning contrary stopped us 4 hours on an Island which we left at 6.15. Crossed the remainder of the Lake & put up at 8.15; yet Strong wind.
**30 August 1836.** A Strong boisterous head wind with Showers of rain. Got up the Portage la Puise. Proceeded untill 11.30 when the wind was too Strong to get on untill 5 P.M. when we pushed off & put up at 7.30 near Tetreau's Rapid.
**31 August 1836, Wednesday.** Got above Tetreau's rapid at 7.30 and reached the Shawernon portage at 8.30. Some Indians brought a few Pieces meat & 3 fishes which were

39 But the boats could be tracked up empty.
40 Pinehouse Lake.
41 Grosse Tete (FNB A, p. 7)
42 I.e. the York boats were sailing close to the wind.

traded. Killed 15 ducks in course of the day.[43] Got over the Portage and the rapids just above it & put up for the night about half way between them & the last rapids to get to Isle a la Crosse Lake. The morning was Cold, but the day fine and Warm.
**1 September 1836, Thursday.** Had a Sail Wind though rather Close[44] in the Lake, which took us to the Fort of Isle a la Crosse a little after sunsett. Three of the Boats were ordered to Encamp at the Point. Mr Deschambault[45] in Charge; All well. Rained in the Evening.
**2 September 1836.** Got 7 Bags Pemican made up. 18 were made of which 2 were found to be Spoiled & unfit for use. The boats took Each 5 bags Except the Guide's & Mine, which take but 4, having some remaining from Cumd.[46] Left the Fort at 5 P.M. & put up for the night at the Presqu'ile. The wind Strong, ahead.
**3 September 1836.** Had Rain in Course of the night. Strong wind kept us ashore untill 8 A.M. when we put off. A Chipewyan young man[47] follows in a Small Canoe to Join some of his relations, somewhere in Methy River. Put up near the narrows or Entrance to Clear Lake. The Indian Killed 6 Ducks for us. Found an Old Free Iroquois[48] with some Indians who Joined us to return to Athabasca, from whence he came in Spring.
**4 September 1836.** The wind favoured us to get to the narrows of Buffalo Lake[49] but was too Strong to attempt Crossing the Lake. Passed the day.[50]
**5 September 1836.** This morning Sailed with a Quarterly wind to the Point of Old Fort where the wind being rather ahead from Land, pulled the oars to the next Point where we were obliged to put up at 9 A.M. for the remainder of the day. Wind increased to a heavy gale.
**6 September 1836, Tuesday.** The wind Continued with the same violence all day. Had rain during the night and the weather Cold & uncomfortable today.
**7 September 1836.** About Midday the wind began to moderate of which we took advantage and reached the Methy River at 6 P.M. Found a Party of Indians[51] Camped near the Lake. Proceeded 2 Points above where the Boats generally leave their masts & yards, for outgoing in Spring. The Indians Joined us and traded some ducks & a little Dry meat.
**8 September 1836.** The water tolerably favourable. Took breakfast at the Portage above the Barrier & the boats took a Half Cargo up to the Grave, while the remainder was Carried to McLeod's Encampment, where our Tents were pitched for the night.

---

43 They also saw flocks of "bustards," i.e. Canada geese *(Branta canadensis)* (FNB A, p. 8).
44 I.e. they had to sail close-hauled.
45 See biography, p. 302.
46 Cumberland House.
47 Named Ignace (FNB A, p. 9)
48 I.e. one whose contract with the Company had expired but who had remained in the West.
49 Now Buffalo Narrows, between Churchill Lake and Peter Pond Lake.
50 In the evening a Cree Indian arrived and provided a bladder of grease and some meat (FNB A, p. 9).
51 Chipewyans (FNB A, p. 10).

**9 September 1836.** At Breakfast time the boats were up with all their Cargoes, having brought the last half Cargo this morning. Proceeded and got to Fir Point not very distant from Pine Portage.

**10 September 1836.** Pine Portage was over at 9 A.M. Proceeded to Methy Lake,[52] which the boats reached at 1.40 P.M. The wind too boisterous to attempt to Cross the Lake; weather Cloudy. Passed the day. Some Indians followed & gave us 10 fishes they took with their netts. A few Ducks were killed & Geese seen flying past to Southward in flocks.[53]

**11 September 1836.** A Continuation of the same Cold boisterous weather made us remove down the River to a Spot more Sheltered, where the day passed without appearance of change in the weather.

**12 September 1836.** Still detained by the violence of the wind which was rendered more unpleasant today, by occasional Showers of rain.

**13 September 1836.** The weather began to moderate about 1 P.M. At 2 we entered on the Lake pulling against a head wind, and got to the little River of Portage La Loche at 8 P.M. where 3 Athabasca Indians met us.

**14 September 1836, Wednesday.** The boats were brought to the usual place [and] drawn up for the season; the Cargoes divided among the Crews.[54] Rain in the afternoon prevented the men from making a move with part of their pieces. The Indians Soliciting employ from the men. 4 of the new hands unfit to pass their shares over this long distance[55] having from 6 to 7 loads each to Carry across.

**15 September 1836.** The day's work was to render the pieces to the rest W. of little River, but some advanced to Petite Verille where our Tents were pitched. The Guide, Cadotte with P. Lalonde, J. McKay & the 2 boat builders[56] were sent to the W. End[57] to have the boats, oars &c. ready by the arrival of the general brigade with their pieces. Thos. Hassal with Sylvester were sent off by C.F. Smith to go direct in a Small Canoe to Athabasca[58] that they may be informed of the time we Expect to get there. We are informed that 5 more Indians have come to the End[59] and are on their way to meet us. They have hunted in Clear Water River and are now on their way back towards Isle a la Crosse. Got 2 Bustards, and 3 ducks sent us from there by Hassal.

---

52 Now Lac La Loche, *methye* and *la loche* being the Cree and French names respectively for the burbot (*Lota lota*).

53 These were again "bustards" or Canada geese (FNB A, p. 11).

54 This was the start of Methye Portage or Portage La Loche, across the major divide between the Churchill and Athabasca drainage basins. Boats were not normally portaged across it. One set of boats was left at the south end and another set, left at the north end, would be used for the continuation of the trip down the Clearwater and Athabasca rivers. For the details of the portage and the impression it made on travellers see Kupsch, "A Valley View in Verdant Prose."

55 The total length of the portage was 20 kilometres.

56 John Ritch and George Flett (FNB A, p. 12).

57 The west (more correctly, north) end of Methye Portage.

58 Fort Chipewyan.

59 Of the portage.

**16 September 1836.** Our Servants with a few of the men got on to the Stoney Springs. The 5 Indians mentioned Came & were employed to carry for such of the men as are unable to carry all their pieces.
**17 September 1836.** Today with some of the ablest men we reached the Little Lake[60] and the property taken across it in an Old Canoe here 2 years for the purpose but now nearly useless. Put up at W. side for the night.
**18 September 1836.** Got to the W. End with the Servants and our Baggage. The general camp are Expected to come near the Little Lake to day.
**19 September 1836.** Employed some Indians to go and relieve some of the weakest men.
**20 September 1836.** All but 5 men came down with their pieces; they have theirs yet above the hills.[61] With Assistance will be down in the morning.
**21 September 1836, Wednesday.** All was prepared, Indians settled with, and 2 trips Carried to the river[62] where we Embarked at 2.30 P.M. A Strong head wind & Cold Cloudy weather. Put up near the White Mud Portage.
**22 September 1836.** Passed the White Mud; boats launched. 4 trips made over Pine Portage; 3 trips at Grosse Roche and 4 over Portage la Bonne remain to be Carried in the morning. The Boats & remainder brought down the rapids Safe. Tents pitched at lower End. Cold windy day.
**23 September 1836.** Fine day; wind ahead. 3 trips made at Cascade Rapids. Put up at 6.30.[63]
**24 September 1836.** Found a mark above the Rapids of Pembina River and a note mentioning that Thos. Hassal had killed a Buffalo Bull[64] on the 19th indicating where it would be found. On going to the spot we discovered that Some Bears had forestalled us, and helped themselves so liberally that but very little remained for us. Afternoon another mark was Seen, where we found a Buck Moose Deer put up, and had not been touched since they left it. Put up for the night at the Junction of Clear Water & Elk Rivers.[65]
**25 September 1836.** Started at an Early hour, met an Indian at the Tar Springs[66]

---

60 Rendezvous Lake, partway across the portage.

61 The portage route drops very steeply into the Clearwater valley.

62 The Clearwater River.

63 About two-thirds of the way between Sulphur Portage and the Pembina River (FNB A, p. 13).

64 Wood bison (*Bison bison athabascae*).

65 Elk River: Athabasca River.

66 A site on the banks of the Athabasca River with conspicuous surface showings of tar, regularly exploited by the Hudson's Bay Company for pitch for caulking boats.

They were first described by Alexander Mackenzie: "At about twenty-four miles from the Fork [the Clearwater/Athabasca confluence] are some bituminous fountains, into which a pole twenty feet long may be inserted without the least resistance. The bitumen is in a fluid state, and when mixed with gum, or the resinous substance collected from the spruce fir, serves to gum the canoes. In its heated state it emits a smell like that of sea-coal. The banks of the river, which are there very elevated, discover veins of the same bituminous quality" (Mackenzie, *Voyages*, lxxxvii). The bituminous sands of this section of the Athabasca River were carefully studied and described in detail by who was sent with one of the boats to shew the men where they may get more than can be collected at the usual place at this Season. We proceeded with the other boats to the Camp, where some Provisions were received, and Tents pitched for the night. The Boat Joined with 5 Kegs Tar.

**26 September 1836.** This morning found 2 young men[67] who had just Cut up a Doe and young moose, which with another sent for a Short distance were taken in the Boats & we proceeded to Poplar Point & put up.

**27 September 1836.** Our usual route took us to the Channel of Grand Remou, where we put up. One of the Fort Hunters[68] came to us, & gave us a Grey Goose.

**28 September 1836, Wednesday.** Started this morning at 1 A.M. and got to Fort Chipewyan at 1 P.M. C.T. Campbell[69] from Peace River has been here since the 18th. Mr. Butcher[70] who is in charge, and all the people Safe and well, with a tolerable good Summers' trade. The boats' Cargoes delivered, and preparing to make up the Peace River & Slave Lake Outfits as soon as possible.

• • •

## *2. Extract from letter from Thomas Simpson to Donald Ross, Norway House, 7 August 1836* [71]

*Private*

Donald Ross Esquire — Forks Red River, 7th Augt. 1836

My Dear Sir,

McNab handed me your favor of 22nd Ulto. after arrival here … By the next trip of the Vessels I will send you my watch, according to the arrangement you were so kind as make with me.

Since my arrival here (29th Ulto) Mr. Taylor has unfortunately been quite laid up with the gout, but I have been practising observations, and find my old mathematics much more easily brushed up than I expected.

• • •

Dr John Richardson on his way north to start his coastal search for Franklin in the summer of 1848 (Richardson, *Arctic Searching Expedition,* 1:122–4). On his way upriver in July 1851, Commander W.J.S. Pullen also investigated and described the site to which Dease refers (Pullen, ed. *The Pullen Expedition in Search of Sir John Franklin*).

These are the surface showings of the Athabasca Tar Sands, in recent years the object of massive exploitation.

67 Above Rivière aux Halliers (FNB A, p. 14).

68 Named Bougon, from Fort Chipewyan (FNB A, p. 14).

69 See biography, p. 300.

70 See biography, p. 300.

71 BCA ADD.MSS 625.

## *3. Letter from Thomas Simpson to Donald Ross, Norway House, 8 August 1836*[72]

Red River, 8 Augt. 1836

Donald Ross Esquire

Dear Sir,

You are aware of our arrangement with Governor Simpson, before his departure, that a fifth boat should be sent to Portage la Loche next season, with the goods & provisions for the Arctic expedition. On comparing notes with Mr. Christie on this subject, I find that he will be able to send out next spring a boat and agrets[73] sufficient for that purpose, except oilcloths, of which you will have the goodness to order two new ones from the Factory. This boat and its crew will belong to the Freight establishment, which will make the regular charge for transport. The lading for the expedition will consist of about 60 pieces from Norway House (say 30 pieces goods etc. ordered from York this fall), 15 bags Flour, and 15 bags of Grease. This arrangement you will be pleased to consider as final, unless you hear further from Mr. Dease or myself on the subject.

With respect I remain,
Dear Sir,
Your most obedt. hble servt
Thomas Simpson

P.S. If Mr. Dease engage Rodk. McLeod or any other retiring servant from McKenzie's river on his way out, and that such engagé come on to Norway House, please to forward him hither by the fall boats. T.S.

• • •

## *4. Extracts from letter from Thomas Simpson to Donald Ross, Norway House, 31 August 1836*[74]

*Private*

New Fort Garry, 31 Augt 1836

My Dear Sir,

I was duly favored with your esteemed epistles of 14th & 17th Inst. … I vacated my quarters at the Forks … and came down here last week with Mr. Taylor, to prosecute our starry Studies in the tranquil seclusion of the Stone hermitage, whose position on the face of this little globe we have settled to a second.

If the deserter Anderson is really anxious to accompany the Expedition, I will still accept of his services and have written him a note to that effect, as I am more than doubtful of Mr. Dease's procuring a good fisherman in the North; in which

72 BCA ADD.MSS 625.

73 Agrets: the boat's accessories (oars, mast, sails, etc.).

74 BCA ADD.MSS 625.

case he must come in with the portage boats, as I have now made up my mind to leave here early in December.

I am much obliged for what you say regarding the watch, ... and have no doubt you will be as good as your word – I now send out my watch, in Cox's charge; it is pretty well regulated at present, and will I hope serve you, till I have the pleasure (D.V.) of restoring you your own ...

Believe me ever, My Dear Sir,
Yours faithfully,
Thomas Simpson

Donald Ross Esquire.

P.S. 1 Septr. I had just finished the foregoing, and gone to bed, when McNab, with McDermot's people, cast up about midnight; and from him I received yours of the 23rd, inclosing Mr. Dease's undirected favor. The contents of the latter are very satisfactory, and I am glad to find that he has hired a fisherman, John Norquay, out of the Portage brigade, and taken him to Athabasca. We are therefore clear of Anderson, which I think is fortunate ... Adieu!

• • •

### *5. Extracts from letter from Thomas Simpson to Donald Ross, Norway House, 27 November 1836*[75]

*Private*

Red River, 27th Novemr. 1836

My Dear Sir,

On the 13th Ulto. I had the pleasure of receiving by Mr. Ballenden your last line letter, to which I can now but briefly respond, being on the eve of my departure (via Manitobah) on my long and unknown journey ... I have been hard pressed to remain for ten days longer, to witness my friend Ballenden's nuptials but private wishes must yield to public duties ...

What has become of the ship? Is she at Charlton or returned to England, or lost? ... I only wish I had my *pistols* – not to shoot the birds – but they were in the ship, and might be of use to wing the Esquimaux. I feel much obliged by what you say regarding your watch, though I fear I cannot now profit by your kindness. I have nothing additional to say regarding the Expedition goods & provisions – except that besides the cargo mentioned in my former letter, there will be a Cassette[76] of mine from hence, which you will have the goodness to get safely forwarded. You will see by the Albion, which has London dates to 11th June, that Captn Back was to come out this fall in HMS "Terror" to Repulse Bay, with the view of prosecuting from thence by boats the survey to Point Turnagain. If *he* has made his way through the ice to his anchorage, and that we are not disgraced by a recall, we shall probably meet

---

75 BCA ADD.MSS 625.

76 A personal trunk.

during our second summer in the debated ground of Ross's Strait or Isthmus – but of this hereafter …

Believe me to be, My Dear Sir,
Every yours sincerely,
Thomas Simpson

Donald Ross Esquire.

• • •

## *6. Extract from letter from Thomas Simpson to Alexander Simpson, 30 November 1836*[77]

Red River, November 30th, 1836.

I myself would have been ere now thoroughly disgusted with this service, had not a ray of light shone, all of a sudden, upon my path, in the shape of a mission to the Arctic regions, where, having shaken off the trammels of pounds, shillings, and pence, I hope to achieve something worthy of the education and talents bestowed upon me. In previous letters, I described to you our plan of operations; and tomorrow I depart on my long and uncertain journey. My only apprehension now is that Captain Back's proposed expedition in H.M.S. 'Terror' to Repulse Bay (which I saw noticed in one of the last *Albions*) may interfere with our eastern voyage; but I am uncharitable enough to wish that the same ice which has prevented the Prince Rupert from reaching York, may have also shut out the gallant captain, in which case we shall endeavour to save him all further trouble.

Since my return from Norway House, my time has been chiefly devoted to astronomy, surveying, and chart drawing; my old mathematics were more readily polished up than I anticipated; and I have all those branches now at my fingers' ends. I have, likewise, read a good deal, with the hope of getting rid of the stiff and ungraceful style of Rupert's Land correspondence, which is still jarring in my ears, and marring the freedom of the pen that may ere long have to figure before the public.

For our poor mother's sake, even more than my own, I pray for success in my present enterprise. But I know, that if the worst happens to me, you will cherish her; for she will then doubly need your affectionate support.

• • •

77 Simpson, *Narrative*, 115–16.

# Chapter Three
# The Winter at Fort Chipewyan

## INTRODUCTION

Once the boats with the outfits for Peace River, Great Slave Lake, and the Mackenzie District had departed, Dease and his men settled down as guests of Chief Factor Smith at Fort Chipewyan for the winter. The major activity for the winter (apart from normal fur trading) was boat building, since two boats were to be built for the Athabasca District in addition to the expedition's boats. For the first few months, work focused on the selecting and felling of the necessary timber and transporting it to the post. The keel of the first boat (for the Athabasca District) was laid on 1 December 1836 and the second (also for Athabasca) on 13 January 1837.

On 1 February 1837 Thomas Simpson arrived from Red River. Since the ground was still not snow-covered when he left Fort Garry on 1 December, Simpson had started his journey on horseback; his baggage was transported in horse-drawn carts while his carriole and three sledges, pulled by dogs, travelled empty across the snow-free ground. At Lake Manitoba, which the party reached on the afternoon of the 3rd, and which was ice-covered, the loads were transferred to the sledges, the horses and carts were sent back to Fort Garry, and the party continued across the lake ice. Simpson used his carriole for transporting his books and instruments while he himself walked. He and his companions wore iron "crampets" to give traction on the lake ice. They passed through the Narrows of Lake Manitoba around noon on the 7th and on the afternoon of the 9th crossed the narrow isthmus (near the present settlement of Meadow Portage) to Lake Winnipegosis. On the evening of the 11th they camped among spruces for the first time, on the shores of Duck Bay, and next morning set off overland towards Swan River; there was still practically no snow, and Simpson's carriole was badly ripped and torn by rocks and deadfalls. However, Simpson managed to shoot numbers of ruffed and sharp-tailed grouse.

On 13 December they crossed Swan Lake in drifting snow; here thin ice, only about two-and-a-half inches thick, meant some tense moments. That evening they camped on Swan River; excitement was created by dry grass catching fire, fortunately without causing any injuries. Their route next lay southwest up Swan River, past the site of the present town of that name, with Duck Mountain visible to the south and the Porcupine Hills to the north. After sunset on the 17th they continued travelling by moonlight and at 8 p.m. reached Fort Pelly on the Assiniboine River. Mr Setter, in charge of the Company post, made them warmly welcome.

Simpson and his party spent the Sunday (the 18th) resting, but on the 19th, despite a temperature of -44°F, they set out, following the Assiniboine and camping where they crossed it, where it turned north; that evening Simpson "burned" his fingers on his sextant while taking his observations. On the evening of the 20th they camped at the foot of the Nut Hills (near the present town of Nut Mountain). Soon thereafter they left the shelter of the trees, and temperatures dropped to below -40°; combined with persistent winds, these made the wide expanses of prairie almost unbearable. Christmas Day was spent on the trail, with no real celebration to mark it. They reached the South Saskatchewan River in the vicinity of Batoche, and were surprised to find some patches of water kept open by the fast current. This provided them with a pleasant change of good drinking water, after the foul-tasting water they had sampled from some of the saline lakes along their route. They camped at Duck Lake and reached Fort Carlton, on the North Saskatchewan, on the 30th. Here Chief Factor Pruden had little difficulty in persuading them to stay to celebrate the New Year.

Mr Pruden held a dance in the hall on 2 January, in which Simpson and his companions joined with gusto. They set off again on the 4th with fresh guides and dogs, crossed the North Saskatchewan, and headed northwest on the Green Lake trail. Travelling through mixed forest, they camped near the present site of Big River on the evening of the 5th. Continuing northwest, they crossed several small lakes in the forest, reached Green Lake at mid-morning on the 7th, travelled north down that lake, and spent the night at the post at its north end. From here their route lay north down the Beaver River, and in the early afternoon of the 10th they reached the Hudson's Bay Company post at Ile-à-la-Crosse, to be warmly welcomed by Chief Factor Mackenzie.

Expecting an express from Dease at Fort Chipewyan, possibly with requests for provisions or equipment, Simpson decided to delay at Ile-à-la-Crosse until it arrived. On the 12th he arranged a dance for the men from Carlton before they started for home; one of his companions, an excellent fiddler, provided the music. In the meantime Simpson wrote dispatches to York Factory, Norway House, and Red River. He also took observations and determined the longitude to be 107° 54' 30"W, only 6" different from that determined by Franklin in 1825.

The packet from Dease arrived on the 20th, and that evening Simpson and party started north again. Since they were now following the standard fur-trade route, which was quite well marked, much of the travelling from here on was done at night – in part to avoid the tedium of the long, snowy vistas on the ice-covered rivers and lakes, in part to reduce the risk of snow-blindness. By the evening of the 21st they had reached Buffalo Lake, and by the 22nd Methye River. On the afternoon of the 23rd they camped on a peninsula projecting into Methye Lake, and on the 24th they started across Methye Portage (Portage La Loche), where they found the snow particularly deep.

As they descended the Clearwater River, on the night of the 25th, it rained; apparently this had also happened earlier in the month, resulting in an ice crust which was now covered with new snow. The ice crust lacerated the dogs' feet while the deep, new snow made for heavy hauling. An additional hazard was the occurrence of ice-free holes maintained by the strong current in some of the rapids.

On the evening of the 26th they camped at the mouth of the Pembina, and on the 27th reached the confluence of the Clearwater and the Athabasca; from here on the snow was less deep and hence progress was easier, at least for a while. The route from here lay north down the Athabasca, but water had overflowed the ice to produce extensive areas of slush, while a return to deep snow hampered progress even more. For the night of the 28th the party camped below the "tar springs," the natural seepages of tar in the river banks. On the 29th in bitter cold and with a strong north wind they pushed on past Pierre au Calumet; three of the dogs collapsed and were unhitched, to follow as best they could. The next two days represented tough, unpleasant slogging in bitter cold and with a relentless, northerly headwind. On 1 February the party got under way at 2 a.m., in relatively mild temperatures but with a clear sky and brilliant aurora; using a "portage" to the Embarras they made use of this short cut to Lake Mamawi and at 3 p.m. reached Fort Chipewyan on the north shore of Lake Athabasca.

By Simpson's calculations he had covered a distance of 1277 miles (2043 kilometres) in sixty-three days, or forty-five days of actual travelling, discounting the stops at Fort Pelly, Fort Carlton, and Ile-à-la-Crosse. It was a performance of which Simpson could well be proud.[1]

After the two Athabasca boats were built, the first of the expedition's three boats came off the stocks on 11 March; this was a large boat (named *Goliath*) intended for the notoriously stormy waters of Great Bear Lake, where the expedition's winter quarters were to be built. In early April a party of men was sent out to collect spruce gum for caulking the boats. The second of the smaller boats designed for the coastwise expedition came off the stocks on 27 April. These boats (named *Castor* and *Pollux*) were clinker built, 24 feet in length with a beam of 6 feet, and with a washboard rising above the gunwales with sliding ports to accommodate the oars.

The first swans, geese, and ducks were seen flying north on 21 April, and thereafter these birds provided a pleasant change to the menu for both post personnel and the expedition members. On 24 May a group of Chipewyan arrived from Fort Resolution on Great Slave Lake and five of them (with their families) were persuaded to join the expedition to act as hunters from the base camp on Great Bear Lake. In late May there was enough open water in the lake near the post that the boats could

1 Simpson, *Narrative*, 26–66.

be taken out for their "sea trials" and fully met all expectations. The expedition was now ready to head north.

## DOCUMENTS

### *1. Peter Dease's journal, continued*[2]

**29 September 1836.** Preparations for sending off the outfits as soon as ready.
**30 September 1836.** The Peace River Outfits ready & Mr. C.T. Campbell prepared to leave this in the morning. James McKay Guide & Mr. McDonald, Expn. men are to go with them to Dunveygan, whence they are to be sent as soon as opportunity affords across to Edmonton, and assist if required to transport 1000 lb Grease[3] ordered for the Expedition next season, to Dunvegan whence it will be brought down in Spring to be made into Pemican here, by which means we can rely on having it good through the season.
**1 October 1836, Saturday.** C.T. Campbell with 2 Boats for Peace River Started at an Early hour. The Slave Lake Outfit prepared to be sent off Tomorrow P.M. A. McBeath with one man for McKenzie's River goes with them as far as Ft Resolution from whence 2 men of this Place are to assist them on to Big Island, entrance of McK R.[4] or beyond if required. Blowing fresh from NW.
**2 October 1836.** The S[lave] Lake Outfit left this in 2 large and half Size Canoes. The wind turned to SE. The men settled for lodgings in the different houses.
**3 October 1836.** Many Indians about the House. Some of them began & took their wants for the Winter. Servants[5] appointed to different duties. Some Geese brought in. A party of men were sent off to cut Timbers, logs &c for Building Boats,[6] As 2 are to be made for this District & those for the Expedition also. There being no Blacksmith at Slave Lake and the provisions to be taken from this Place, renders it necessary that the Departure in Spring Should be from hence, which Can be at a sufficiently early season to get down the River[7] before the Lake[8] can afford a free passage to McKenzie's River. The men are therefore placed at the Command of C.F. Smith, to assist in doing the work that may be required of them either for the Establishment or Expedition, I shall therefore confine this Journal to occasional remarks as may be thought most interesting during the Winter.
**to the 12 [October 1836].** The weather Fine & mild. The Party sent for Boat wood did not find a sufficient Quantity for their purpose & came back. Yesterday they were

---

2 MCM M214.
3 Probably buffalo fat.
4 Mackenzie River.
5 The Company employees.
6 According to the Fort Chipewyan post journal, the timber-cutting party (Cadotte, the two carpenters, and seven men) left on the 11th (HBCA B.39/a/33, f. 9d).
7 Slave River.
8 Great Slave Lake.

sent again with Ritch, the Boat builder.[9] The natives are all off for the Season, in the direction of their different hunting grounds. The wild fowl has also left these parts; few now to be seen.

**20 October 1836.** The Boat wood was Cut & brought except a few logs which the change of weather since the 15th has prevented them from getting with boats, as they report that Ice is formed over the River Qui Evite le Lac,[10] along which it was cut.[11] The winds have been from S to SE and SW. The people appointed to the Fisheries left this on the 18th. 2 men appointed to saw.

**26 October 1836.** The wind changed to N and NE. The **27 [October]** had rain & snow with a Change of wind to NW. Weather became mild to the end of the month.

**10 November 1836.** On the **5th** Swans were seen flying to the Southward, on the **6 & 7th** had Snow. The weather yet mild for the Season.

**16 November 1836.** The Lake frozen over between this and the Fishery at Old Fort.[12] Some men were Sent for fish.

**21 November 1836.** I had news from the Fort Hunters, who have killed 4 moose for the Fort. They are along the Athabasca or Elk River, and Complain of want of Snow to hunt.

**25 November 1836.** Two Indians from the Barren Grounds came in from a Party who in Course of the Summer fell in with a Camp of Esquimaux, about 50 men they say, near the Falls of the Thellon Dessy[13] with whom they remained some days, in a friendly manner and traded some articles with them for the Furs they had, which consist chiefly of Foxes, principally the White Species.[14] These can give very little

9 The Fort Chipewyan journal reported that on the 12th "Cadotte and La Lond came in for a supplie of provisions for the Timber Men; they are in Riviere qui evite le Lac. Wood plenty and fine" (HBCA B.39/a/33, f. 9d).

10 Embarras River and its continuation, the most westerly distributary of the Athabasca Delta, which, as the name implies, almost completely avoids the open waters of Lake Athabasca for canoes or boats heading for Fort Chipewyan.

11 Entries in the post journal for this period, pertaining to the boat-building operation, are as follows:

15th. Cadotte and party return with two Boat loads of Timber for Boat Building.

17th. Men depart for reminder of Boat Timber; put back, could not pass the Quatre Fourche, the River being full of Ice. This is a disappointment, Our Wants being beyond the Means we possess to accomplish what we would require.

24th. … Rich and Geo. Flett with two Men begin to prepare the Boat Store to work in over winter, and erect a saw pitt.

1st. November … Men employed in collecting the timber still wanted for our Boats – say Logs for planking, Keels, Gunwalls, etc., etc. Tournangeau finishes the Coals and in Carting Home the Logs for Boat timber sprained his Knee, and is unable to walk …

15th. Men go and come to Rivière qui évite le Lac for some Boat timber squared their last Month – such as stem and stern Posts, timbers etc. (HBCA B.39/a/33, ff. 9d–10d)

12 The previous site of Fort Chipewyan, on the south shore of Lake Athabasca about eight miles east of the Athabasca Delta. Founded by Roderick Mackenzie of the North West Company in 1788, the fort was moved to its present site on the north shore of the lake in 1804.

13 Thelon River.

14 Arctic fox (*Alopex lagopus*).

information regarding that part of the Country generally, being the first time their Excursions have Extended so far.

**27 November 1836.** The men sent to accompany P.M.[15] McBeath to Great Slave Lake arrived, having proceeded as far as Big Island, head of McKenzie's River, where they found a party from Ft. Simpson sent to fish there. By these men we have no other news of MK River than verbal reports, which state the Failure of a party, sent in Spring to Form a trading Post up the West branch[16] at Dease's Lake under the management of Mr. J. Hutchison.[17]

**1 December 1836.** The Keel of a Boat for this District was laid. The usual work about the Establishment affords nothing new,[18] untill the **13[th]** that 2 men arrived from Peace River, bringing dispatches from Forts Dunvegan & Vermilion. C. Trader Campbell got to the former place with his Outfit & people safe the 24th Oct. From thence the Expedition men, J. McKay & W. McDonald, were sent on to the Saskatchewan on the 27th. The Object for which they were Sent[19] I am glad to hear, has been already Effected. C. Factor Rowand has Sent it across with the Lesser Slave Lake Outfit & from there it was rendered to Peace River with horses on the 18th October.

**21 December 1836.** On Considering the scarcity of men in P.R.[20] Chief Factor Smith sent back the men come down from there without waiting for our news daily Expected from the Southward. Indeed the Ship letters should have been here sooner, if nothing has occurred to Stop the usual Express conveying them from York Factory to the Interior Posts. By these men's return wrote to our friends in Peace River. This is the Coldest day we have had this Season yet; the Thermometer this morning was 46 below 0.[21] Weather Clear.

**28 December 1836.** A great change for some days past in the weather which is much milder since the 25th than for some days previous. C.F. A.R. McLeod[22] from Great Slave Lake, arrived this morning with letters from our friends in McKenzie's River, confirming what was reported of the Failure of the Party under Mr Hutchinson to form an Establishment up the West Branch. That Gentleman and the men that were with him, arrived at Fort Simpson before the return of the Brigade with the Outfit from Portage la Loche. It is Stated that they were making a Portage above Ft. Halket 2 days Journey when they were informed by Scouts of a large party of natives seen at the upper End of it preparing their arms to come and attack them, which with many threatening reports Spread by these people previous to the departure from

---

15 Postmaster.

16 Liard River.

17 See biography, p. 303.

18 The post journal noted that on 2 December Cadotte was "sent to square timbers for Oars" (HBCA B.39/a/33, f. 11d), and that on the 12th the "remainder of the Boat Timbers brought Home from Riviere qui evite le Lac."

19 They were sent for buffalo fat.

20 Peace River.

21 -46°F or -43°C.

22 See biography, p. 304.

the Establishment of Mr. H. Caused him to retreat leaving part of the property he had behind him. After having got in their Canoes & pushed off to a Short distance, the Enemy to the number of at least 60 men, armed, made their appearance on the Bank, but could not injure them.

**5 January 1837.** The new Season commenced with fine mild weather which continues. On this day the General Express to the Southward was Sent off for Isle a la Crosse by 2 men, the dispatches from MKs R., Great Slave Lake & Peace River forwarded with our letters. It is rather Surprising that we have no news yet from the Southward.

**10 January 1837.** C.F. McLeod left this on his return to G.S. Lake.[23] This Gentleman has very handsomely volunteered his Services to cooperate with our Expedition to the Eastward of the Copper Mine River summer 1838 by leading a party overland to the Thlooedessy[24] by which route, if we Should Succeed in reaching it by the Coast, to form a Junction & Secure a retreat that way, in the event of not having time to return to GB Lake[25] by the Coast. This proposal has been mentioned to the Governor & Council in my dispatch, and submitted for their Consideration. Should any party be appointed to such an undertaking, I know no one more Capable of leading it than Mr. McLeod, independent of his Knowledge of that Route, since his having Joined Captain Backs' Expedition, his unwearied activity, and Experience with the natives of the Country would render such a Security highly desirable to our party. He takes letters to forward to our Friends in MKs River, Stating what assistance the Expedition may require from them in the course of next Summer, for rendering the outfit Expected from York Factory as well as to forward the party that may be appointed to form a Winter Establishment at the NE end of G B Lake somewhere about Dease's River, being the nearest part of that Lake to the Copper Mine River, for convenience of Winter transport.

**13 January 1837.** The Keel of the 2nd Boat laid.[26] The first seems to be of superior workmanship, which leads me to hope that those he may make for the Sea Coast, will be good.[27] The Weather continues very mild for the Season with no great depth of Snow.

• • •

23 Great Slave Lake.

24 Thlewechodesse, or Back River.

25 Great Bear Lake.

26 The first boat for the Athabasca district (begun on 1 December) had come off the stocks on 12 January (HBCA B.39/a/33, f. 13).

27 Thomas Simpson (*Narrative*, 68) claimed that the decision to use two boats on the coastal trips, rather than one, was made after his arrival at Fort Chipewyan on 1 February; clearly, however, from Dease's remark here, he had already arrived at this decision over two weeks earlier.

## *2. Letter from Thomas Simpson to Donald Ross, Norway House, 19 January 1837*[28]

*Private*

Isle a la Crosse, 19th Jany. 1837

My Dear Sir,

Here have I been for the last nine days, impatiently waiting for the lazy knaves from Athabasca. There may be something in my letters from Mr. Dease, respecting the arrangements for the Expedition, that may require my attention here, or may render it necessary for me to write to yourself or Mr. Christie by the express – and this is the cause of my long halt. I mentioned to you from Red River that the lading for the Expedition by the portage brigade by this ensuing spring from Norway House, would consist of:

our YF Requisition – from 30 to 40 pieces
15 bags Flour ea. 1 cwt.
15 " Grease " 100 lbs.
1 Cassette of mine; & any parcels belonging either to Mr. Dease or myself.

While at Carlton I arranged with Mr. Pruden that he is to prepare and deliver at *Cumberland*, under the ALE[29] mark, the 15 bags of Grease; I shall write Mr. Christie to forward the flour, similarly distinguished, from Red River, and you will be pleased to instruct the Guide to embark the Grease at Cumberland, and to charge him particularly to have a care of our property, any damage to which will not be passed over in silence (as has long been the case with the McK's R. outfit), and, if serious, might entirely defeat the objects of our enterprise. So much on business, and if any change occurs, I shall scratch you a line or two before putting on my snowshoes for Fort Chipewyan.

I have had a right pleasant jaunt so far, very little snow, and the weather (excepting our desperate week between Fort Pelly & Carlton) mild for an Arctic winter. We enjoyed, however, our quantum of "misere", during the eight days of our passage across the Manitobah & Winipegos Lakes. The slippery surface of the young ice obliged us constantly to walk on iron "crampets" (you have no doubt seen these instruments of torture) which reduced our unfortunate feet to the consistency of a jelly. Then followed a long and *tearing* (as my poor cariole found) woody portage of 32 miles, to Swan Lake, which occupied two days more, a pitiful quarter of an inch of snow having fallen like manna to help us through this otherwise impracticable part of our journey. But I had capital shooting during these two days, likewise the two of our cart procession from Fort Garry to Manitobah, a distance of 60 miles – the rabbits were swarming in every bush & hummock, and we literally loaded our vehicles with them. From Swan Lake three days took us to Fort Pelly, where I rested three days more with honest George Setter, whose regular and oecomical management of the place does him infinite credit. The prospects of trade, both there and at Carlton, are by no means flattering, and notwithstanding all J.R. McKay's palaver & extravagant demands for

28 BCA ADD.MSS 835, f. 202.
29 Arctic Land Expedition.

Beaver Creek, he writes that the Assiniboines have again all gone to Fort Union. Goulet winters at Shoal river, and with all due respect for Dr. Todd's discernment, I think he has ben deceived in that man. Richard at Manitobah told me that he is a "loup", "un flatteur", and that seems to be the general opinion of him entertained of him in the District. During my short stay at Fort Pelly, the thermr. fell to -44°, and to this extreme cold was superadded a stiff northwester, which assailed us when we emerged into the wide & unsheltered plains, and compelled us to muffle up all save our eyes; yet, even with this precaution, two of my men got bitten, and as if we had not enough of it, my guide Pierre led us a whole day's march out of our way, through wild "hilly plains" (if I may use the phrase), which looked as if the vast Atlantic had been suddenly ice-bound, while rolling its huge waves in a storm. The prairies were all burnt, and in consequence we fell in with no buffalo, but we saw a few straggling tracks of Bulls and of Moose, when we regained the woods, or rather open bluffs, which form the country along the Bow River.[30] On the banks of this noble stream, one bright morning we espied four splendid red deer,[31] inhaling the fresh breeze, but the beautiful creatures took the alarm when we endeavored to creep towards them through a woody hollow, and darting down the eminence on which they stood, with the speed of light disappeared from our view. The whole journey from Fort Pelly to Carlton occupied ten days – the encampments were all bad and we considered ourselves fortunate when, at the close of day we could find a tolerable poplar hummock, with enough of dry grass to lay our blankets on. My Christmas was spread on willow branches but to make amends my new year was spent at Carlton, tripping it on the light fantastic toe (still innocent of snow shoes) with the celebrated Miss Pruden. I staid six days partaking of her father's hospitality; and six days more of rapid marching brought me to this place, where I was cordially welcomed, and have since been most kindly entertained, by C.F. McKenzie. The total distance from Fort Garry is about 900 English miles. I have travelled so far entirely on foot – have regularly surveyed the whole route, and am highly satisfied with the perfect accuracy of my astronomical observations, which afforded me amusement in many a tedious encampment – for excepting from Carlton hither – we scarcely ever could start before daylight. This is a nice quiet place and the appearances[32] are favorable. A small party of Chipewyans left here yesterday; they brought in a good many furs. I could not help admiring their careful and honest conduct – so strikingly contrasted with that of the worthless tribes which surround Red River.

Your correspondents in the North will give you all the Atha. & McK's R. news; and with kind regards to Mrs. Ross,

Believe me to be, unalterably,
My Dear Sir,
Most sincerely Yours,
Thomas Simpson

P.S. If Dr. Todd be still with you, pray remember me to him; I hope he is convales-

30 The South Saskatchewan River.

31 Wapiti, or elk (*Cervus elaphus*).

32 The prospects for furs.

cent. I got Mr. McKenzie to kill his wild bull the other day, and what with fresh fish & moose meat this is quite a land of plenty, and I am the feasted pilgrim. Adieu.

• • •

### *3. Letter from Thomas Simpson to Roderick McKenzie, Ile-à-la-Crosse, 20/21 January 1837*[33]

Pointe a l'épinette, 20/21 Jany 1837
Midnight

My Dear Sir,

I find that it was owing to the disgraceful practise of expressmen hunting or pretending to hunt on their road that our Athabasca letters were so long en route. I regret that you did not send me a perusal of the General letter to which Mr Dease *refers* in his letter to me without mentioning what that document said on the subject of provisions for the Expedition. I must therefore leave all further arrangements respecting these provisions to Mr. Ross & yourself and beg you will write Mr. Ross what you see fit on the subject. I thought that my own arrangement with Mr. Pruden to deliver at Cumberland 15 bags strained Grease ea. 100 lbs. for the Expedition would have sufficed, and from the favorable provisions trade in Athabasca I think so still, but what is stated in the General Letter – if different – must of course be acted on in preference to any opinions or arrangements of mine. I send a few lines for C.F. Charles which you will please forward. And with kindest wishes,

Believe me to be,
My Dear Sir,
Yours most sincerely
Thomas Simpson

p.s. Have the goodness to send a copy of this scrawl (under a private cover) to Mr Ross. – Farewell!

• • •

### *4. Peter Dease's journal, continued*[34]

**20 January 1837.** Nothing of any particular interest has transpired. The weather continues Fine & mild. All hands occupied at their different Duties.

**1 February 1837.** Since last notice we have nothing particular to insert. This afternoon we had the Satisfaction of Greeting the Safe arrival of Mr. Thos Simpson, after his long travels from Red River, which he left on the 1st Decr. He brought letters from our friends, with the melancholy tidings of the Death of Mrs Jones[35] at Red River, a loss to that Settlement very much felt. Her Amiable Character had rendered [her] deservedly Esteemed and respected by all Classes & particularly by those Chil-

---

33 BCA ADD.MSS 635.

34 MCM M214.

35 Mrs Jones had run a boarding school at Red River, attended by the children of many of the Company's personnel.

dren who had the advantage of being under her Care. The privation of such a benefactoress is more deeply felt by them as well as their parents. We have as yet no news of the arrivals of the Ships at YFactory[36] up to the 24th Sept last at which date the last RR[37] boats left it.[38] Severe Summer frosts have destroyed the Crops in RR to Conclude the unfavourable season there. Fire raging over the Plains towards the Fall, has Consumed great part of the Hay made for their Cattle, while some of the Cattle have also fallen a prey to the devouring Element. In consequence of these unfortunate Circumstances, it is to be feared great distress will be felt among the Settlers in that Colony, particularly in the upper Parts among the Half Breed population, whose restless habits makes them neglect agriculture, and all Economy.

**20 February 1837.**[39] The Weather Continued Mild. This day arrived the Express from the Southward. By our Letters we are informed that the Ships for York Factory reached that place, the Prince Rupert and the Eagle and Esquimaux brigs, the former 25th Sept., the others after. The first had not Crossed the Bar, remaining in the offing. After landing the passengers, and a few Bales Goods, with the Gunpowder, was taken by a violent Storm, her boats Stove, Cable broken & driven out to Sea, since which no tidings have been heard of her. The Esquimaux was sent with the Churchill Outfit for that Place, was also driven by a Storm to the Eastward, among Ice and rendered unmanageable by the Quantity of Ice formed about the rigging &c. In that State C.F. Lewis, with the Crew being in the utmost danger, abandoned her, taking the Boat to get to land which was Effected with great difficulty & risk of their lives. The Eagle got Safe to the River at YF and was frozen in after having landed the Passengers & Cargo. These disasters were fortunately not attended with any loss of life as far as we know of yet, but will Cause much inconvenience to the trade in Some parts of the Country.[40] The Different Districts afford nothing beyond the common routine prospects of the Season [which] so far are favourable enough.

• • •

36 York Factory.

37 Red River.

38 On this topic the post journal noted: "He is the Bearer of bad News. 24th last September the Ship from England had not arrived and all Hopes of Her getting in this season was over. The Bay being at the time mentioned covered with Ice" (HBCA B.39/a/33, f. 14).

39 On 6 February George Sinclair was sent to Point la Brie with the expedition's dogs, presumably to support them by fishing (HBCA B.39/a/33, f. 14). On the 16th the second Athabasca boat came off the stocks and the keel of the first of the expedition's boats was laid. This latter was the large boat which the expedition would use on Great Bear Lake.

40 1836 was indeed a disastrous year for the Company's ships trying to reach the posts in the Bay. *Prince Rupert* (Captain John Grave) was beset in the ice in the Bay from 23 August until 19 September. She did then manage to reach York Factory, but, having landed only his passengers and a few items of cargo, her captain decided in early October to head for home with his cargo largely still intact. He reached London on 12 November. *Eagle* (Captain Charles Humphreys) ran aground on the Button Islands while outward bound (27–28 July). She reached York Factory on 7 September, too late to attempt the homeward voyage, and wintered in Hayes River. *Esquimaux* (Captain Butterwick) was badly damaged by the ice in the Bay but reached York Factory on 24 September (the day the Red River boats left). It was decided that she should winter at Churchill, but she was forced to turn back by ice and was driven ashore on 16 October. The ship was abandoned and her crew walked back to York Factory (Cooke and Holland, *The Exploration of Northern Canada 500 to 1920*, 163).

### *5. Letter from George Simpson to Peter Dease and Thomas Simpson, undated (winter 1836–37)*[41]

Messrs. P.W. Dease and Thos. Simpson
Dear Sirs,

The Govr. & Committee have been pleased to approve all that has been done in reference to the Expedition on which you are at present employed & I am instructed to convey to you their best wishes for the welfare of yourselves and party and for the success of the arduous enterprise in which you are engaged.

The unexpected return of the Prince Rupert with her outward Cargo which you will have heard of before this can get to hand, will occasion a great scarcity of Goods next year. I have therefore to beg that you economise your supplies so as to render your demands on the Dept. next year for the Expedition to the Eastward of Copper Mine River to be undertaken Spring 1838, as moderate as possible.

My instructions to you on the subject of both these Expeditions were so full and explicit that the Govr. & Committee have nothing further to suggest thereon. By their Honors' desire a Copy of Capt. Back's narrative of his late Expedition[42] will be forwarded to you from Montreal Dept., which I trust may be useful & interesting to you. That Officer sailed from hence in the month of July last in the Terror bound for Wager Bay with the view of prosecuting the survey of the Coast to the West. These operations, however, are in no wise to alter your plans and sincerely hoping that the honor of completing these interesting discoveries remains for the Hudson's Bay Company through your exertions.

I remain. Dear Sirs, etc., etc.
Geo. Simpson

• • •

### *6. Peter Dease's journal, continued*[43]

**1 March 1837.**[44] An Express was Sent off with the Letters for Great Slave Lake and McKenzie's River Districts, by which we wrote to the Gentlemen in Charge of both these regarding the assistance we might require from them in course of the next Summer to enable us to carry on our operations, and have the property Ordered

41 HBCA D.4/22, p. 161.

42 Captain George Back's account of his journey from Fort Reliance on Great Slave Lake down the Great Fish River [Back River] to the sea and back in the summer of 1834, namely *Narrative of the Arctic Land Expedition to the Mouth of the Great Fish River, and along the Shores of the Arctic Ocean, in the Years 1833, 1834, and 1835.*

43 MCM M214.

44 On 27 February (according to the post journal) Dease and Simpson first informed Chief Factor Smith that they would require two boats for the coastal voyage, rather than one. As a result, on the 28th, "Men out in quest of Timber for Gunnels, Keel, etc., etc. for Second Sea Boat" (HBCA B.39/a/33, f. 15). They brought the necessary timber home on 1 March.

from York Factory rendered to the Establishment to be formed at NE End of Great Bear Lake, and the Assistance of Some Natives to Conduct our Party there in Summer, as none of them are acquainted with the country.

**March 1837.** Our Boat builder has finished the 2 boats for the District and is working at the one intended for the Expedition to proceed to the NE end of Great Bear Lake[45] With that part of it that are to Erect the buildings for a Winter Establishment at the Entrance or near to Dease's River. The Various duties of the place going on in a very Satisfactory manner. Such of the natives as have made their appearance from the North & Eastward inform us that some of the Chipewyans who have gone to their lands the last and preceding Summers have met (on both occasions) a party of Esquimaux near the Falls or last rapids of the Thellon Desseh (or River of Barren Land Indians) which by all the information we can obtain from them regarding that Stream falls into Chesterfield Inlet,[46] the Charts they have drawn of the Country, leaving no room to Doubt that it is the same with that mentioned in Captain Back's narrative as the Thellew Desseh receiving the Stream flowing from Hearne's Doobaunt Lake in its course to the Ocean.[47] The Falls mentioned are near the Extreme verge of the Woody Country, on which account the Esquimaux ascend it that far to fish & provide materials for making their Kayacks Oomiaks, Bows and arrows &c. During these meetings they have Kept up a friendly intercourse and Bartering their different Commodities with one another. To facilitate their future intercourse a plan had been proposed and assented to that 2 men of each part Should remain with the others, in order to acquire the Language. This present plan was however frustrated by the unwillingness of the wives of the Esqx. to leave their own tribe for so long a time even when accompanied by their own husbands.[48]

**30 April 1837.**[49] The fine Season coming on, our boats made, and all preparations for our Voyage advancing as well as can be wished. We have had 2 Sea Boats built of

---

45 This was the boat which would be named *Goliath*. It came off the stocks on 11 March and the keel of the first sea boat was laid on the 13th (HBCA B.39/a/33, f. 16).

46 This early report about the Thelon River is, of course, accurate, but was not confirmed until David Hanbury ascended the Thelon from Chesterfield Inlet, then travelled via the Hanbury River to Clinton-Colden and Artillery lakes, then south to Great Slave Lake in the summer of 1899 (Hanbury, *Sport and Travel in the Northland of Canada*).

47 The fact that the Dubawnt flows into the Thelon was not confirmed until J.B. Tyrrell and J.W. Tyrrell descended the Dubawnt while surveying and exploring for the Geological Survey of Canada in the summer of 1893 (Tyrrell, *Across the Sub-Arctics of Canada*).

48 Simpson (*Narrative*, 71) noted that the same Indians also reported that in 1832 a mixed group of Athabasca and Churchill Chipewyan had made rendezvous with a group of Inuit at Yathkyed Lake on the Kazan River.

49 Late March and April saw a great deal of activity in connection with building and fitting out the boats. On 31 March the post journal recorded: "They [the men] have been employed squaring Timber for 35 Oars and spars for masts etc. and bringing the same Home to the Boat Store … Sawyer stoped since the 25th when the quantity of Timber wanted for our Boat Building was completed … Pawse given to Black smith to assist in preparing the Iron Works required for the Four Boats. In the Article of Iron we are scarce owing to the quantity required for the Boats" (HBCA B.39/a/33, f. 16d).

24 ft. Keel [gap] ft. broad in thaft[50] and [gap] ft deep from Gunwales with a Wash board over them fixed, allowing Spaces for the Oars with Sliding ports when not pulling, to Close them. In the Course of this month most of the Fur hunters made their appearance, with pretty good hunts. On the 19. we heard of Swans being Seen by Some of them, and Subsequently were Seen flying about the Channels of the River in numerous flocks, Geese, Ducks, Swans, &., &c.[51] The Indians have approached the Establishment & Some of them Tented by the House. Our Expedition men have been Working occasionally at making two new Sails for one of the Boats & repairing the Old Oars.

**May 1837.**[52] The weather has been rather backward this month,[53] the Water remarkably low. Consequently the Ice confined to the Channels Cannot break up by the Wind as usual when the Water is high and affords it Space to drive about. On the **19.** the Peace River Craft made their appearance but had to remain at the English Island, the Channel being Still covered with Ice. The Returns from that Quarter are Satisfactory and I am pleased to find that the District[54] will have the means of Supplying all the Wants of the Expedition without any Embarrassment to the General business. Our preparations for departure hence on the first proxo.[55] will be complete before that time. On the 24th C.F. A.R. McLeod in a halfsized Canoe manned by Beaulieu[56] & Some of his followers arrived from Great Slave Lake. Spoke to the

---

An important task tackled in mid-April was the gathering of spruce gum for caulking the boats. The post journal recorded:

> 10th April ... 7 men provisioned for Four days departed to gather Pine gum, the clear gum intended for the sea Boats, the other a rough Gum, when cleaned, is to mix with the country Tar for our Larger Craft.
>
> 11th. Sylvester and Monette came home with the quantity of gum asked of them, 90 lbs.
>
> 13th. Jeanveine, John McKay, Fr. Fremand and Bt. St. Cyre returned with 258 lbs Gum, part of which (nearly all) is clear Gum intended for Sea Boats.
>
> 17th. The two Carpenters and assistant are getting on smartly with the last sea Boat.
>
> 27th. Last sea Boat off the stocks today. (HBCA B.39/a/33, ff. 16d–18)

50 *Thaft* or *thoft*: a rower's bench and hence, as here, a measurement: the beam of a boat.

51 Simpson (*Narrative*, 72) reported that the first swans were seen on the 17th on the south side of Lake Athabasca; the first waterfowl flew past the post on the 21st.

52 Work on the boats and equipment continued into early May, as revealed by entries in the post journal:

> 1 May. Geo. Flett and Norquay set to making Nets and Boat Sails for expedition.
>
> 3rd. Preparing Pitch for the Boats ... Black Smith off duty with a sore hand. This is a bad slap we did not expect. His work is far from being finished and some little jobs is yet wanting to compleat the Boats. (HBCA B.39/a/33, ff. 18d–19)

53 But according to Simpson (*Narrative*, 79) the weather was fine, and even sultry latterly. There were two thunderstorms in mid-month. Swallows appeared at Fort Chipewyan on the 19th and flocks of geese passed through on their northward migration throughout the month; these were snow, Canada, white-fronted, and Hutchin's geese. They provided both sport and food for the expedition members, the post personnel and the local Indians.

54 Athabasca District.

55 1 June.

56 See biography, p. 299.

Indians to procure hunters for the Expedition to accompany us down, and hunt for us at G. Bear Lake. After some Consultations among themselves a party Consisting of Rabbits Eyes (an Elderly man) & family – Nahlhal, Eengheen, Toozeyaze & Egoothe, with their families agreed to join us and were accordingly advertised to prepare to take their departure on the 30th, as they do not travel as Expeditiously as we will. As Beaulieu is to return to G.S. Lake, they were Settled to go down the Slave River in Company with him. The latter Expected to have been employed on this Service but having too many useless followers We did not Consider him on that account as a proper person to employ in that Quarter. Our hunters were according to these arrangements provisioned for the voyage to Slave Lake and on the 30th left the Fort. A Space beyond the Islands being Cleared of Ice, Our Ice Boats were taken out for a trial, under Sail and although rather leaky from their being very Dry, gave much Satisfaction as to their Sailing and Stowage. Our Letters for the Govr. & Council were given to C. Factor Smith, with those for our Friends. C.F. McLeod[57] does us the pleasure to take a passage with us to Fort Resolution. And every thing being in readiness to commence our Voyage to the Coast, The Crews of the Boats temporarily appointed untill our separation at Bear Lake River, where or at Ft. Norman the final Settlements will be made. We hailed with pleasure the approaching moment when we would be under Way on a voyage to unknown Coasts to us, but not the less interesting for being so. Our Little Boats were named after the twin Constellations, Castor & Pollux & the large baggage Boat the Goliath.[58]

• • •

## 7. *Letter from Thomas Simpson to George Simpson, 31 May 1837*[59]

Fort Chipewyan, May 31st, 1837

Dear Sir,

As we are on the eve of our departure for the polar regions, I seize the opportunity of addressing you, in the hope that you have by this time safely reached the shores of Lake Winnipeg.

The winter was mild with little snow, and spring has been pleasant though lingering, and the waters are very low.

---

57 Chief Factor McLeod volunteered to lead a group of Chipewyan down the Back River to meet Dease and Simpson at its mouth in August 1838 (Simpson, *Narrative*, 80). This offer was declined, which was just as well, since Dease and Simpson did not reach the mouth of the Back River until the summer of 1839.

58 Simpson (*Narrative*, 80–1) described the boats as follows: "Our sea-boats … were light clinker-built craft, of twenty-four feet keel and six feet beam, furnished with wash-boards and carrying each two lug-sails. They were expressly adapted for a shallow navigation, by their small draught of water; were payed with a mixture of clear pine-resin, which gave them a light and elegant appearance, and with the coloured earth of the country we manufactured paints for their further decoration. So perfectly alike and admirable were they, that they were honoured with the classical appellations of the twins Castor and Pollux; while the more capacious bateau for Great Bear Lake gloried in the redoubtable name of Goliath. Each of the sea-boats was provided with a small, oiled, canvass canoe, and portable wooden frame, which proved highly serviceable in the sequel."

59 Alexander Simpson, *Life and Travels*, 222–5.

I have now been four months here, but have not spent that long interval unprofitably; and my astronomical observations, I am glad to find, correspond exactly with Captain Franklin's. My walk from Red River has invigorated me in body and mind; and the mere "office man" is laid aside for a sphere of action more congenial to both my taste and my acquirements. *Though deprived of the command after it had first been held out to me, and was, I think, my due,*[60] yet I enter on this enterprise with confidence and alacrity. It holds out to my imagination the prospect of realizing some, at least, of the romantic aspirations which first led me to the New World, and disappointment has given place to ardent hope. There is but one painful thought which mingles with and clouds that feeling – it is that if it be my fate to perish on this expedition, my mother will be left in comparative indigence; yet I trust that in such an event the Company, with their usual high sense of honour, would make an adequate provision for her. I know not whether my brother still remains in the service or has embraced some more promising line of life; but wherever his lot may be cast I feel satisfied, from my high opinion of him (which detraction has only served to enhance), that he will do well.

We have received Captain Back's journal by the winter express. It contains, indeed, little thought, with no small portion of French sentimentality and self-admiration; but, altogether, I think that he has made the most of his subject, which was not a fertile one.

Mr. Christie has forwarded to Mr. Dease an extract of a letter from Governor Pelly, stating that Captain Back's present voyage of discovery was not to arrest or interfere with any expedition which you might have set on foot for the same object. This is most gratifying to us, and our utmost energies shall be put forth to reach the "Pole" first; or at least to win a share of the prize.[61] I hope that you have, this spring, ordered out Captain Ross' Journal,[62] and the additional instruments etc., requested by me, as they are all essentially necessary.

I have procured from two intelligent Chipewyans a minute chart of the routes by the "Thelewdessey"[63] and the Kasandesse,[64] or White Partridge River, from Athabasca Lake to Chesterfield Inlet, into which, beyond a doubt, they empty their waters, after flowing respectively through the Doobaunt and Yath-kyed Lakes of

60 It is hard to believe that George Simpson would at any point have offered Thomas sole command of the expedition; there is certainly no surviving documentary evidence of such an offer. One suspects, rather, that Thomas had misconstrued George Simpson's proposal and had falsely assumed that the command was to be his.

61 From a remark in another letter (see p. 59), it is clear that Thomas Simpson is referring to the North Magnetic Pole (which James Ross had found on the west coast of Boothia Peninsula in 1831). However, it lay well beyond where either expedition could realistically expect to reach.

62 Captain John Ross's account of his voyage to the Gulf of Boothia in *Victory* in 1829–33, namely *Narrative of a Second Voyage in Search of a North-West Passage, and of a Residence in the Arctic Regions during the Years 1829, 1830, 1831, 1832, 1833*.

63 The Thelon River.

64 The Kazan River. This information demonstrates that the Chipewyan were familiar with the major rivers of the Barren Grounds. The Thelon and Kazan rivers were first mapped by Joseph Tyrrell in 1892 and 1893 respectively. For further details see Tyrrell, *Across the Sub-Arctics*.

Hearne. The latter stream is rapid, and navigable only for Indian canoes; but the former is described as broad, deep, and interrupted by only one cataract, below which the Chipewyans had their friendly rencontre with the Esquimaux last summer. There is said to be but one considerable portage of about two miles in the whole route – which were it properly explored, might at some future time prove the most eligible channel of communication between the Northern Districts and Hudson's Bay.

I think it fortunate that Mr. Dease determined on wintering at this place, which possessed a blacksmith and many other means for the building and complete equipment of our boats, and the supply of other wantages, which Great Slave Lake could not have afforded. The abundant fishery, too, enabled Mr. Smith to keep the people concentrated; and much useful work has in consequence been got through, which could not have been done had they been scattered about searching for subsistence. Besides it is here, of course, that our whole stock of pemican for this summer can be embarked. The requisite arrangements for the transport to Great Bear Lake of our supplies for 1838 are detailed in the public letters; and the substitution of two small boats for a larger one this season is a measure that will materially contribute both to the safety and to the success of the expedition.

With sincere best wishes,
I have the honour to be, dear Sir,
Ever yours faithfully,
Thomas Simpson

• • •

## *8. Extract from letter from Thomas Simpson to Alexander Simpson, 31 May 1837*[65]

Fort Chipewyan, May 31st, 1837

The waters are very low, which has retarded the final disappearance of the ice; but we are now ready for a start tomorrow. We have two excellent boats for the coast – light, and of light draught, each adapted for six men; besides a third of larger dimensions for Great Bear Lake. We are thoroughly provided with provisions, clothing and every necessary; confident in ourselves, and in our arrangements, we look up to the Ruler of the elements for a successful issue to our undertaking. Captain Back's present "terrific"[66] voyage is not to interfere with ours; and we are in high hopes of reaching the Pole first, perhaps dining there together. His book is a painted bauble, all ornament and conceit, and no substance.

Hurrah for a *Husky* wife! I have got the portrait of mine at full length in Captain Franklin's last voyage.[67] Our worthy mother favoured me with some lengthy

65 Alexander Simpson, *Life and Travels*, 227–8.

66 A pun on the name of Back's ship, HMS *Terror*.

67 The reference is to the Inuit woman, an engraving of whose portrait (painted by Lieutenant Back) was included in Franklin's narrative of his second expedition (Franklin, *Narrative of a Second Expedition*, facing p. 118).

strictures respecting Indian connexions. What would she say to see me figuring by-and-by with a young Esquimaux wife, and a pair of urchins in her boots?

• • •

## *9. Letter from Thomas Simpson to Donald Ross, Norway House, 31 May 1837* [68]

*Private* Fort Chipewyan 31 May 1837

My Dear Sir,

My last to you from Ile a la Crosse was I believe long enough, but I shall perhaps be somewhat briefer now for the best of all possible reasons – that I have little to say, unless I were to storm the bright realms of fancy, which would disparage my dawning credit as a Traveller. But let me first acknowledge your very welcome epistles of 23 Novemr. & 27th Decr. with your watch (of which more anon), both of which reached me here on the 20th February. Why did you not send on the Northern packet at once to Carlton, after I had written you that I purposed leaving Red River on the 1st December. It would have saved the expense of a second express – and your humble Servant some delay to the bargain. Perhaps I was to blame for being in such a hurry, but it was the Governor's wish, and moreover Mr. Charles told Mr. Christie that his letters would be at Fort Garry *before* the 1st of December. As for the sequel of my journey, I breakfasted at the north end of the celebrated Portage la Loche my fourth morning from Lac la Crosse, and reached here my 12th day, without losing a dog, though with much trouble in preserving the poor lame brutes from the jaws of a pack of hungy wolves which followed us. I received, as I expected, a cordial welcome from our worthy friends, C.F.s Smith & Dease, and have spent the long interval both profitably & agreeably, – in study and in exploits among the partridges & geese. Fort Chipewyan is a paradise compared with York Factory or Red River. But what hell's game has been going on among the shipping at the former place. Alas, alas, for the wisdom of the wise! "like the leaves of the summer their banners are flown, like the leaves of the autumn those banners are shown";[69] 'tis well if "the angel of death" have not completed the picture. To descend to *solid* thoughts, what a deplorable sweep was made there of *money*!, the "elixir vitae", as it may well be called, of men who waste both body & mind in the pursuit of a scanty competence. The hands of our servants are stained with innocent blood! expiate *that*, and *then* pray to the great Ruler of the Storm to avert his wrath from our ill-starred coast. The trade of this season, judging by that of the *North*, and by what I learn from Mr. Rowand via Peace river, promises to shew well, and well may it sell, to replenish in some measure your lightened purses.

---

68 BCA ADD.MSS 635.

69 A badly garbled quotation from Lord Byron's "The Destruction of Sennacherib." Byron's version reads:

Like the leaves of the forest when Summer is green,
That host with their banners at sunset were seen:
Like the leaves of the forest when Autumn hath blown,
That host on the morrow lay wither'd and strown.

I am glad to hear that, notwithstanding the failure of your fishery, you had still somewhat wherewith to entertain our friend the Doctor, and the numerous passants whom last fall's mishap have thrown upon your hospitality. They have had a pleasant little society at Fort Garry this spring, indeed what was said of Cicero may with equal truth be applied to our invaluable friend Mr. Christie, that wherever he goes happiness is found. I trust that Ballenden will be fortunate in his matrimonial choice; he deserves to be so, for he is warm hearted & honorable, but "entre nous" I have misgivings on the subject. Physically speaking I think the poor fellow sadly wanted a wife, and that the *application* will have quite an opposite effect from that anticipated by that rum D.D. *Cockran*.

I have perused Capt. Back's journal; the style of the book is showy, and of the writing often amusing and sometimes elegant; but it is "parvum in multo", which is the sum & substance of my criticism. You need not fear that your humble servant, if we do aught, will prove a deletante in that line; you shall have the whole in small enough compass. I was telling Mr. Dease that I wish we had carrier pigeons to bear a message to the gallant Captain, inviting him to dine with us some day next summer at Ross's Magnetic pole. If both parties prosper it is indeed not improbable that we may meet somewhere in that parallel.

I cannot tell you how much obliged I am by your generously depriving yourself to send me your watch, which by frequent observations I find to be a good timekeeper. The Committee *might* have sent a chronometer, but "a friend in need is a friend indeed". Mr. Smith will deliver you my watch, which though scarcely fit for astronomical purposes, now goes very regularly, and if not exposed to cold, will keep your civil reckoning well enough. It seems as if I would end my letter in *Time*, but time shall not efface your kind wishes from my memory, while my heart warmly reciprocates them. Our two boats are as fine craft as ever swam the Arctic Sea, and C.F. McLeod is our fellow passenger in them to Great Slave Lake.

May God be with you & yours, And believe me to be,

My Dear Sir,
Yours most faithfuly
Thomas Simpson

Your hint on the Title of our Expedn. is taken.
I too *had* property in the Prince Rupert.

• • •

## *10. Letter from Thomas Simpson to Donald Ross, Norway House, 1 June 1837*[70]

*Private*

Fort Chipewyan, 1st June 1837

My Dear Sir,

I send you a line, almost at the twelfth hour (and while Mr. Smith's Artillery man is sounding the note of preparation for a Salute to the future Discoveries) to tell you

70 BCA ADD.MSS 635.

that your watch, which had gone with exemplary regularity all the spring, has stopped several times yesterday & today, while I was arranging our cargoes, but without her having received any blow; nor can I with a magnifier discover any thing in the works to authorise such a freak. This caper has shaken my dependance on her for astronomical purposes, and I therefore return her to you, equally indebted to your kindness as if I had retained her according to your intentions. In this dilemma Mr. Smith has handsomely lent me his gold watch which is the best time-keeper I ever handled, of those not called chronometers; I even think that it may sometimes be of value to me in the latter capacity.

No traveller I believe ever started on a voyage in higher spirits than your humble servant; how light & exhilirating is the free air of the North, contrasted to the penthouses of York Factory & Red River. It shall go very hard with me if I ever turn *copyist* again. Seven years of that work well nigh made a fool of me, but I now feel the mind expanding again with an elasticity commensurate to the grievous pressure to which it was so long subjected. In short, My Dear Friend, I am happy, most happy; and trusting that you are no less so, though your cares are weightier far than mine,

I ever am
Yours most faithfully
Thomas Simpson

Mr. McLeod & I had a boat regatta the other day; Our sea craft pull well and sail to admiration, lying within 3 points of the wind in a stiff breeze. The Canvass "Safeguards" are oiled, and tolerably staunch. Farewell.

• • •

# Chapter Four
# Down the Slave and the Mackenzie

## INTRODUCTION

The party set off down the Slave River on 1 June. They portaged past its major rapids (near the present town of Fort Smith) over the period 3–5 June and by the morning of the 5th had reached Salt River. Simpson made a brief side-trip by canoe up this tributary to the Salt Plains to obtain a supply of salt for the expedition's use and for the various posts of the Mackenzie River District. Continuing down the Slave, the party reached Great Slave Lake on the 10th to find it still ice-bound, but managed to reach Fort Resolution on the same day. Here they were forced to wait, but on the 21st got under way again, working their way westwards along a shore lead along the south shore of the lake, although severely hampered by ice. On the 25th, at the head of the Mackenzie River, they met the Mackenzie boats heading for Portage La Loche. Simpson took the opportunity to send a letter to his friend Donald Ross at Norway House.

On 26 June, Dease, Simpson, and party started down the Mackenzie; in his first reference to mosquitoes, Dease noted that they were so persistent that the men were glad to get out on the river away from them. To avoid the mosquitoes that night they did not land, but slept in the boats, drifting with the current all night, and reached Fort Simpson on the morning of the 28th.

Having picked up more supplies, they continued down the river on the evening of the 29th. By travelling all night they made excellent time down the Mackenzie, reaching Fort Norman late on 1 July.

## DOCUMENTS

### *1. Peter Dease's journal, continued*[1]

**1 June 1837.** The Weather rather Cloudy, had a little rain, but not such as to be any hindrance to our Departure. The forepart of the day was passed in arranging & Storing the loadings in the Boats. the Ice appearing at no distance in the Lake. Our Friend Mr. Smith proposed to leave this for Norway House with the 2 last boats on the 3rd. Two of them were sent off on the 30th for the same Place. At 5 P.M. we took leave of our Friends of Fort Chipewyan, whose Kindness and pleasant Society throughout the long winter Season has been unvaried and unremitting as it was gratefully felt.[2] And with mutual Kind feelings and Wishes we pushed off from the Rocks, and proceeded but about 5 miles when we Stopped for the night on an Island between Dog Head and the Entrance of Rocky River.[3] A light Shower of Rain Came on. The Small boats rather leaky.

**2 June 1837.** At 3 A M we left our Station; the Weather Cleared up. We soon got to the Stream of R River,[4] which in Consequence of the low State of the Waters of Peace River, assisted us down its Channel with a Steady Current untill midday when we got to its Junction [with] the latter, where it presents quite a different view, turning from a narrow Smooth Channel into a large rapid Stream Studded with numerous Islands Clothed with verdure, and a beautiful Surrounding Scenery. Our men pulling their Oars, all in the highest Spirits, and occasionally Singing one of their boat songs to the time with very pleasing Effect. Vegetation Although Commenced has not yet entirely Effaced the Idea of the Winter Season; [on] some Northern Exposures leaves hardly beginning to bud yet.

**2 June 1837.** Our progress was retarded by adverse wind, and we put up for the night at 8.30' near a Sand hill where the River Contracts its breadth and runs along a lime Stone Cliff on the South side, out of which Issues at two places a sort of Coal Bitumen, but not in any Quantity. The Sand bank opposite, from its height and fine level Surface on the top Commanding a beautifull view of an Extent of Surrounding Country, has been named by the Canadian Voyagers la Jolie Butte, or the Handsome Hill.

**3 June 1837.** Left our last night's encampment at 3 A.M. and took breakfast at the Commencement of a Range of Islands Called the Rein Deer Islands, and got to the Rapid of Dog's River at Midday. The Boats although heavily laden were run down Safely. Below this rapid immediately, lies the numerous Islands and rocks which, Crossing the bed of the River, form the rapids where the Cargoes are Carried over

---

1 MCM M214.

2 The relevant entry in the post journal reads: "About 6 o'clock C.F. Dease and Mr. Thomas Simpson of the Northn. Expedition took their departure in three Boats, 14 Servants & 2 Boys. Mr. McLeod accompanied them on his return to Fort Resolution" (HBCA B.39/a/33, f. 20).

3 Also Rivière des Rochers, the outflow of Lake Athabasca; after its confluence with the Peace River it becomes the Slave River.

4 Rocky River.

the Portages, the first of which is named the Cassette. The Weather, Cloudy all day previous, turned to rain while the people were Carrying the property, which made us Erect our Tents & pass the remainder of the day from 4 P.M. The rain Continued untill evening.

**4 June 1837.** On account of the Rain & vicinity of the next Portages we left the Cassette at 7 A.M. to allow the Rain to be shaken off the leaves by the wind in the next portages and avoid wetting the property Carrying it across them. We got over the Embarras or Island Portage, two next Rocky Portages and the 2 Points of the Mountain Portage, below which we put up for the night. By their traces it appears that our Hunters with Beaulieu's party passed here yesterday.

**5 June 1837.** We got over the Pellican[5] and Portage of the Drowned by 2 P.M and proceeded to Salt River about 18 miles distant from the last Portage where we landed at 5 P.M. Our Hunters with Beaulieu's party waiting our arrival brought us 2 bustards Killed in course of the day. Some of them have been up the River[6] hunting and tell us that the Water in the River is so low it will be impracticable for us to go up with a Boat for Salt,[7] an article we must procure for our Winter. We therefore made an arrangement with them for the loan of 3 of their Canoes to go up the River, as they will be more easily got over Shallow parts than a Boat can.

**6 June 1837.** Mr Thomas Simpson with 5 men & 3 Indians left us to go for Salt. Our Sea Boats were Examined, Caulked & pitched as they have been leaky Since we left Athabasca. Several small leaks were found & Stopt. They were not finished quite today. The Goliath does not require any Examination being tight. Our hunters have 2 netts down and gave us a mess of fish of the Inconnu Species.[8] Several Geese were Killed today.

**7 June 1837.** The party came back at Midday from the Salt Springs with 30 Bags of Salt about a Bushel each, 16 of which being Deposited in Security for the McKenzie River Boats. We got the Boats Loaded and left the River at 4 P.M., the hunters having pushed off about 2 hours previously. Fine weather. We Continued to 9 P.M. and put up on an Island at the great Bend of the River named Grand Dêtour. Passed the Indians.

**8 June 1837.** A Contrary Wind retarded our progress much today. We could only get by 8 P.M. near the head of Big Island, below a long bend of the River round a narrow Isthmus or Point.

**9 June 1837.** The same Cause as yesterday retarded our progress so much that it was 9 P.M. when we got to the Channel branching off to the Northward called John's

---

5 Pelican Portage. Simpson (*Narrative*, 84) comments on the large numbers of white pelicans (*Pelecanus erythorhynchos*) resting on the islands in the Pelican Rapids. These birds still frequent the area, nesting on the islands and fishing in the rapids; this is the most northerly nesting colony of this species.

6 The Salt River.

7 Natural surface accumulations of salt occur in the valley of the Salt River and were used by the Hudson's Bay Company as a source of salt for the Athabasca and Mackenzie river districts.

8 Inconnu (or "connie") (*Stenodus leucichthys*), in North America largely confined to the Mackenzie and Yukon basins and to northern Alaska. A very popular food fish, it supports a commercial fishery in Great Slave Lake (Scott and Crossman, *Freshwater Fishes of Canada*, 295–298).

River or Channel leading to Slave Lake eastward of Stony Island. Here we put up for the night.

**10 June 1837.** At 8 A M we got to the Lake which presented a view of unbroken Ice at no great distance from the Channels of the River. We had however a Clear Space in the Bay to Enable us to hoist our Sails to a free wind which took us to Moose Deer Island, but endeavouring to get round the W End our advance was Stopped by the Ice adhering yet to the Islands and apparently solid as far as our view Could Extend. We turned therefore to pass the Narrows between the E end & mainland, but there was too little water to pass the Boats with their Cargoes. With the Assistance of Small Canoes the 2 Sea Boats were got to the Establishment of Fort Providence[9] and the Goliath remains untill a passage may Clear to bring her round the Island. There is no prospect of our leaving this for some time yet. The Ice has not yet made any move beyond the Shallow Sands between this and the Island, where the heat of the Sun soon dissolves it. We must therefore wait patiently for a passage over the Lake.

**12 June 1837.** The Goliath was got round the Island and Stowed her goods in the Store, to prevent any depredations on our Provisions by Dogs or Indians.[10]

**18 June 1837.**[11] A Clearance along Shore was formed as far as Buffalo Creek. We Sent off our hunters ahead with a nett & amunition to fish and hunt as far as the Ice will allow them to get on. Beaulieu & party also left this for the north side of the Lake.

**20 June 1837.** The wind yesterday Changed from a NW direction to ENE which broke up the borders of the Ice Considerably and today Changed to SE which drives it out from the Southern Shore, but it is yet in Sight.

**21 June 1837.** A Continuation of wind from the Same Quarter removing the Ice, we took leave from Friend Mr McLeod & his people at 11 A.M.[12] The wind rather near ahead. Plied the oars for some time to pull in the Bay, and hoisted our Sails which took us on at a Good rate to some Miles Westward of Island of the Dead when the breeze died away, but not untill we had got in the Bay of Large Burnt Point, near the Ice, where we found the Hunters, who cannot proceed. Put up near them and gave them a Supply of provisions.

**22 June 1837.** Although the Ice appears Solid outside we pushed off at 5.30 and by Keeping along Shore, pushing among broken Ice we got to the Sulphur Springs[13]

---

9 Actually Fort Resolution; this slip of the pen is inexplicable. There had earlier been a Fort Providence on the east side of the North Arm of Great Slave Lake, but it had been abandoned over a decade earlier, and the present Fort Providence (at the entrance to the Mackenzie River) would not be established for another few decades.

10 That evening a dance and supper were arranged for the expedition members and the personnel and Indians of Fort Resolution (Simpson, *Narrative*, 87).

11 During this ten-day enforced sojourn at Fort Resolution, Peter Dease vaccinated all the children, Indian and Métis, against smallpox, something he had had also done earlier at Fort Chipewyan (Simpson, *Narrative*, 88).

12 *Goliath*'s cargo included twenty-one sled dogs, most of which Simpson had brought from Red River (Simpson, *Narrative*, 89).

13 Simpson describes these springs as follows: "They are large, clear, and leave a deep sulphureous deposit wherever their waters run, as well as in the fountains themselves" (Simpson, *Narrative*, 90).

at Big Burnt Point. Then we waited for the Indians, who Exchanged a Canoe of theirs for a larger one left here last autumn, the repairing & pitching which Afforded us time to take breakfast. Their arrangement of the Canoe not being completed, we Continued on as far as Buffalo Lake River, which being a Secure harbour and good fishing place in the Event of being Stopped by the Ice, we put ashore & waited 4 hours for them. A light Rain and Cloudy Weather came on. The Indians joined us, and taking advantage of every hour to get forward, pushed off at 6 P.M. At 9 the Indians Stopped behind, while we continued untill we got to Hay River which we Entered at 5 A.M. of

**23 June 1837.** The wind turned to N W which brought the Ice in Shore and Stopped us all day.[14] The Hunters Came up at 11 A.M. We sent them on to the Channel West of the Large Island that is in the mouth of the River, dividing its Stream for about 3 miles from where it falls into the Lake. We were much annoyed by Swarms of musquitoes. From a Cloudy & Rainy morning, the Weather turned fine & Clear untill 5 P.M when the wind, to our Joy, turned to E by S and rather Cloudy weather. This was

**24 June 1837** followed by a thick fog, but the Ice by the East wind being a little detached from the Land, we got under way at 4 A.M, and were some time working among the broken pieces of Ice, the fog so thick that although the boats were not 200 yards distance Could not be seen. About two hours were Spent in this manner, when the weather Cleared & the Bay towards Stoney Point appeared quite coverd with Ice Close in to Shore. Knowing, however, that the bay near Land was Shallow and a probability that the Ice would be broken over these places, we pushed through, and Succeeded after much labour in getting to Stoney Point. There again we Experienced another delay of 3 hours. A violent Shower of rain accompanied by a Strong Gust of wind brought the Ice upon the Extremity of the Point for some time. We got round after the wind Ceased & proceeded to the Entrance of the River opposite the first large Island[15] where we put up for the night.

**25 June 1837.** At an Early hour this morning we were under way. At 8 A.M., being at the Commencement or head of the first Little Lake, met the McKenzie River boats[16] on their way to meet their Outfits, taking the Returns of the District.[17] Put ashore & had the pleasure of Shaking hands with Messrs. McPherson[18] and Hutchison, and passed the day with them. Wind ahead; too Strong for us to proceed.

• • •

---

14 This delay was put to good use: a net set in the mouth of a stream produced a good haul of dory (walleye, or pickerel, *Stizostedion vitreum*) and inconnu, while wild onions were harvested on its banks (Simpson, *Narrative*, 90).

15 Big Island in the mouth of the Mackenzie River.

16 From Fort Simpson.

17 "Outfits" refers to the incoming trade goods, provisions, and supplies coming mainly from London via York Factory; the "Returns" refers to the outgoing bales of furs and other country produce, destined for the London market.

18 See biography, p. 304.

## *2. Letter from Thomas Simpson to Donald Ross, Norway House, 25 June 1837*[19]

*Private*

D. Ross Esquire — McKenzie's River

My Dear Sir, — 25th June, 1837

I write you a line from the region of unfading day, and alas! of undying musquitoes. These "enfans du diable" tickle one's catastrophe so confoundedly (to use the words of Falstaff and our own "Bob logic") that my ideas, like an ill taught rope dancer, are every moment ready to topple over from the thread of placid discourse. However it was not for an harangue I began now, having only two minutes to spare, in which I epistolise not a soul but yourself, begging you to inform or not – as you choose – our friends in general or (which is the same thing) our general friends, that we were alive & kicking at this place and date. We reached Fort Resolution on the 10th, having halted two days at Salt River, to bring down a supply of its useful & curious product for Atha., McK's River, & ourselves. I made the trip with our hunters' canoes, and was much gratified with the beautiful scenery of the Salt plains, and by an inspection of the singular formation which the saline annually deposit in such quantity under the influence of a scorching sun.

The relentless ice detained us eleven days at Fort Resolution. Mr. McLeod did his utmost to make the hours pass pleasantly, but the Look Out and Glass in hand often testified my impatience of so early & tedious a check. Mr. Dease being an old stager took it more cooly, and the truth is we shall be quite soon enough on the coast. We met Mr. McPherson today near the entrance of this gigantic stream and had the pleasure of forming my first personal acquaintance with a friend already well known to me by correspondence. We have just breakfasted together, and are about separating to pursue our separate destinies. We are glad to learn from him that tranquillity pervades the district, which has surpassed the handsome returns of Athabasca by several hundred pounds. I forgot to mention that we had four days tough work to effect our passage through the ice of Great Slave Lake, a fair initiation into the operations of the summer. By the 10th of next month, at the latest we hope to be on our ocean theatre; but whether to enact a comedy or a tragedy, the fates must determine.

Meanwhile Believe me to be
My Dear Sir,
Ever your sincere friend
Thomas Simpson

• • •

19 BCA ADD.MSS 635.

### *3. Peter Dease's journal, continued*[20]

**26 June 1837.** At an early hour this morning Our Friends took leave of us. The wind being fair for them, and directly contrary for us, Kept us ashore untill Midday, when we got afloat working against a strong wind and heavy Swells to & near the head of the Rapids where we took supper. During our Short Stay for that purpose, we were so much tormented by our Enemies, the Musquitoes, that all hands preferred being in the boats afloat than enduring such a torment on land. According with our own wishes to advance we did not hesitate to grant their desire. Continued our Route on the 27th. We took our Supper at the Contraction of the River, where the first Establishment was Erected in McKenzie's River Called Livingston's Old Fort,[21] From thence, driving down the Stream all night[22] and an occasional advance with the Oars brought us to Fort Simpson on the **28 June 1837** at 8 A.M.[23] Our hunters did not come up today. Got all our wants from Mr McBeath who is in Charge, and some work done by the Blacksmith which was not quite finished this day.

• • •

### *4. Fort Simpson journal, 28–9 June 1837*[24]

**Wednesday 28th June 1837.** In the forenoon we were much gratified by the arrival of Msrs. Dease and Simpson on the Expedition with three Boats …
**Thursday 29th.** Fine warm weather, very calm. The Expedition people remained here the whole day arranging themselves and at 9 P.M. took their departure …

• • •

### *5. Peter Dease's journal, continued*[25]

**29 June 1837.** Our Hunters Came this morning, and having everything we wanted in Course of the day we left the Fort at 9 P.M. and continued our usual routine of Travelling to Ft Norman which we got to at 11. PM. on the second day after.[26]

• • •

20 MCM M214.

21 The first trading post on the Mackenzie River, established by Duncan Livingston of the North West Company. It was abandoned in 1800 following Livingston's murder in 1799. Dease and Simpson found a Dogrib camp here; to Simpson, the attitude and behaviour of the Dogribs were much more appealing than "the sullen indifference of the Chipewyan" (Simpson, *Narrative*, 92).

22 Dease means that the boats were lashed together and allowed to drift with the current while most of the party slept (Simpson, *Narrative*, 92).

23 Simpson (*Narrative*, 93) noted that while the crops looked good, the fields of barley were suffering from the depredations of the passenger pigeon (*Ectopistes migratorius*), an interesting comment about a bird which is now extinct, the last specimen dying in Cincinnati Zoo in 1914. Fort Simpson must have been close to the northern limit of its range.

24 HBCA B.200/a/19, f. 3d.

25 MCM M214.

26 They had not stopped for the night of 30 June/1 July but had simply drifted with the current for

## *6. Letter from George Simpson to Peter Dease and Thomas Simpson, 30 June 1837*[27]

Norway House, 30th June 1837

Messrs. Dease & Simpson,
Gentlemen,

In a private letter I had the pleasure of receiving from Mr. Dease, dated 3rd. January, he refers to a public letter written by him which I am sorry to say never reached me. Mr. Ross says he saw it and forwarded it to Red River; had it been in Duplicate, I should have found one of the copies here. Mr. Ross informs me that he has attended to the request made in that Letter respecting provisions. That Gentleman likewise says the Letter in question contained some suggestions in reference to an Esquimaux Interpreter, and to other points upon which it did not appear to him, the Council could, at this Distance, and with the limited information we possess in regard to the Country, offer an opinion or advice.

All we can now say in regard to the Expedition is, that the Government, the Company, and the Council feel the most lively interest in its success, and that a failure would be an everlasting source of vexation & disappointment to every member of the Fur Trade.

The objects in view are distinctly laid down in the letter of Instructions I wrote you from this place last year. In regard to supplies, you have a carte blanche; our depôts are open to your demands, and you are authorised to call on the Districts of Athabasca & McKenzie's River for any facility or assistance in men, Goods, Provisions, Indians, Craft, etc., etc. you may require. It rests with you to apply for and employ those means as you may find necessary; but we have no further instructions to give than to entreat you will use your best endavours to accomplish the great object in view by any means, or in any way you may determine upon. The season has hitherto been unusually cold; even here we are now rarely without fires in the Sitting Rooms, and to your mission I fear it is very unfavourable. When you started from hence, it was expected the objects of the Expedition could be completed in two years, but should the unfavorable state of the season prevent your accomplishing the western survey this summer, you had better make another attempt next year, and defer the Eastern survey until the following, i.e. take three summers instead of two. In short, we are more anxious to accomplish this important and interesting object than I am well able to describe & are willing to incur any expense or inconvenience to the service to that end. Capt. Back's narrative was sent you last fall and Capt. Ross's and

five hours. In this fashion they covered the impressive distance of 250 miles (400 kilometres) in forty-eight hours (Simpson, *Narrative*, 95).

27 HBCA D.4/23, pp. 86–7.

Mr. King's[28] are now forwarded. With warmest and best wishes for the welfare & success of the Expedition, I Remain, Gentlemen,

Your very obed. servt, Geo. Simpson.

• • •

## 7. *Fort Norman journal, 1–2 July 1837*[29]

**Saturday 1st July 1837.** At a leate our the Expedition Brigade arrived, three Boats and two Canoes. Mr. Chief Factor Dese and Mr. Thomas Simpson put up their Tents for the night. They came from Fort Simpson to this Place in two days. the Boats was anchored out from the Land, and the officers went to bed as well as their Soldiers …

**Sunday 2nd.** … thos two Indians (Etatille and Cute Hand) will accompany the Boat up the Bear's River. the gentlemen was very busily employed making out their outfit for the sea and sent off John Rich to Bears Lake with a boat, wher he is to Establish. the whol Party went off this after nune.

• • •

28 Dr Richard King's account of his journey down the Back River with Captain Back in the summer of 1834, namely *Narrative of a Journey to the Shores of the Arctic Ocean in 1833, 1834 and 1835, under the Command of Capt. Back, R.N.*

29 HBCA B.152/a/16, f. 4d.

# Chapter Five
# West to Point Barrow and Back

## Introduction

At Fort Norman the expedition split into two groups. A party of four men, led by John Ritch and accompanied by Indian guides and hunters, headed up the Great Bear River with instructions to establish winter quarters for the entire expedition near the mouth of the Dease River at the northeast corner of Great Bear Lake. The remaining members of the expedition (fourteen men), led by Dease and Simpson, set off down the Mackenzie in the two sea boats, *Castor* and *Pollux*, on 2 July. Having called briefly at Fort Good Hope, they reached the sea at the west side of the Mackenzie Delta on the 9th. Heading west past Herschel Island, they found themselves battling a baffling combination of fog, sea ice, spits, bars, and shoals. Small groups of Inuit whom they encountered were invariably friendly, communicative, and keen to trade. They passed Return Reef, Franklin's farthest point, on 23 July. The frustrations of fog, sea ice, and shoal waters continued; on one occasion (23–24 July), because of the shallowness of the water, they were out of sight of land for twelve hours; that "day's" travel lasted for twenty-five hours!

On 31 July a solid barrier of ice farther west brought boat-travel to a halt, some fifty-two miles (eighty-three kilometres) in a straight line from Point Barrow. Hence it was decided that Simpson and five men would attempt to cover this final lap on foot, taking a collapsible canvas canoe in order to cross any rivers they might encounter. Dease and the remainder of the party stayed to guard the boats. Simpson and his men made good progress westwards but were halted by Dease Inlet, which would have been impossible to circumvent on foot in the available time. But by good luck here they encountered a group of Inuit from whom they negotiated the loan of an *umiak*, and using this craft they reached Point Barrow on 4 August. There they found a large Inuit settlement, whose inhabitants were found to be very friendly and eager to trade. This was in marked contrast to the reception they gave to the Russian party led by A.F. Kashevarov, who reached Point Barrow from the west the following year (1838) and was forced to turn back only a short distance beyond it because of the threatening attitude of the local inhabitants,[1] and to the the officers and men of HMS *Plover*, under the command of Captain Rochfort Maguire, who wintered here in 1852–54, and who found the local people extremely uncooperative and often openly hostile.[2]

Still in their borrowed *umiak*, Simpson and party rejoined Dease and his men on 6 August. The reunited party then started back east in their own boats and re-entered the Mackenzie Delta on 18 August. After a brief stop at Fort Good Hope (28–30 August), they reached Fort Norman on 4 September. Having taken delivery of supplies and mail which arrived from the south, and having spent several days writing reports and private letters, Dease and Simpson set off for their winter quarters on 10 September. They started up Great Bear River that afternoon and, tracking their boats, reached Great Bear Lake on the 14th. Despite strong winds and heavy seas, which kept them ashore on several days, they reached their destination on 25 September. As the site for the winter quarters (named Fort Confidence), John Ritch had selected a location overlooking the narrow channel between Ritch Island and the mainland, some seven kilometres west of the mouth of the Dease River. By this stage only a storehouse had been completed; the walls and roof of the dwelling house had been erected but none of the fireplaces, chimneys, or internal fittings had been installed. Ritch and party had been delayed severely by a solid ice cover on Great Bear Lake and had reached the site only on 16 August; in the light of that fact, he had made quite admirable progress in erecting the buildings, especially since the expedition hunters and the local Indians had contracted some serious illness. Having written and sent off a further batch of mail with one of the boats returning to Fort Norman, Dease, Simpson, and party now set about serious preparations for the wintering.

Dease and Simpson had good reason to be satisfied with their achievements in their first season. They had done precisely what George Simpson's instructions had specified – "closed the gap" between Franklin's farthest at Return Reef and Elson's farthest at Point Barrow. On the basis of the regular observations which Simpson had made, he was able to produce an impressively accurate map of the arctic coast of Alaska (fig. 7).

The only jarring note is to be found in Simpson's selfish claim that this achievement was *solely* due to his efforts. Thus in a letter to his brother Alexander, he wrote: "for I and I *alone*, have the well-earned honour of uniting the Arctic to the great Western Ocean, and of unfurling the British flag on Point Barrow."[3] It never appears to dawn on Simpson that Dease made a major contribution by provisioning, encouraging, and reassuring the voyageurs in an environment which was almost totally alien, travelling along an ice-infested sea coast with a barren tundra landscape devoid of

1 See VanStone, ed., *A.F. Kashevarov's Coastal Explorations in Northwest Alaska, 1838*.

2 Bockstoce, ed., *The Journal of Rochfort Maguire 1852–1854*.

3 Alexander Simpson, *Life and Travels*, 257.

trees. Unfortunately, as the story of the expedition unfolds, Thomas Simpson's overweening egotism will be repeatedly revealed.

## DOCUMENTS

### *1. Peter Dease's journal, continued*[4]

**2 July 1837.** Our Cargoes were Seperated, and a division of the Property intended for Great Bear Lake & what is wanted for the Coast.[5] Our hunters got some advances; arms and amunition given out to the men. The Crews of the Sea Boats, viz. Jas. McKay, Steersman, François Felix, bowman, Charles Begg, Hector Morrison, John McKey, Frans. Boucher (boy), Mdn. of the Castor, P.W Dease Offr.; Geo Sinclair, Steersn., Pierre Morin bowsn., Peter Taylor, Geo Flett, Wm. McDonald & Ferdd. Wentzel (boy), Midn.[6] Pollux, Mr Thos. Simpson, Offr. John Ritch in Charge, J. Norquay, François Framand & Laurent Cartier, the men for Great Bear Lake, with the Hunters & 3 Indians of this Post to assist and Guide them to Dease's River where or near they are to erect buildings for the Party to pass the Season.[7] All our arrangements occupied the day.[8] At 11 P.M. we left Fort Norman & drove down with the Stream all night. Landed on the Banks along the Coal mines[9] at 6 A.M. of **3 July 1837** where we took Some White Clay for White washing the Houses. At 8 We Seperated from our Bear Lake party[10] and pulled at the oars all day. Landed for supper below Blue fish River[11] & got a few fishes from 2 Indians that we saw at that Place.

4 MCM M214.

5 From this point a party under John Ritch in the large boat, *Goliath*, was to head up Great Bear River and across Great Bear Lake to erect winter quarters on Dease Arm, near the mouth of the Dease River, while the remainder of the party, in the two sea boats, *Castor* and *Pollux*, continued down the Mackenzie to start their exploration westwards towards Point Barrow.

6 Middlemen. Ferdinand Wentzel may have been a young son of Willard Ferdinand Wentzel, who spent most of his long career in the Mackenzie District, first with the North West Company and then with the Hudson's Bay Company (Keith, *North of Athabasca*, 157–65), or possibly a son of Willard's son Alexander.

7 They were also to establish a fishery, and the hunters were to lay in a stock of caribou and muskox meat for the winter (Simpson, *Narrative*, 96).

8 This included selecting trade goods for the Inuit they might encounter on the coast, and provisions for the trip: 30 pieces of pemmican and 10 hundredweight (1120 pounds) of flour (Simpson, *Narrative*, 96).

9 Dease is referring to seams of lignite which outcrop a few miles above the confluence with Great Bear River. Alexander Mackenzie reported on this phenomenon in 1789 as follows: "in our progress [we] experienced a very sulpherous smell, and at length discovered that the whole bank was on fire for a very considerable distance. It proved to be a coal mine, to which the fire had communicated from an old Indian encampment" (Mackenzie, *Voyages*, 96). They were also noted by Franklin, who also found them on fire in 1826 and took a sample for experiment but found it unsuitable for blacksmith's work (Franklin, *Narrative of a Second Expedition*, 91). While heading down the Mackenzie to initiate his search for Franklin in 1848, Dr John Richardson took the opportunity to examine the lignite beds in detail and reported on them and the fossils they contain (Richardson, *Arctic Searching Expedition*, 1:186–9).

10 At the mouth of the Great Bear River.

11 Possibly the Carcajou or Mountain River.

**4 July 1837.** To avoid the Torment of Musquitoes as usual we passed the night in our Boats, gliding down the Smooth Stream untill we came to the Rapids, which is no ways dangerous on the W side, without meeting any Indians. We got to the Ramparts[12] where Joyful Shouts were heard echoing among the Rocks from side to Side by Small parties Stationed near the Eddies of the Points to fish at this Season. In a short time we were Joined by a large escort of small Canoes, who Continued down to Fort Good Hope[13] where we landed at 10 P.M. Mr. Bell[14] and his people all well. 5 men of the Loucheux[15] tribe here. Accounts of a murder having been committed by a party of Esquimaux upon their tribe, was brought by these men. This happened only about 20 days ago. 3 men were killed and one wounded.[16] The relatives talk of going to retaliate upon them. Another Cruel murder was committed upon a family of Hare Indians[17] in Feby. last about 2 days Journey hence. A man, his wife & 2 Children were killed, supposed to be by his own Son in law, an Indian from Ft. Norman.

• • •

## *2. Fort Good Hope journal, 4–5 July 1837*[18]

**Tuesday 4 July 1837.** A fine warm day. At 8 A.M. The Expedition Gentlemen, Messrs. Dease and Simpson arrived in two Boats after a passage of five days from Fort Simpson all well.

**Wednesday 5.** A fine day. At ½ past 2 P.M. precisely the Expedition Gentln. and crews took their departure for the Sea Coast in high spirits, after giving us three hearty Cheers, which we cordially returned with our fervent Prayers for their safety and success ...

• • •

## *3. Peter Dease's journal, continued*[19]

**5 July 1837.** At 1 P.M We left our friends at F.G. Hope. Gave a passage to one of the Loucheux to Join his party at the Narrows. Saw some Hare Indians along the River but did not Stop. They complain of fish being very Scarce this year. Put ashore to

---

12 At its confluence with the Ramparts River, the Mackenzie flows over the Ramparts Rapids, formed by a series of limestone ledges, then constricts to about four hundred metres in width as it enters the Ramparts. In this section, some ten kilometres in length, the river flows between vertical limestone cliffs, thirty to sixty-five metres high, at speeds of ten to twelve kilometers per hour.

13 Relocated high above the river on the right bank the previous fall. Previously it had been on nearby Upper Manitou Island, but it had been totally destroyed by a surging mass of ice which overwhelmed the island during breakup in June 1836 (Simpson, *Narrative*, 99).

14 See biography, pp. 299–300.

15 Later commonly called Kuchin and now by their own appellation, Gwich'in.

16 Hostility, suspicion, and violence had marked relations between the Inuit and the Gwich'in for centuries. This put paid to the plans of Dease and Simpson to take two Loucheux from here as Inuktitut interpreters, although the Indians concerned were still keen to go (Simpson, *Narrative*, 100).

17 Now known by their own appellation of Sahtu Dene.

18 HBCA B.80/a/14, f. 3.

19 MCM M214.

take our Supper and drove down the rest of the night as usual. On the 6th we saw a few of the Loucheux tribe along the Banks of the River. Passed the Old Lower Fort Good Hope[20] at 6 P.M. after landing to take an observation. The day was very warm. The Thermo. Stood at 107 at 1 P.M. Exposed to the Sun.[21] Usual route the **7 [July]**. At 5 P.M got to the Narrows a little above Red River.[22] We saw the Chief of the Loucheux with a Small party, where we left our passenger and got to Point Separation of Captn. Franklin[23] at 8.30. Weather very Sultry. Musquitoes innumerable made us follow our usual plan & Keep on the Water. Came in sight of the first range of Mountains[24] before morning **(8 July)**. Continued on all day without anything occurring, untill passing along the Islands off the mouths of the Channels of Peels River, we Saw 2 Moose & 3 R Deers[25] at different times. Below that River Some distance, another apparently Considerably large Stream falls in. The Channel widens and about 4 miles from the mouth of the last we put ashore on a small Island to take our Supper. Here we found a Deposit of Esquimaux property, upon a Scaffold, Consisting of various articles, as lines, netts, Spears, Sleds, Skins and some Dried fish.[26] We did not disturb it, but put up a Small package for the Owners when they might come for it, of beads, Rings, a file, Hatchet, awls &c with a Hierogliphic letter on Birch Bark intimating that it was left for them by the White people who were Come with two boats & proceeding down to the Coast. As usual after taking our Supper we got afloat again.

**9 July 1837.** At 7 A.M. we got below the Pine or fir wood,[27] the Islands having no other than the Willow and a few Small poplar trees. Wind E by S helped us forward with the Sail. On leaving one of the Channels, falling into a larger, before us about 200 yards distance we saw an Esquimaux Oomiak[28] or boat with one man and 4

---

20 Located on the right bank of the river just upstream of where it makes a marked bend to begin the east-west section which takes it to the head of the Delta. This post was occupied from 1823 until 1826.

21 And 77°F in the shade (Simpson, *Narrative*, 102).

22 Arctic Red River.

23 The head of the Mackenzie Delta, so named by Sir John Franklin on 4 July 1826 since it was here that his party, heading west to explore the coast towards Point Barrow, separated from that of Dr John Richardson, which was bound east to explore the coast to the mouth of the Coppermine (Franklin, *Narrative of a Second Expedition*, 95.) Since Point Separation lies north of the Arctic Circle, Dease and Simpson's party observed the midnight sun here for the first time (Simpson, *Narrative*, 104).

24 The Richardson Mountains, which lie west of the Delta; the party was following the West or Peel Channel on the west side of the Mackenzie Delta (Simpson, *Narrative*, 105).

25 Caribou (*Rangifer tarandus Groenlandicus*), which Dease referred to as reindeer.

26 This cache of materials used in winter would have been left here for the summer by the Inuit, who at this time would be at the sea coast, fishing and whaling. It was placed on a scaffold to keep it out of the reach of foxes, wolverines, etc.

27 I.e. they had passed the treeline, which in the Mackenzie Delta reaches almost its most northerly position in North America.

28 The *umiak*, or women's boat, was a capacious and seaworthy boat, capable of hauling a surprising amount of cargo. It consisted of a skin of walrus or bearded-seal hide stretched over a light, flexible wooden frame, and was propelled by oars and/or by a sail.

women Stemming the Current towards us, but the Sight of the Strange boats & Sails caused a great alarm. Seeing by turning round they could not escape they drew their boat a little up and run for their lives, no ways endangered by us. As we could not speak their language, it was useless for us to attempt bringing them to a meeting. None of our repetitions of the words Teymak,[29] Kabloonak,[30] &c. had any effect. At 10 A.M. we landed to breakfast. While preparing this meal, we were hailed in a very rough manner by a Single Esquimaux coming down, apparently after us, in his Small Canoe or Kayak, making many threatening gestures with his long Knife, being quite naked. As he perhaps did not consider it safe to pass below our Boats, the Channel being not more than from 20 to 30 Yards wide, he landed. I then walked a little distance from the Fire without arms and invited him over, more by signs than words. He at length put on his Cloathes, Shirt & Trousers, & came over to our Fire. He was, a tall, active, and well-shaped man, with a very pleasing Countenance. As far as I could make out his meaning he said that he had seen the Whites who went that way before[31] & that he did all in his power to prevent his tribe from pillage & stealing the Whites' property. Notwithstanding these favourable traits which he gave of his own Efforts to Assist on that occasion, it does not appear that he considered the violation of the 8th Command as a very heinous offence. While drawing our attention to what he was saying he was actually endeavouring to conceal a Small Tin pan which he had been allowed the use of to drink with. Some victuals were given him on a plate. The latter after having emptied of the Contents he took to his Canoe & put it away. A Knife & Fork was the next trial. Our vigilance detected all his attempts and the articles taken back, telling him as well as we could it was bad to Steal. He was however presented with a few useful articles which seemed to please him much. After leaving the place he followed the Boats about 1½ mile & left us. Those that had seen us previously were of his Family & Camp. We continued with a fair breeze untill 4 P.M. when we took a Western Channel which turned off near an Esquimaux hut of not a very old date. This we found too Shallow. It Spread into a Small Lake, which was not of sufficient depth to float our Boats. We retraced our way & taking the principal Channel that continued on, at 6 P.M. got to the Sea Coast about 3 or 4 Miles to Westward of Pillage Point[32] which was Shewn us by Frans. Felix.[33] We landed for an observation taken by Mr. Simpson, and got Clear of the Shallows near Shore when the breeze favouring, hoisted our Sails. Tent Island[34] in view. We were

29 *Teymak* or *Chimo* was the standard greeting in Inuktitut.

30 *Kabloonak* or *Qablunaq* – the Inuktitut name for a white man.

31 This reference is to an incident involving Sir John Franklin's party at Pillage Point on 7 July 1826 as they headed west from the Delta along the arctic coast to Return Reef. Franklin, *Narrative of a Second Expedition*, 105–6; Davis, ed., *Sir John Franklin's Journals and Correspondence: The Second Arctic Land Expedition 1825–1827*, 201–5.

32 So named by Franklin because of the incident of 7 July 1826 (Franklin, *Narrative of a Second Expedition*, 193–204; Davis, *Sir John Franklin's Journals and Correspondence*, 201–5).

33 Who had been a member of the crew of Lieutenant Back's boat *Reliance* on Franklin's expedition (Franklin, *Narrative of a Second Expedition*, 96).

34 The most northwesterly island in the Mackenzie Delta.

not long Clear from the Shoals, when we saw some Kayacks out from the Island coming after us. Four first appeared, but as we advanced the number increased latterly to above 20. We did not Stop for them as the lightness of these small craft enabled them Easily to Keep up with us. They were allowed to come alternately near our boats and each got a Small present without allowing any of them to hold by our boats or oars. This rendered them Clamorous to barter whatever they had, Bows, arrows, &c. We bought a few, and then desired them to leave us & return to their Camp. This they understood perfectly well, but said they would Cross the Bay and land with us to pass the night. In Short they seemed so intent upon it, that in order to show them we were not to be trifled with I showed them we had firearms, which not yet having the desired Effect, Fired a Couple Shots over their Heads. They all dodged down their heads, vociferating the Words Kaw, Kaw, Kaw, with Signs to us to desist. They were told to return which they did and left the Boats free. The Wind turning to S of West, we had a long and fatiguing pull at the oars to get to Shore near Escape Reef[35] at 11 P.M., where after taking our Supper, we were so much tormented by musquitoes, that we remained but 4 hours.[36] The wind had veered to NW and was working the Boats too much. We pushed off.

**10 July 1837.** Had a partial sail wind but the wind again turned and came ahead after we had passed a Reef Point[37] with many huts on it not long abandoned. Large piles of Drift Wood have been Collected near them. The oar was plied untill we got to Gravel Reef[38] which, affording a Secure Shelter for our Boats at Same time Safety from any Surprise from the Natives, we put ashore and looking out from the Top of the bank about 80 feet could see that the Ice was at no distance a head of us. Many huts and frames of boats were on the Point near us. Some property also left in one of the Houses. Several Bones of Whales were lying near us but these appeared to be of an Old Date as they were partly rotten & decayed. The Keel or bottom piece of one of their Boat frames[39] measured 29 feet 4 Ins. length. The longest thaft 5 ft. 4, depth 2 ft. 2 in. to the Ribbon or thaft but the Gunwale is from 8 to 10 Inches above that.

**11 July 1837.** The wind moderated in course of the night. At 4 A.M. we pushed off, and soon got to the Ice. After trying several places, we were obliged to go along Shore, and pushing through,[40] succeeded in bringing the Boats to a good harbour at Point

35 So named by Franklin on 8 July 1826 because of his party's escape from the unwanted attentions of the Inuit (Franklin, *Narrative of a Second Expedition*, 111).

36 Here large numbers of burbot (*Lota lota*) came in on the rising tide and some were speared (Simpson, *Narrative*, 111). The party's departure, according to Simpson, was dictated by a heavy swell which began pounding the beach.

37 Like Franklin before him, Dease used the term "reef" for the shingle spits, which are such a conspicuous feature of this part of the arctic coast.

38 Shingle Point.

39 This was an *umiak* frame.

40 Progress was also hampered by fog (Simpson, *Narrative*, 112).

Kay,[41] a reef running across part of the Bay (Phillips)[42] and several narrow Channels cutting through it. We attempted inside the Reef but not depth enough for our Boats. A Couple Esquimaux huts were on a Reef about a mile from us seperated by a Channel. Outside the Ice Close packed over the Whole Bay and no open Water appeared out to Seaward. About an hour After we had landed 2 men came to the small Channel that Cut through the Reef, and began an oration, which we could make nothing of. I told them to bring us fish if they had any, being provided with a Short Vocabulary from Captn. Parry's voyage.[43] They went to their tents and soon came with their whole families in a Boat, the party Consisting of 4 men, 4 women & 6 Youths & Children.[44] They brought a few fishes of the *Bear Lake herring Species*[45] and a few pieces whalebone. Each of them were gratified with a Small present and their fish paid for. They did not Show that prying disposition that I Expected, or attempt to purloin anything. True, they were too few to act out with any boldness. After remaining about two hours they left us for the day. Mr Simpson got an observation. I collected a few flowers in course of the day. The Seals were very numerous out on the Ice; could see I think near a hundred at a time.[46] A watch was Sett for the night.

**12 July 1837.** We had a visit from our Esquimaux friends who brought 4 Small fishes. Their visit was not a long one. The Weather very thick & foggy. The Ice has been driven in Closer to the Shore. The tide rose at 1 P.M. 9 Ins. inside the Reef where the Boats were Stationed. A few flowers were Collected but the Weather being Wet and but little wood to Keep the damp from the Paper in putting them up,[47] did not go far from the **[13 July]** Camp. The Wind Changed a little in the afternoon to SE and towards evening the Ice moving a little. The tide at 1 A.M. much the Same as yesterday; the fog Still Continues. Seals very numerous out on the Ice. Some ducks flying over but none killed.

**14 July 1837.** The Fog Still Continues. At 8 A.M. we pushed off along the Bay through the narrow passages formed between the Pieces of moving Ice, by running out and in as the Channels permitted us to get on. We put ashore for breakfast at Point Stokes.[48] Here we saw Huts of the Esquimaux with their property in them, Closed up untill

---

41 Named by Franklin after some relatives on 15 July 1826; Emily Porden, the sister of his first wife, Eleanor, had married Henry Kay (Davis, *Sir John Franklin's Journals and Correspondence*, 220, n. 860).

42 Named by Franklin on 15 July 1826 after Mr Phillips, professor of painting at the Royal Academy (Franklin, *Narrative of a Second Expedition*, 125); he had painted portraits of both Franklin and Richardson before they started on their expedition (Davis, *Sir John Franklin's Journals and Correspondence,* 221, n. 866).

43 Captain William Edward Parry's account of his second arctic voyage to Igloolik, which includes an Inuktitut vocabulary (Parry, *Journal of a Second Voyage*, 559–71). Dease had carefully copied this vocabulary into one of his field notebooks.

44 Three men, five women, two boys, and several children, according to Simpson (*Narrative*, 113–14).

45 Bear Lake herring: cisco, or lake herring (*Coregonus artedii*) (Scott and Crossman, *Freshwater Fishes*, 236–43).

46 These were ringed seals (*Phoca hispida*) or bearded seals (*Erignathus barbatus*); both species spend hours basking on the ice near leads or holes at this time of year.

47 I.e. there was little firewood to dry the paper in the plant press.

48 Named by Franklin after Mr Charles Stokes, naturalist and amateur geologist, on 16 July 1826 (Franklin, *Narrative of a Second Expedition*, 126; Davis, *Sir John Franklin's Journals and Correspondence*, 222, n. 867).

the return of the inhabitants, who it appears are off Somewhere. Pushing off from there we found the next Bay more free of Ice than we Expected. Passing along the Point after rounding P. Stokes we saw two Esqx. families, consisting of 3 Old men,[49] 2 young, 4 women and 3 Children. Landed and got some Bear Lake herring & 2 fine Salmon Trout[50] fresh. The Weather Cleared up. A large Seal was Sporting about near the Tent, and being apparently tamed by them, we did not attempt to molest it. Gave a Small present to the natives & proceeded. Between Herschel Island[51] & Shore the Ice was thick but the wind had seperated it sufficiently to afford us a passage under easy Sail.[52] We met 2 Boats with families in them and gave a small present to each & continued, they having nothing to trade. The Ice appears Close packed to Northward or outside of the Island. Having an Easterly breeze we got through the Channel. Saw some Tents on the Point of Mountain Indian River.[53] but did not land untill we got to the End of the Reef attached to the Island Westward, on which we took supper. Saw many traces of Esqx. having lately been here, and apparently boiling down blubber, as the Gravel & Stones near the place were Stained with it. We saw the Bones of whales in Several places but of an Old Date. One particularly of the head Seperated and the part seen although not quite half the breadth measured near 5 feet, by which it appears the whole breadth of the head bone would have Exceeded 10 feet.[54] We pushed off at 11 P.M., wind favouring, and a Space along the land clear **[15 July]** Enough to allow us to profit by the opportunity, took us to Demarcation Point[55] for breakfast, after which pushed off again. Saw a large Herd of Rein Deers[56] along Beaufort Bay,[57] and a Camp of natives on the Reefs. The wind being fair & brisk we did not Land. Some of them attempted to come after us in their Kayacks but abandoned after pulling Some time without gaining any thing upon us. At Point Humphries took our Supper & there being a Space along Shore pushed off again, which soon closed. After proceeding but ½ hour from Point Griffin,[58] we landed.

---

49 One of whom was blind (FNB B, p. 1).

50 Arctic char (*Salvelinus alpinus*). The Inuit had caught these fish in a net made of strips of baleen (Simpson, *Narrative*, 113).

51 Named by Franklin after the astronomer Sir John Frederick William Herschel on 15 July 1826 (Franklin, *Narrative of a Second Expedition*, 126; Davis, *Sir John Franklin's Journals and Correspondence*, 221).

52 According to Simpson (*Narrative*, 115), the strait between Herschel Island and the mainland was open.

53 Named by Franklin on 17 July 1826 after Indians from the interior (Gwich'in) who used this route to come north to the coast to trade with the Inuit. Now the Firth River, named after John Firth, patriarch of the Firth clan of Fort McPherson.

54 This would have been the skull of a bowhead whale (*Balaena mysticetus*).

55 Named by Franklin on 31 July 1826 as lying on the boundary with Russian America (141°W) (Franklin, *Narrative of a Second Expedition*, 142; Davis, *Sir John Franklin's Journals and Correspondence*, 242, n. 931).

56 The North Slope, i.e. the lowlands between the British and Brooks ranges and the Arctic Ocean, are still the calving and summering grounds of the Porcupine caribou herd.

57 Named by Franklin on 3 August 1826 after Sir Francis Beaufort, hydrographer to the British Admiralty (Franklin, *Narrative of a Second Expedition*, 144). Early in the twentieth century, Beaufort's name was transferred to the Beaufort Sea, the marine area north of the Mackenzie Delta and Herschel Island and west of Banks Island.

58 Named by Franklin on 3 August 1826 after his wife's family (Franklin, *Narrative of a Second Expedition*, 145).

**16 July 1837.** The Tide rose 8 Ins. at 1 A M., the Wind NE by E. Cleared the way to enable us to push off at Midday with a breeze that favoured us untill we got near Point Manning. The Ice packed on the Reef. We landed, and afternoon got the Boats & Cargoes Carried across the Reef. A fine view of the Romanzoff Chain of Mountains[59] from the Reef about 20 miles distance, their Tops covered with Snow. Finding water enough inside the Reef, with a favourable wind we Steered between the Reef and a Point bearing N of West, with an Island Eastward of it, which answers to Barter Island[60] of Sir John Franklin; the N Side composed of high mud banks for about 2 miles. Passed on. Saw a Solitary Esquimaux lodge on the Reef west of it, whose inmates were apparently enjoying their repose about 1 A.M., guarded only by a Dog laying near the door of the Hut, apparently as unconscious of Strangers being within 60 Yards as the Inmates. Nor did we disturb their peacefull slumbers. We proceeded untill 5 A.M. of the **17 [July]** when we found the Ice so packed that we had to land near the point of a Reef on which we could see some Esquimaux Tents, who no sooner perceived us landing than in great agitation they made preparations either, I presume, for attack or defence. The party who came Shortly to hail us Consisted of 22 men, about as many women and the same number of children. When about 80 Yds from our Tents, which had been erected, I advanced to them with friendly signs & told them they must not come nearer. A fire was made, round which, when all seated, a trifle was given to each as a present.[61] Some, however, to double their portion, endeavoured Slily after getting the present, to remove to the end of the Circle, in hopes of Escaping detection, but laughing when pointed out. After having distributed our presents a few Pairs Sealskin Boots [that] were brought by them, I purchased.[62] They are very keen traders always offering the least valuable articles first & holding out to get as much as they can for them. Several of their mouth ornaments[63] they willingly parted with, as well as Bows & arrows. From what we could make out from their signs, with the few words we used, they are apparently on their way to the Westward to meet a party from that Quarter to barter with them, but at what particular Spot we could not learn. One of them who had a bold impudent appearance, after returning some article he got for a pair of mouth ornaments & wishing for an additional payment which was refused him, to shew his dissatisfaction Seized a Tin wash hand basin & was walking off with it very deliberately untill I took it from him. He seemed to hesitate how to Act or rather whether he should not endeavour to keep it against our will, but gave up the article. Another very boldly turned, Shewing his long Knife to Mr. Simpson who was sending him from near the Tents to their

59 Named by Franklin after Count Romanzoff, the Russian Chancellor (Franklin, *Narrative of a Second Expedition*, 145).

60 So named by Franklin on 4 August 1826 for a large camp of Mackenzie Inuit who had come here to trade with Inuit fom the west (Franklin, *Narrative of a Second Expedition*, 147; Davis, *Sir John Franklin's Journals and Correspondence*, 249, n. 956).

61 These gifts were knives, fire steels, awls, files, beads, rings, and needles (FNB B, p. 2).

62 Enough for the entire party, according to Simpson (*Narrative*, 119).

63 Labrets, in this case made of ivory and blue beads, worn by the men in the corners of their lower lips.

prescribed limits.[64] The morning was very Cold, and the day, although the Weather Cleared a little, Continued uncomfortable, raw & chilly. The Wind Continuing and an opening appearing a little distance out from Land, we pushed off about 3 P.M. to endeavour to get through. The Ice moving in large masses retarded our progress much. After a laborious Struggle to reach an open Space about 3 miles from Land, we found our retreat Closed up & that the Ice, driven both by wind and Tide, was Closing fast upon us. Our Boats much in danger of being Crushed. After several escapes by Jumping out and rising the boats as much as we could, the Pollux was Completely Squeezed. The Cargo Carried over the broken pieces upon the Oars, Serving as a Bridge, & the Boat also dragged over, by both Crews, in imminent danger of falling through. After two hours hard work we got both boats aside and drew them upon a large floe as far as we could, and in this Situation to pass the night as cold and uncomfortable as can be imagined. The Ice moving but not to allow us a passage **[18 July]** untill 9 A.M., when by Pushing through broken pieces, and after a long and fatiguing trial, we made the land about 3 miles to the Eastward of where we landed yesterday, while a warm breakfast was a most acceptable meal, after having been in the Boats 12 hours most of which we could not keep ourselves Warm enough to prevent Shivering, & the teeth Chattering. Our Friends the Esquimaux it seems did not see us. None of them came from the Summit of the Banks which near our Camp are about 90 to 100 feet high. The Ice appears packed as far as we can see. The Wind NE & Weather hazy. A fine White Owl[65] was seen by Mr. Simpson but too Shy to get a Chance of firing at it. To the E of us a Small river Channel is now nearly Dry, come from the S. and E. The Channel about 70 Yds. wide at high water. A Small Stream to the Westward running from a Small Lake.

**19 July 1837.** At 10 A.M. two of the Esquimaux who have been out hunting, apparently, Saw our Tents. Shortly after 11 of them Came with Sealskin Boots & some other trifles to trade. To avoid any more difficulties a line was drawn for them about 60 yards from the Tents & none allowed to pass it. At first they sett high prices on their articles, but when disregarded came down to whatever they could get.[66] Some of them remained all day. The day mild and Foggy. Tide rose 8½ Ins. The Ice moving with the Tide, an opening forming.

**20 July 1837.** At 30' past midnight the wind being still from the same favourable Quarter, after pulling out for about 2 miles we got a Space Clear enough to hoist our Sails which took us on at about 5 or 6 knots, and we landed for Breakfast at Point Bullen. From there we hoisted our Sails again but the Ice ahead seems packed in Shore in the Bay and as far out to Sea as we can distinguish any thing. We therefore had to land in the Bay.[67] The wind increased to a Storm. Got the Tents Erected

64 This limit was set at 6 yards (5.4 metres) from the tents (FNB B, p. 2).

65 Snowy owl (*Nyctea scandiaca*). This species breeds on the tundra throughout the Arctic, nesting, of necessity, on the ground.

66 They wanted knives and daggers, but settled for awls, beads, rings, buttons, etc. (FNB B, p. 4).

67 Foggy Bay (Simpson, *Narrative*, 124); Franklin had been detained here by fog on what he called Foggy Island in August 1826 (Davis, *Sir John Franklin's Journals and Correspondence*, 257, n. 981).

& collected a few Plants. No variety found in these low Swampy grounds. Not a Shrub Except the Creeping Willow[68] to be Seen & that does not attain the size of ½ an inch diameter. We have a range of mountains yet in view. The Foggy Island of Sr. John Franklin in sight. Rein Deer tracks very numerous, but none seen. The traces of Natives out hunting of a recent date are also here. Some of the men were out, they say, about 4 miles, but the weather too thick to see any distance.

**21 July 1837.** The Same Weather as yesterday Kept us to the Same Spot. McKay & Sinclair went out again but could see no opening near land or out at Sea. A quantity of Sea Wrack[69] has been brought ashore since yesterday, and some small pieces of branch Sponge. A Bustard and 4 brown Ducks were killed to day.[70] The Lat. of this Place by Observation 70° 10'N.

**22 July 1837.** At 7 A.M., the wind seeming to have moderated, got the Boats loaded & pushed off. Weather thick & foggy. After being out 2 hours wind too near and increasing, bearing us to leeward on the Ice which we narrowly Escaped. After making Several tacks, got to Shore about 3 miles Westward of our Yesterday's Station & passed the day.

**23 July 1837.** At 8 A.M. the wind had moderated to allow loading the Boats. We accordingly Started and put up the Sails. A Broad Space of open Water out to Seaward but much Ice packed in Shore. This obliged us to Keep out from 5 to 7 miles from land. Passed Foggy Island, Point Anxiety & Chandos. The Stream Issuing from Yarborough's Inlet formed a Channel through the Ice, which enabled us to approach nearer the Land, and we had the Satisfaction of taking our Supper on Return Reef of Captn. Franklin.[71] The Ice packed outside the Reef but the Inside Channel free of Ice. We therefore took the Inside & pushed off at 10 P.M. with a light fair wind; weather Clear. Passed Point Back & Point Beechy.[72] The Reefs Continuing outside almost parrallel with the Coast, running rather out **[24 July]** at their western Extremity and 2 Islands appear at the termination of them from Point Berens[73] covered with verdure, but not very high. Still Keeping the Same Course. Several mounds, as if Artificial, appear along the Coast but on approaching, are caused by the pressure of the Ice against the Banks, either in the Fall of the year, or Breaking of the Ice. About 10 A.M. the wind became stronger, the fog more dense. The Bay[74] along which we are coasting Shallow. Had to Keep out Several miles, occasionally taking ground, without seeing the land. Wind increasing, had to keep out, & fell in with reef, Without being able to Land. On pulling out from the Shoals a distance we got among Pieces

---

68 Arctic willow (*Salix arctica*).

69 Seaweed.

70 Large numbers of seals were seen in Foggy Bay (Simpson, *Narrative*, 125).

71 The farthest point reached by Franklin and his party on 16–18 August 1826 (Franklin, *Narrative of a Second Expedition*, 159–66).

72 Seen in the distance by Franklin from Return Reef and named after Captain George Back and Captain Frederick Beechey respectively.

73 Named by Dease and Simpson after Henry Berens, a member of the Committee of the Hudson Bay Company (Simpson, *Narrative*, 129).

74 "Named Harrison's Bay in honour of the Deputy Governor" (Dease's original note).

of Ice and soon to the borders of field Ice which appeared as if it had not been broken yet. Sailed along it for some time when, finding ourselves in a narrow Stream of Clear water with the Ice Close on each side, obliged us to back round & endeavour to make our way to the inside of the Ice. With some difficulty we got round a reef and, the Boats for some time in a dangerous position, to get over a Shoal Covered with pieces of floating Ice, much agitated by a Strong wind which prevented us from managing the Boats as we wished, and with difficulty we got them over the Shoal which was our only alternative, otherwise they must have been broken among the heavy Ice, having no room to Sail Clear. The weather very Cold and unpleasant, the thick fog Keeping us Constantly wet. At 10 P.M. after being looking out a long time, we thought we descried Something like a land fog. And Shortly After had the Sight of Land which we made for and got ashore at 11 P.M. after 25 hours labour, of which we had been 12 without seeing land. The Point was Called, from the gratifying enjoyment of a warm meal after landing, Point Comfort.[75] It forms a projecting point with a large opening to Westward, where from the freshness of the Water, we are inclined to think a River of Considerable magnitude falls in.[76] Fuel rather Scarce, but traces of Esquimaux having been here Since Spring were visible. The wind turned round to Northward, which is directly against us, and kept us here **[25 July]** all day. Although the Fog Continued, partial Clearing enabled Mr. Simpson to get observations, which place this point in Lat. 70° 43'N. Long. W. 152° 14', varn. 43°E. The Channel W of us is that of a River, the water quite fresh with a Stream.[77] The tide rose only 11 Inches. Several of the men looking about saw vestiges of Esquimaux but not very recent. The partial Clearance of the Fog Shews a Point out to NE of us about 5 miles from us; and to Eastward saw the Mountains, or a range of them.[78] The wind and fog Continued all day & Kept us ashore where fuel is rather scarce; what is found mostly Saturated, difficult to burn.

**26 July 1837.** The weather Cleared up at 7 A.M.; wind north. Had to pull against it for 1½ hours untill we reached a point of an Island, where we landed for Breakfast.[79] Here again are vestiges of Esquimaux having worked wood & Killed Rein Deer. Some pieces of the Antler found about appear to have been sawed off with a regular Saw. The Ice close in Shore but got a passage through Small Channels to get round, when the land trending more to the Westward. Passed 3 Points & got to a large opening, water fresh, appearance of a River of Considerable breadth.[80] Kept

75 Simpson (*Narrative*, 131) stressed that they would all have been even more uncomfortable, had it not been for the waterproof boots they had traded from the Inuit.

76 "Named the Garry in honor of Ns. Garry Esqr." (Dease's original note).

77 This river (the Garry) was found to be 1.6 kilometres wide, and its banks thickly covered with driftwood; the water off its mouth tasted fresh for several miles (Simpson, *Narrative*, 133).

78 "Named the Pelly in honour of the Govr." (Dease's original note).

79 "The point named from the Cape of the Mainland Cape Halkett, Lat. 70° 48'N, Long. 151° 55'W" (Dease's original note). This cape was named after one of the Company directors (Simpson, *Narrative*, 133); the large bay extending from here east to Point Berens was Harrison Bay.

80 "Named after Wm. Smith Esqr." (Dease's original note). Smith was the secretary of the Company (Simpson, *Narrative*, 134).

the northern side along the Ice which continued at no distance from Shore, the distance about 35 miles which brought us to a Stop,[81] where the Ice seemed fixed ashore. In wishing to follow an outer lead of water the head of the rudder of the Pollux was broken in wishing to avoid a piece of Ice, the boat going fast before the Wind. We had to pull ashore against a heavy Swell direct ahead. Wind Strong from NE. Cold raw weather.[82] A deer Pound near where we landed, with heaps of moss raised so as to resemble a man in a sitting posture.[83] Some deers were seen at a little distance. Peter Taylor & James McKay went out but could get no chance to fire at them.

**27 July 1837.** The repairs of the Rudder brought us to 1 P.M. before we embarked when, by following an opening formed since last night, took us into a Bay[84] defended from the Ice outwards[85] by a range of Shoals which we followed under Sail. Occasionally taking ground & Crossing the Ends of them that seem to run parrallel, or rather approaching Shore in the Bay,[86] we Continued with a good breeze most of the time untill 11 P.M. when we put up on a high and projecting Point where the Ice has been heaped very high in the Fall & fixed ashore but for a narrow lane between these ramparts of Ice & Land by the dissolving of the Ice & the land Waters. The Weather all day tolerably Clear but very Cold. At this Point are the remains of an Encampment of last Autumn.

**28 July 1837.** This morning at 7 A.M. we pushed off with a Strong Wind but only advanced about 3 miles when we found the Ice so packed & Close in Shore that we were obliged to land, & from the highest Spots we Could find no open Water to be seen. Walked about 4 miles along the Coast, but without any prospect of advancing for some time or untill Some Clearance is made. The Country low, Swampy, with many Marshes or Small Lakes; the Soil perfectly frozen at the Depth of 4 inches from the Surface.[87] Esquimaux traces of last Season yet to be seen all along. In Some parts where they have travelled with their Dogs & Sleds the marks appear plainly.

**29 July 1837.** The wind Continues the Same. A thick mist prevented any observation being taken to day. A Great many ducks passing by large flocks and generally near the land afforded some Sport by taking a Seat along the edge of the Ice. 30 were killed and were found pretty good Eating but rather a Strong Oily taste. They are large birds and I believe called the King Duck.[88] A Seal was Shot to day by the men but sunk too suddenly to get it.[89]

---

81 "Point Pitt, named after one of the members of the Hon. Committee" (Dease's original note).

82 The boats were encrusted with ice (Simpson, *Narrative*, 135).

83 The two sides of the pound converged inland to a lake where the driven caribou could be speared while swimming (Simpson, *Narrative*, 136).

84 "From Pt. Drew named after one of the Hon. Committee, Richd. Drew Esqr." (Dease's original note).

85 I.e. on the seaward side.

86 "Named after Geo. Simpson Esq., our worthy Governor" (Dease's original note).

87 The entire North Slope is underlain by continuous permafrost; a depth of only four inches for the active layer, i.e. the seasonally thawed layer, is remarkably shallow for the time of year, but is quite feasible in a peaty area.

88 King eider (*Somateria spectabilis*).

89 Whether ringed seals float or sink depends on the thickness of the blubber and the salinity of the water.

**30 July 1837.** No Alteration in the appearance of the Ice. Observations were obtained both for time and Longitude.[90] The weather Cold; a thick Mist or fog most of the day which very soon wet the clothes through when Exposed to the Air. No Specimens could be Collected on account of the Wet. Fuel very Scarce, what is found at Some distance being merely enough to cook our victuals & that quite Saturated. A number of Ducks killed today.

**31 July 1837.** We made a move by pushing among the broken pieces of Ice along Shore, but only advanced about 1½ miles or 2 when we were under the necessity of Landing again. The Weather Continues the same as yesterday. Seeing no prospect of the Ice Clearing soon to afford us a passage with our Boats, the Distance from our present position to Point Barrow being but 52 miles in a direct line, we came to the conclusion[91] that it would be saving much time, and the object of our Mission equally Satisfactorily effected by a party performing the remaining distance on foot along the coast. Accordingly 5 men were appointed to accompany Mr. Ths. Simpson, viz James McKay, Peter Taylor, Fr. Felix, Pierre Morin and Hector Morrison. One of the Canvas boats was put upon the Frame to be carried that way. Provisions for 10 days with some articles for Presents or trade in the event of Seeing any natives. Every thing prepared for an Early Departure in the morning tomorrow. All the men were equally anxious to be of the party, but as the Boats & property remaining requires to be well guarded we cannot all go, however desirous. The evening was passed in talking of the probable time may be taken to accomplish the voyage, all in high Spirits. The Canvas Canoe seems to be Sufficiently tight to perform any Short distances such as Crossing a River or a couple miles without a heavy load, or to have no more than the weight of 3 or 4 men in at once. The Country behind much covered with Lakes or marshes; no larger plant than the dwarf Creeping willow, and Several Mosses.

**1 August 1837, Tuesday.** The party mentioned left us at 8 A.M. today. The weather Cold with a dense wet fog; Wind NE. George Sinclair & George Flett were Sent to accompany them untill tomorrow when they will come back & report the progress, and accompanied by our most hearty wishes for success. The Boats and baggage were put in Safety and under Cover of the Oil Cloths. At 1 P.M. the tide rose 15 Ins. The Thermo. at +35°. The Country about us affords no hope of any variety of plants or change of Scenery. All is alike: low Swampy, wet grounds, neither adapted for Exercise or amusement.

**2 August 1837.** The Weather at 6 AM turned out fine & Clear and continued so untill 8 P.M. The Thermo. rose to 46° in the Shade and for the first time Since we are on the Coast Butterflies were Seen. At 9 P.M. Sinclair & Flett came back; left Mr Simpson and party this morning 18 Geo. miles from here. They crossed 2 Rivers yesterday where the Canvas Canoe was of great Service to them. They got beyond a

90 "Boat Extreme is in 71° 3'N Lat., 154° 26'W Long., varn. 42E" (Dease's original note); but Simpson records the position as 71° 1' 44'N; 154° 22' 53"W, var. 42° 36'18"E (Simpson, *Narrative*, 140).

91 According to Simpson (*Narrative*, 141), he was responsible for the idea: "I therefore lost no time in imparting to Mr. Dease my desire of exploring the remainder of the coast to Point Barrow on foot."

range of reefs & hope to get on better now but may take 4 days more, including this one, to reach Point Barrow. They saw fresh vestiges of Esquimaux and Expect to see some Soon. These two men Saw Several Deers today but could not get near enough to fire at them. They brought a White Owl Killed by Sinclair but too much injured to serve as a Specimen. The tide rose the Same as yesterday.

**3 August 1837.** Although the sun appeared this morning, the mist was dense and did not Clear away untill 2 PM. when the weather became fine & Clear. The Thermo Stood at +39° at 1 P.M. The Tide the same as yesterday. The Sun disappeared below the horizon at 11.20 P.M. This is the first time we could notice the Setting sun since it dips below the horizon.[92] The Ice has broken considerably along Shore and Clearing outside.

**4 August 1837.** The day was fine & Clear. At 1 P.M the Tide was 15 Ins and was Steady near an hour. The Thermo. at 2 P.M. was +55°. All our things were Spread out today to dry, as Every thing has got very Wet and Damp in the Weather we have had.

**5 August 1837.** This morning at 6 A.M. 2 Oomiaks, with a family of Esquimaux in Each, hailed us from a large piece of Ice they stopped at about 400 Yards from Shore. They were answered by the friendly Word Teymah, and desired to Come ashore which they did. The party consisted of 2 middle aged men, 2 young lads, 2 Women and 4 boys & girls. They had a few fishes of the Bear Lake Herring Species, of which they gave us readily a Share, and were paid for with Beads, needles and awls. They used a Substitute for Tobacco & had pipes, and seemed very fond of Smoking, but seldom inhaling more than six or seven draughts of the Smoak which is retained and all drawn into the lungs, when it produces Coughing and a Kind of Sickness which appears to be the Chief Enjoyment and Object with them in using it. The Women were Employed mending Sealskin Boots & Shoes for the men, untill about 5 P.M. afternoon when they left us to proceed on their Journey to the Westward, from whence they Came & have been to the Eastward to barter with those of that Quarter. About 2 hours after their Departure it was discovered that they had committed a theft out of my travelling Case (which my Servant had neglected to lock as usual[)]. This Consisted of 3 Tea Spoons and part of a Gun rod which Screwed in 3 pieces. The Articles not being of much value I did not think it worth while pursuing them for. The Tide rose the Same as yesterday. The Thermo was at +57°. The day was tolerably Clear throughout.

**6 August 1837.** Between 5 and 6 A.M. this morning we were agreeably surprised by the distant Sound of a Boat Song which in a few moments we Knew to proceed from the party gone with Mr. Thos Simpson and we rejoiced in seeing a Flag Hoisted in the stern of an Esquimaux Boat and all our friends land in Safety. Particularly have we reason to be thankful to the Almighty for his mercies in having granted us the Complete accomplishment of our Object, Point Barrow having been attained on the morning of the 4th. The Party, as Expected, on the afternoon of the

92 This was probably not the first time the sun had dipped below the horizon, but the first time the clouds had allowed them to see it (FNB B, p. 11).

2nd found a family of natives at the beginning of a Deep Bay,[93] which by the account of these people would require several days to make the Circuit of by Land. Therefore Mr Simpson proposed to them to lend their Boat for the voyage which they agreed to for a hatchet as payment. By this means they were enabled to proceed more Expeditiously and, as Stated above, got Early in the morning of the 4th to the Point Barrow, where the Ice was fast ashore to Some distance in the Bay Eastward. A Camp of natives were found there, Consisting of about 40 to 50 men & families, who received our Friends in a very friendly manner, and willingly disposed of any article they wished to purchase from them, and not so Extravagant in their demands as those we saw to the Eastward of this. Seem to Know the use of Tobacco and are very fond of it. An Elderly man among them took charge of a letter which he said would be conveyed to the Whites by the Interior. The party remained at the Camp about 10 hours. Mr. Simpson having got Observations and ascertained the position, left Point Barrow afternoon of the 4th and travelled since with but little stoppage, Yesterday they took up 4 Esquimaux who were to come with them and take back the Boat, but meeting with the 2 families that were here, who no doubt informed them of the Theft they had committed yesterday, rendered the others reluctant to continue, perhaps fearing that they might be punished instead of the others. They therefore landed & wished to Keep the boat, but were not allowed.[94] The men were permitted to take a few hours rest, after their hard labour and very little rest since the 3rd. In mean time we drew up the Oomiak with its paddles, oars &c., all safe, that the owners may not be disappointed in finding it. At 1 P.M. Loaded our Boats & pushed off from Boat Extreme on our return. A favourable breeze from NW rose, which Continued untill 9 P.M. when it changed to head wind. We land at 10h 30' for Supper. Some Rain fell. Our landing Place was the Entrance of a Small River,[95] where plenty drift wood Enabled us to have a comfortable fire & Warm Supper. The Wind Changed again in our favour, of which we took advantage **[7 August]** & pushed off at 1 A.M., Continuing under Easy Sail untill 11.30 A.M. when we landed on the Island we took breakfast upon the 26th Ulto., named [Cape Halkett]. Weather Cloudy but mild. Wind Ceased; rained a little. The position was ascertained by observations & compared with former ones, courses &c. We pushed off at 4.30' P.M. Wind very near. Kept out along the edge of the Ice[96] but wind falling, plied the oars from 7 PM untill midnight, when the wind having altered a little **[8 August 1837]** the Sails were hoisted and Carried easy untill 4 when the wind increased & turned to Westward with light Showers of Rain. We passed the mouth of a large River, with an Island among its Western Channels; the Middle Channel deep and appears, as far as we could judge, to be about or near 2 miles wide. The water quite fresh a consid-

---

93 Named Dease Inlet by Simpson.

94 This is a simplification of the events. The Inuit tried to drag the *umiak* ashore. Simpson levelled his gun at them. The Inuit drew their bows and there was a tense stand-off till the Inuit backed down (Simpson, *Narrative*, 166).

95 Boat Creek, behind Point McPherson (Simpson, *Narrative*, 169).

96 On a direct course across Harrison Bay towards Point Berens (Simpson, *Narrative*, 169).

erable distance out. This River was named Colville River.[97] The Gale continued increasing and our Boats Stood it well. We had to reef and at last to go under the foresail only, which brought us to a Small River opposite Return Reef named Fawn River, from a Young Fawn, while we were there, having nearly Entered one of the tents. We landed at 3.30' PM. being 23 hours since our last landing. Wood was plenty but the Wind too Strong to Enjoy our Fire. A Warm meal was however very invigorating and rest much required by the men. In order to get observations[98] we **[9 August]** did not leave our Campment untill 1 P.M. when a light wind helped us on to a point Eastward of Yarborough Inlet at 9 P.M. From the Bay Southward a Stream of fresh water makes its way some distance from Shore, which leads us to Suppose a River of Some Size falls into the Bay.[99] We took supper and pushed off again at 11 P.M. with a light NW **[10 August]** wind untill we passed Foggy Island. From there the wind increased again to a heavy gale from WNW, which obliged us to take in Reef. The Seas very heavy, but our little Boats Stood the Gale admirably, without Shipping any water. We passed Flaxman Island;[100] the Ice packed on the north side of it. We passed through much Ice to the East of it, and landed on a reef to take Breakfast in Camden Bay.[101] We pushed off again & Continued to the Reef, West end of Barter Island and landed to take our Supper. Large pieces of Ice along Shore. Cold Cloudy weather. After taking a Warm Supper we **[11 August]** pushed off and Crossed Beaufort Bay.[102] On the Reefs E Side we saw 5 Esquimaux Tents. Being under Sail with a fine breeze did not Stop although many Signals were made to us & Calling out. The Ice, thick and heavy, Kept us near Shore and after having passed about 5 miles Eastward of the Camp we were obliged to Land as no passage was afforded us beyond. We landed near a deep bay inside the Reef we were upon,[103] where some huts Contain the Baggage of the Inhabitants and several deposits of Oil & Provisions were made, buried in the Gravel & Covered with the Same over logs of Drift Wood. While we were preparing to Embark after having taken Breakfast 5 Kayacks Came up, being some of the Party we saw on the 17 & 19th Ulto. The moving Ice enabled us to get through some narrow passages, and although an open Space appeared about 2 miles outside of us, we could not find any opening to get outside the line of Ice which has been driven in by the late Gales of wind. We crossed the Bay & Landed. Our Friends the Esquimaux came up to us in the Evening with their

97 "In honor of one of the members of the Hon. Committee of the Hudson's Bay Coy" (Dease's original note). Simpson (*Narrative*, 173) suspected the Colville was identical with the Lewes (as the Upper Yukon was then known).

98 And to let the cargoes dry, since the sun was now shining (Simpson, *Narrative*, 174).

99 Dease was correct; the river is the Sagavanirktok, entering the sea just east of Prudhoe Bay.

100 Named by Franklin after the recently deceased sculptor, on 6 August 1826 (Franklin, *Narrative of a Second Expedition*, 151).

101 Named by Franklin after the Marquis Camden on 4 August 1826 (Franklin, *Narrative of a Second Expedition*, 147).

102 It was now bitterly cold with several light snow showers; the British Mountains to the south were snow-covered (Simpson, *Narrative*, 176).

103 This was at Demarcation Point (Simpson, *Narrative*, 177).

whole Camp, which they erected on a Reef about ¾ of a mile from us.[104] All came to see us & brought some fresh fish & a few Ducks; the latter we did not take. A few Pairs Boots were purchased from them. They were very desirous of selling any thing they could dispose of. The Ladies appeared no less so of attracting our notice by the most bewitching and significant Smiles and gestures. Whether any of them were Successfull in their Endeavours I know not. Some of them passed the night near the Tents with their Husbands at various Amusements, the females occasionally Singing & the men leaping & playing at games in use among them.[105]

**12 August 1837.** After breakfast this morning an opening formed among the Ice which Enabled us to pull out. The wind being from NE we made but little progress with our oars, over very heavy dead Swells from N & W. We continued along the Ice 4 hours when the Wind freshening & the Swell increasing (being abreast of the British Chain of Mountains) we took advantage of a Narrow Channel to get to Shore where the remainder of the day was passed. Our landing was in a fine Basin at the entrance of a Small River coming from the mountains & Sheltered from all Winds being inside a Sand bar which leaves but a passage of about 10 feet broad to get in & then Extends about 200 yards across. Some Deer were seen, but none Killed.

**13 August 1837.** Passed the day here again. Several of the men were out looking for Deer but no one has been Successfull, unless it may be George Sinclair who has not returned.

**14 August 1837.** This morning George Sinclair brought the meat of a Doe Deer Killed yesterday, which afforded an Agreeable change for Breakfast. The wind Continues direct ahead; a long line of Ice running out between us & Herschel Island, which is in Sight from this place. John McKay Killed & brought another deer today.

**15 August 1837.** The wind fell in the night. The Ice loosened enough for us to push off at 5 A M. The wind ahead yet, what there is of it.[106] Put ashore at 9 to go take our Breakfast & pushed off again. The Ice very thick as we drew near the Channel of Herschel Island but we found it Clear. Entered it at 4 P.M. quite free of Ice as far as we could see. Saw Several Tents of Natives but did not land. 2 Kayacks & 2 boats Came out to us but had nothing to trade. The afternoon was fine, but the Swells against us, as we could not use the Sails. We Continued on[107] untill we got to Point Stokes at Midnight & Landed not far from 2 tents of natives, although not aware they were inhabited untill the Old blind man who we saw near here before Came to us led by a Boy, and accompanied by a young man, 5 Women and as many Children. The men, they Say, are out Hunting in the Interior & will be back in 4 nights. After having taken a Warm meal and some water on board, we Embarked to

---

104 They were travelling in five *umiaks* and pitched five tents (FNB B, p. 16).

105 The Inuit men competed with the men of the expedition at jumping, presumably long-jumping (Simpson, *Narrative*, 178).

106 There was a striking mirage that morning, with ice floes and the hills of Herschel Island severely refracted (Simpson, *Narrative*, 180).

107 The moon rose at 10 p.m., the first time they had seen it since leaving Fort Chipewyan (Simpson, *Narrative*, 181).

**[16 August]** land at Point Kay which we got to at 6 A.M. Landed on the Same Place we were at the 13th Ulto. and allowed the men 6 hours rest.[108] In mean time a Couple of natives were again seen on the Reef in Philips Bay, and they came to us, the party consisting of 2 men, 8 women & 8 children. The men of this Camp are also out hunting Rein Deer, they say. We pushed off at 1 P.M.; fine weather. Near Point King we saw 4 White Whales,[109] the only ones we have seen yet. Seals numerous every where. Saw Several Esquimaux huts but did not wish to land before we would get to gravel Reef[110] which affords a Secure harbour, and an open Place at Same time. Many Kayacks & Oomiaks came out to us inviting us to go ashore and Sleep which we declined. Among them we recognized some of those seen on the 9th Crossing the Bay of Shoal Water. They seemed to wish to give us some information which we could not comprehend. We therefore Continued on to gravelly reef & landed at midnight. We had not **[17 August]** been an hour ashore before an Oomiak with 6 men & 3 women came from behind a point where we had not seen them. Shortly after a party from another Camp ahead of us came & appeared much pleased to see us. Brought us a few fresh fishes & wished to Dispose of their Bows & Arrows which we declined purchasing. The party was augmented to 12 men with a number of women & Children. After allowing our men 4 hours rest,[111] we prepared to embark when it was perceived that a frying pan had been Stolen out of one of the Boats. At this moment a great Stir was making with the Boats to push off, but they were Stopped and accused of the theft, & Shewn another pan with intimation that the Article must be restored before any of them would be allowed to depart. This produced some loud words among them, when an Elderly man went to near where the fire had been made, and taking up some pieces of Drift Wood & removing the gravel underneath produced the Pan, at Same time pointing out the thief who, upon a motion of Mr Simpson towards him Stood on the Defensive with a long Knife in his hand. Others preparing to take their weapons,[112] no doubt Expecting that we intended to inflict punishment on the delinquent, but as our property was returned we did not wish to draw on a serious Quarrel. Therefore, giving them to understand as well as we could that we disliked thieves, allowed them to embark as we did also. They followed us as far as their huts & left us. Several other houses were Seen, some of the natives wishing apparently to communicate some information regarding the entrance of the River, which I took to be that a party of the Loucheux tribe had come down to retaliate for the murder of their friends last Spring & that most of those seen in the Bay on our way to the Westward had gone off to the Eastward. If that was the Subject of their information the Idea of not seeing any more of them was far from being displeasing to us, having but few Small articles of trade

108 The temperature rose to 54°F (12.2°C) and Simpson at least went for a swim (Simpson, *Narrative*, 181).

109 Beluga (*Delphinaptera leucas*).

110 Shingle Point.

111 While Dease and Simpson stood guard on the beach (Simpson, *Narrative*, 182).

112 One man drew his bow, the arrow aiming at Simpson's back (Simpson, *Narrative*, 182–3).

remaining & they having little to offer worth purchasing. We had to pull against a head wind to get to the Channel we descended by,[113] and being at Ebb of the tide had to drag our Boats about an hour before entering the Channel where we put up for the night. Tent Island seems to be deserted as we could see no one through a Glass on it.

**18 August 1837.** We Embarked at 3h 30' A.M.; fine Calm & Warm weather. Had musquitoes during the night. Mr Simpson & myself Keeping watch to allow the men rest during the night.[114] About Midday a light breeze in our favour helped us on. We got to the Large Channel we descended by; our distance about 30 miles from the Coast.[115] Put up at 8.30 P.M. above the first Pines[116] on the way up.

**19 August 1837.** The weather fine & Calm. Pulled at the oar all day. Our progress not very great although the Current is not Strong. Passed at the Deposit Island[117] & found the property had been taken since we passed. Passed the Entrance of the River Coming from the mountains,[118] but whether a branch of Peels River I cannot say, but not the Same marked by Sir John Franklin. The Current seems to be Strong and Sand Shoals formed at its Junction with the McKenzie. By the appearance of the land I think it flows from the West of South through the nearest ranges of mountains. Put up at 8.30'.

**20 August 1837.** The Same fine weather, but a light head wind did not help us. Put up at the usual hour above the lowest Channel of Peels River as Stated in Franklin's Narrative.

**21 August 1837.** We find the Weather so different from what we had along the Coast that a person may imagine himself in a different country altogether. We passed through a Channel of an Island which was very Shallow for a Short distance but brought us into the main one above Peels River. Continued on untill 9 P.M & put up for the night at Point Separation.

**22 August 1837.** A Warm day; Calm. The tracking line principally used this day.[119] At 1.30 P.M. passed at Red River[120] where a Camp of Loucheux appeared much pleased to see us back again. The Chief this day has gone downward with a Party hunting, but we do not hear that they have been down to retaliate upon the Esquimaux. Gave each of them a Piece of Tobacco & left ½ a yard for the Chief. They complain that the fish has been very Scarce in the River this Summer & fear that it will Cause Starvation in Winter. We put up at 8h 30'. Weather Cloudy; a light Shower of Rain fell.

---

113 The West Channel.

114 The Inuit they had encountered over the previous few days had warned them against a possible ambush, presumably by the Loucheux (Gwich'in)(Simpson, *Narrative*, 184).

115 Progress was almost entirely by tracking (towing along shore), the tracking crews being relieved every hour (Simpson, *Narrative*, 184–5).

116 In reality spruce, probably black spruce (*Picea mariana*), rather than pine.

117 Where the Inuit cache had been found on the way down.

118 The Rat River.

119 The boats were tracked, with only one man remaining in each boat to steer it.

120 Arctic Red River.

**23 August 1837.** My Boat met with an accident to day, Striking on a Stone which Split one of the Bottom Streaks[121] and obliged us to unload & pitch it[122] which occupied us for 1 ½ hour ashore. Put up at usual hour.
**24 August 1837.** The Weather fine & Clear. Passed by a Loucheux family from whom we got 2 fine White fishes. A light wind ahead most of the day. Put up at usual time about 3 miles above Old Fort Good Hope.
**25 August 1837.** This morning met a family of Loucheux going down Stream on a raft.[123] Have been hunting deer above and now returning to their Tents. Afternoon saw a Camp of Hare Indians and with them 2 Loucheux on their way to the Fort with some furs. These last, in 2 Canoes, followed and Encamped with us. The Weather not so warm as yesterday; Some Clouds; Wind SE.
**26 August 1837.** Saw Several Indians to day, who seem to be very poor and Complaining of Want. At 4 P.M. the Boat[124] was Stove upon a Shoal. About 6 feet of the Garvel Streak[125] Split near the Keel. Lost 4 hours in mending the Boat and Drying the property.
**27 August 1837.** A fine Clear day this morning. Overtook a Loucheux & family on his way to the Fort, who Joined his Countrymen to follow us. In course of the day we were much Shocked on finding a Child about 5 years old lying Dead near the Water edge, a Girl of the Hare Indian tribe that must have been left by some of the tribe to perish alone, as there were no marks of violence that we saw. It was left recently, not more than 2 days past.[126] Put up at 8h 30' as usual.
**28 August 1837.** A Continuation of the Same fine Warm Weather. Got to Fort Good Hope at 11 A.M. where Mr. Bell & his people were highly pleased to see our Safe return. The water is very low in the River & has been remarkably so all the Season & the fishery very scanty indeed. To allow our men rest we **[29 August]** passed this day also with Mr. Bell[127] & prepared to start tomorrow.

• • •

## *4. Fort Good Hope journal, 28–30 August 1837*[128]

**Monday 28 August 1837.** A fine warm day. At 11 A.M. was agreeably surprised by the early, and safe return of the Expedition from the Sea Coast in excellent spirits

121 Strakes.
122 I.e. caulk the split with tar.
123 Simpson noted that the raft "was formed of two large logs joined by a cross bar thus, A. On the fore part rested a raised platform, where the pasengers sat; and the men escorted these primitive vessels in their bark canoes, which, when they choose, are conveniently secured between the projecting arms of the after-part of the raft" (Simpson, *Narrative*, 185–6).
124 The *Castor*.
125 The garboard strake – the first plank next to the keel.
126 Simpson indicated that the incidence of infanticide such as this had been greatly reduced by the fur traders' strong denunciation of the practice, and that they had spotted the corpse only because they were well beyond where the traders normally went (Fort Good Hope being the most northerly post), in a location where the Indians responsible would not have expected it to be found (Simpson, *Narrative*, 187).
127 Wet bags of pemmican were opened and dried (Simpson, *Narrative*, 190).
128 HBCA B.80/a/14.

and good health. they have successfully accomplished the object of their voyage to their satisfaction, and were received in a friendly manner by the Esquimaux …
**Wednesday 30.** A fine day. About 12 o'clock Messrs Dease and Simpson took their departure for Bear Lake their winter quarters. they will proceed to Fort Norman, unless they meet the Boats from Fort Simpson at Bear's Lake River.

• • •

## *5. Peter Dease's journal, continued*[129]

**30 August 1837.** We left Ft. Good Hope at 2 P.M.[130] Had to carry at the Rapid most of the Cargoes. Put up at 9 P.M at a little River[131] above.
**31 August 1837.** Started at 5 A.M. and got half up the Rapid of Old Fort where we put up, being too late to continue in the Rapid.
**1 September 1837.** Started at 4 A.M. Weather became Cloudy. Afternoon had a Shower of rain. Put up at 8 P.M. in sight of Mammillary hills.
**2 September 1837.** Had a light Wind which helped us with the Sails afternoon and Stopped for the night in the reach of Bluefish River.
**3 September 1837.** Got to the Entrance of Bear Lake River, where finding no marks of any one having passed, unloaded the Boats in order to proceed to Ft Norman to meet our Outfit and Dispatches.
**4 September 1837.** Left George Flett & Hector Morrison in charge of the property and got to Fort Norman at 6 P.M.[132] All well. Mr. Mowat[133] and his man Expecting the Outfits daily.

• • •

## *6. Fort Norman journal, 4–5 September 1837*[134]

**Monday 4th September 1837.** The Expedition gentelmen arrived from the Sea Coast. they have been fortunate enof to Discover Point barrier [Barrow] …
**Tuesday 5th.** The gentelmen very busy writing their letters, as in a few Days they are to send off their Expreys. their soldiers are enjoying themselves dancing in the Hall. a fine warm day …

• • •

129 MCM M214.
130 Rather strangely, Dease fails to mention that his wife, niece, and granddaughter joined the party at this point (Simpson, *Narrative*, 191).
131 Old Chief's River (FNB B, p. 23).
132 On the way upriver to Fort Norman they killed a black bear (FNB B, 25); it was shot in the water by McKay and the meat was eaten (Simpson, *Narrative*, 192).
133 See biography, p. 305.
134 HBCA B.152/a/16, f. 8d.

## 7. *Letter from Peter Dease and Thomas Simpson to the directors, Hudson's Bay Company, 5 September 1837* [135]

Fort Norman, September 5th, 1837

Honourable Sirs,

We have now the honour to report the complete success of the expedition this summer to westward of Mackenzie's River.

Our arrangements up to the departure of the expedition from Fort Chipewyan were stated in full to Governor Simpson and the Northern Council, who, no doubt, communicated them to your Honours; we shall, therefore, present a brief detail of our subsequent operations.

On the 1st of June we quitted Fort Chipewyan with two small sea-boats, accompanied by a luggage-boat and a party of hunters, for Great Bear Lake; visited the salt plains, and arrived at Great Slave Lake on the 10th, where we were detained by ice until the 21st. The same cause prolonged our passage across that inland sea, and, having been for two days stopped by a strong contrary wind at the head of Mackenzie's River, it was the 1st of July when we reached Fort Norman. Our Indians cast up on the following day, and the crews and cargoes were finally divided and arranged. Our boatbuilder, John Ritch, received his instructions to proceed immediately, with a fisherman, two other labourers, and the hunters, to Great Bear Lake, and at its north-eastern extremity to erect our winter-quarters, and lay in a stock of provisions against our return from the coast. We then took our departure, and on the 4th reached Fort Good Hope. There we found an assemblage of Hare Indians and Loucheux. The latter informed us that three of their tribe had been killed, and a fourth severely wounded by the Esquimaux in the preceding month, which at once put an end to our intention of procuring an interpreter from among them, although several volunteered to accompany us in that capacity. They at the same time earnestly cautioned us to beware of the treacherous arts of their enemies. On the 9th of July we reached the ocean by the most westerly mouth of the Mackenzie, which Sir John Franklin sought for in vain. It is situated in lat. 68° 49 ½'N., long 136° 37'W., and perfectly answers the description which the Esquimaux messengers gave of it to that officer when they came to apprise him of the intended attack by the Mountain Indians.

We had proceeded but a short distance to seaward when a party of nineteen men came off to us from Tent Island. We gave each of them a small present, a practice which we continued throughout the voyage, and employed our vocabularies to the best of our ability, and to their great surprise, to explain the friendly feeling of the whites towards their tribe. Being a lively and communicative people, we in the course of the season acquired some facility in our intercourse with them, and when words failed signs supplied their place, so that we seldom experienced much difficulty in making ourselves understood or in comprehending their meaning. When indulged, however, they always became daring and excessively troublesome, and they were ever

135 Alexander Simpson, *Life and Travels*, 229–54; Dease and Simpson, "An Account of the Recent Arctic Discoveries."

on the look-out for plunder. On this first meeting they made several unsuccessful attempts in that way; and it was no easy matter to induce them to return to their camp after we had finished our business with them. They said, they wished to accompany us to our encampment, where they would have soon been joined by fresh parties; and we had a shoal and dangerous navigation before us that night. We therefore peremptorily ordered them back; but it was not until we fired a ball over their heads that they put about and paddled off. A storm soon after arose; but we made the land in safety the following morning at Shingle Point, in lat. 69°, where we were detained until the 11th. The thermometer had already fallen to 48° (Fahr.), being 30° lower than on the evening we left the Mackenzie River; and, instead of the bright and beautiful weather we enjoyed in our descent of that noble stream, we were now doomed to travel in cold dense fogs, which enveloped us during nearly the whole of our progress along the coast. But, although they perplexed and retarded us, we never allowed them to arrest our course; nor did we ever throughout the voyage encamp but when compelled to do so by ice or contrary winds, to which line of conduct may, under Providence, be ascribed the early and successful accomplishment of our undertaking. In the afternoon of the 11th of July we reached Point Key, where we were detained by a compact body of ice occupying Phillip's Bay, until the 14th. There we were visited by another party of Esquimaux, whose tents were pitched at no great distance from us. They live in the country bordering on Babbage River, and informed us that, except when flooded by the melting of the mountain snows, it is an insignificant stream, not fifty yards in breadth; of this we had ocular proof on a clear day on our return. A handsome flora was collected in this neighbourhood. Having found a passage through the ice in Phillip's Bay, we reached Herschel Island the same evening (14th July), and had intercourse with other parties of the natives who were pretty numerous along this part of the coast. We found on the island the skull of a whale eight feet in breadth; and whalebone[136] is everywhere an article in extensive use among the natives, especially for the making of their nets, and the fastenings of their sledges. We continued our route before an easterly wind, along and through the ice, with very little interruption till two A.M. of the 17th, when an unbroken pack, extending to seaward, made us seek the shore in Camden Bay, near a considerable camp of Esquimaux. As soon as the fears of the latter were removed, an amicable meeting took place; and, having made them the usual presents, we purchased a good many of their mouth ornaments,[137] weapons and other articles, which will be forwarded in due course to the Hudson's Bay House. Three of the men were remarkable for their good looks, upright figures, and a stature of from five feet ten to six feet. They gave us a specimen of their dances, and one of them afterwards won the palm from all our people at leaping. They informed us that they have two sources of trade: the first and most regular with their countrymen, who come annually from the westward; the other with the Mountain Indians, who use fire-arms, and travel a great way across land from the direction of the Russian settlements. They showed

136 Baleen.

137 Labrets.

us the knives, iron-kettles, beads, and other things thus procured, which we have no doubt are of Russian manufacture. Their means of repayment appeared to us very limited, consisting in seal-skins, whalebone, ivory, and a few inferior furs, viz., wolvereens,[138] foxes[139] and muskrats.[140] A pair of indifferent beaver gloves were purchased from them, which they had probably procured from the Mountain Indians, for we saw no other symptom of the existence of that valuable animal near the coast, though it doubtless abounds at some distance up the large wooded rivers which we subsequently discovered. In the afternoon there appeared a narrow lane of water stretching outwards, and we immediately embarked. We had advanced about three miles from the land when the ice suddenly closed upon us before a strong north-east wind; one of the boats got squeezed, and it was only by throwing out the cargo upon the floating masses that she was saved from destruction. By means of portages made from one piece to another, the oars serving as bridges, the cargo was all recovered, and both boats finally hauled up on a large floe, where we passed an inclement and anxious night. Next morning the gale abated, the ice relaxed a little around us, and by a long circuit we regained the shore, about a league to the eastward of our former position. There we were detained until midnight of the 19th, when a favourable wind enabled us to round the body of ice at a distance of four miles from the land; and continuing, carried us on the 20th into Foggy Island Bay. There we were stopped by the ice and a violent north-east wind until the 23rd, having on the preceding day made an ineffectual attempt to weather Point Anxiety, in which we narrowly escaped with a thorough drenching. The latitude ashore was 70° 10'. From this situation we had the satisfaction of discovering, during a clear afternoon, a range of the rocky mountains to the westward of the Romanzoff chain, not seen by Sir John Franklin; but, being within the limit of his survey, we called it the "Franklin Range", as a just tribute to his character and merits. On the 23rd we again set sail, rounded the pack of ice which extended six miles to seaward from Yarborough Inlet, then, abruptly turning in, we supped near Return Reef, and the survey commenced.

Return Reef is one of a chain of reefs and islets which runs for twenty miles parallel to the coast, at the distance of about half a league, affording water enough within for such light craft as ours; the mainland is very low. From "Point Berens" to "Cape Halkett" (named after two members of your Honourable Board) it forms a great bay, fifty miles broad by a third of that depth, which in honour of the Deputy-Governor was named "Harrison's Bay". At the bottom of this bay another picturesque branch of the Rocky Mountain range – the last seen by us – rears its lofty peaks above these flat shores; we called them "Pelly's Mountains", in honour of the Governor of the Company. At their base flows a large river, two miles broad at its mouth, which we named after Andrew Colville, Esq. This river freshens the water for many miles, and its alluvial deposits have rendered Harrison's Bay so shallow that it was not until after a run of twenty-five hours, during which we had repeatedly

138 More correctly "wolverines" (*Gulo gulo*).

139 Most likely arctic foxes (*Alopex lagopus*) but possibly also red foxes (*Vulpes vulpes*).

140 *Ondatra zibethicus*.

to stand well out to seaward, that we could effect a landing on a grounded iceberg,[141] nine miles to the south-west of Cape Halkett. A north-east gale kept us there the whole of the following day. The country extending to the foot of the mountains appeared to consist of plains covered with short grass and moss, the favourite pasture of the reindeer, of which we saw numerous herds. Observations were obtained, determining our position to be lat. 70° 43'N., long. 152° 14'W.; variation of the compass 40°E. Next morning (26th July) the tide rose nearly two feet at six A.M., and enabled us safely to cross the shoals. At no great distance from our encampment we passed the mouth of another large river, one mile broad, whose banks were thickly lined with drift timber. We named it the "Garry", in honour of Nicholas Garry, Esq. Cape Halkett forms the extreme point of a small island, separated from the main shores by a narrow channel too shallow for boats. Its situation was likewise found by observation to be in lat. 70° 48'N., long. 151° 55'W. It appears to be a place of resort to the Esquimaux, for we found a spot where they had been building their baidars[142] last spring. We suppose them to have been part of a very large camp, which we saw in the bay of Staines River, as we sailed past the east end of Flaxman Island on the 20th of July; that this camp consisted of the western traders of that tribe, on their annual journey to meet their eastern brethren at Barter Island; and that we missed them on our return, from the circumstance of their being then dispersed along the rivers, lakes, and in the skirts of the mountains hunting the reindeer.

From thence the coast turned suddenly off to the west-northwest. It presented nothing to the eye but a succession of low banks of frozen mud. The ice was heavy all along, but there were narrow channels close to the shore; the soundings on these averaged one fathom on sandy bottom. In the evening we passed the mouth of a considerable river, which was named after William Smith, Esq. From thence, for about six miles, the coast-line is formed of gravel reefs, near the extremity of which, at "Point Pitt" (called after another member of your Honourable Board), the land tends more to the westward. The ice lay much closer here; numerous masses adhered to the bottom under the water, which obliged us to search for a passage out from the shore. The night was dark and stormy, and we were in considerable danger; one of the rudders gave way, but we at length effected a landing on a place near an immense reindeer pound. This was ingeniously formed by the Esquimaux with double rows of turf, set up on a ridge of ground enclosing a hollow four miles by two, the end farthest from the beach terminating in a lake, into which the unsuspecting animals are driven, and there despatched with spears. The vegetable soil in this vicinity was barely four inches in depth, beneath which the clay was frozen as hard as rock, so that our tent pickets could not be driven home. The men had to go a good mile to find a log or two of drift-wood for fuel, the scarcity of which essential article is doubtless the chief

141 Probably more accurately an ice floe; icebergs, large masses of ice which have calved from glaciers, are an extreme rarity in the Beaufort Sea, although ice islands, very large masses which break away from the Ward Hunt Ice Shelf on northern Ellesmere Island, may be carried here occasionally by the Beaufort Gyre.

142 *Umiaks*: *baidar* was the common name for these skin boats in Siberia and Russian Alaska.

cause of the want of inhabitants along so great a portion of the coast. We were detained at this place till the following afternoon (27th), when the ice opening a little, enabled us to resume our route. It blew a cutting blast from the north-east, and the salt water froze upon the oars and rigging. "Point Drew", called after Richard Drew, Esq., seven miles distant from our last encampment, is the commencement of a bay of considerable size, but extremely shallow and much encumbered with ice, in pushing through which the boats received several blows; and we had on this, as on many other occasions, to admire their excellent workmanship. To seaward the ice was still smooth and solid, as in the depths of a sunless winter. At midnight we reached a narrow projecting point, across which the peaks of some high icebergs appeared and were from a distance mistaken for lodges of the natives. This point we named "Cape George Simpson", as a mark of respect for our worthy governor. It was destined to be the limit of boat navigation, for during the four following days we were only able to advance as many miles. The weather was foggy and dismally cold, the wild fowl passed in long flights to the westward, and there seemed little prospect of our being able to reach Point Barrow by water. "Boat Extreme" is situated in lat. 71° 3'N., long. 154° 26'W.; variation of the compass, 42°E.

Under the above circumstances Mr. Simpson undertook to complete the journey on foot, and accordingly started on the 1st of August with a party of five men. They carried with them their arms, some ammunition, pemican, a small oiled canvas canoe for the crossing of rivers, and necessary astronomical instruments, and a few trinkets for the natives. It was one of the worst days of the whole season, and the fog was so dense that the pedestrians were under the necessity of rigidly following the tortuous outline of the coast, which for twenty miles formed a sort of irregular inland bay (being guarded without by a series of gravel reefs), the shore of which was almost on a level with the water, and intersected innumerable salt creeks through which they waded, besides three considerable rivers or inlets which they traversed in their portable canoe. They found at one place a great many large wooden sledges joined with whalebone and strongly shod with horn. Mr. Simpson conjectures that these vehicles were left there by the western Esquimaux already spoken of, on their eastward journey, to be resumed again on their return when winter sets in. The tracks of reindeer were everywhere numerous. Next day the weather improved, and at noon the latitude 71° 10' was observed. The land now inclined to the south-west, and continued very low, muddy, and as on the preceding day, abounding in salt creeks whose waters were at the freezing temperature. The party had proceeded about ten miles when, to their dismay, the coast suddenly turned off to the southward, forming an inlet extending as far as the eye could reach; at the same moment they descried, at no great distance, a small camp of the western Esquimaux, to which they immediately directed their steps. The men were absent hunting, and the women and children took to their boats in the greatest alarm, leaving behind them an infirm man who was in an agony of fear. A few words of friendship removed his apprehensions, and brought back the fugitives, who were equally surprised and delighted to behold white men. They set before the party fresh reindeer-meat and seal-oil, and besought them for tobacco (Tawaccah), of which, men, women, and even children, are inordinately fond. Mr. Simpson now determined to adopt a more expeditious mode of travelling, and applied for the loan of one of their "oomiaks" or family canoes, to convey the party to Point Barrow, with which, from a chart drawn by one of the most intelligent of the women, it appeared that these people were acquainted. The request was immediately complied with; four oars were fitted with lashings to this strange craft, and the ladies declared that our party were true Esquimaux and not "Kabloonan". Before starting, the hunters arrived, and were likewise gratified with tobacco, awls, buttons and other trifles. "Dease's Inlet" is five miles broad at this place, yet so low is the land that the one shore is just visible from the other in the clearest weather. It now again blew strongly from the north-east, bringing back the cold dense fog, but the traverse was effected by the aid of the compass, the waves ran high, and the skin boat surmounted them with a buoyancy which far surpassed that of our boasted north canoes. The party encamped on the west side of the inlet. The banks there were of frozen mud ten or twelve feet high; the country within was perfectly flat, abounded in small lakes, and produced a very short grass, but nowhere had the thaw penetrated more than two inches beneath the surface, while under water along the shore the bottom was still impenetrably frozen.[143] Not a log of wood was to be found in this land of desolation, but our party followed the example of the natives and made their fire of the roots of the dwarf-willow in a little chimney of turf. Next morning (August 3rd) the fog cleared for awhile, but it was still bitterly cold and the swell beat high on the outside of a heavy line of ice which lay packed on the shore. The latter, after extending five miles to the northward, turned off to the north-west, beyond which the latitude 71° 13' was observed. From this point the coast trended more westerly for ten miles, until the party came to what appeared a large bay, where they stopped for two or three hours to await the dispersion of the fog, not knowing which way to steer. In the evening their wish was gratified, and from that time the weather became comparatively fine. The bay was now ascertained to be only four miles in width; the depth half way across was one fathom and a half on a bottom of sand; that of Dease's Inlet was afterwards found to be two fathoms, muddy bottom, being the greatest depth between Return Reef and Point Barrow, except at ten miles south-east from Cape Halkett, where three fathoms on sand were sounded on our return. After crossing "Mackenzie's Bay", the coast again trended for eight or nine miles to the W.N.W. A compact body of ice extended all along and beyond the reach of vision to seaward; but the party carried their light vessel within that formidable barrier, and made their way through the narrow channels close to the shore. At midnight they passed the mouth of a fine deep river, a quarter of a mile wide, to which Mr. Simpson gave the name of "The Belle Vue", and in less than an hour afterwards the rising sun gratified him with the view of Point Barrow stretching out to the N.N.W. They soon crossed Elson Bay (which in the perfect calm had acquired a coating of young ice), but had no small difficulty

143 This is a very perceptive observation of an unusual phenomenon; submarine permafrost extends out from the coast for a considerable distance, underlying a considerable part of the continental shelf along the North Slope and the Beaufort Sea.

in making their way through a broad and heavy pack that rested upon the shore. On reaching it, and seeing the ocean extending away to the southward on the opposite side of the Point, they hoisted their flag, and with three cheers took possession of their discoveries in his Majesty's name.

Point Barrow is a long low spit composed of gravel and coarse sand, which the pressure of the ice has forced up into numerous mounds, that, viewed from a distance, assume the appearance of huge boulder rocks. At the spot where the party landed it is only a quarter of a mile across, but is broader towards its termination. The first object that presented itself on looking round the landing-place was an immense cemetery. The bodies lay exposed in the most horrible and disgusting manner; and many of them appeared so fresh that some of the men became alarmed that the cholera or some other dreadful disease was raging among the natives. Two considerable camps of the latter stood at no great distance on the Point; but none of the inmates ventured to approach till our party first visited them, and with the customary expressions of friendship dissipated their apprehensions. A brisk traffic then commenced, after which the women formed a circle and danced to a variety of airs, some of which were pleasing to the ear. The whole conduct of these people was friendly in the extreme; they seemed to be well acquainted with the character if not the persons of white men; were passionately fond of tobacco; and when any of the young people were too forward, the seniors restrained them, using the French phrase, "c'est assez," which like "tawaccah," they must have learned from the Russian traders. They designated the latter "Noonatagmun," and a respectable-looking old man readily took charge of a letter addressed by Mr. Simpson to them, or to any other whites on the western coast, containing a brief notice of the proceedings of the expedition. To the northward, enormous icebergs covered the ocean; but on the western side there was a fine open channel, which the Esquimaux assured the party extended all along to the southward; and so inviting was the prospect in that direction, that, had such been his object, Mr. Simpson would not have hesitated a moment to prosecute the voyage to Cook's Inlet in his skin canoe. The natives informed him that whales were numerous to the northward of the Point; and seals were everywhere sporting among the ice. These Esquimaux were well clothed in seal and reindeer skins; the men all used mouth ornaments; and the "tonsure" on the crown of the head was universal among both men and boys; the women had their chins tattooed, but did not wear the lofty top-knots of hair which are fashionable to the eastward. They were very inquisitive about the names of our party, and equally communicative of their own. A number of their words were taken down, some of which are different from the corresponding terms given by Sir Edward Parry, but the greater part are either the same or dissimilar only in their terminations. They lay their dead on the ground, with their heads all turned to the north. There was nothing else either in their manners or their habits remarked as differing from the well-known characteristics of the tribe except an ingenious and novel contrivance for capturing wild fowl. It consists of six small perforated ivory balls, attached to cords of sinew three feet long, the end of which being tied together, an expanding sling is thus formed, which, dexterously thrown up at the birds as they fly past, entangles and brings them to the ground.

Mr. Simpson could not learn that there had been any unusual mortality among this part of the tribe, and is of opinion that the concourse of natives who frequent Point Barrow at different periods of the year, coupled with the coldness of the climate, sufficiently account for the numerous remains already noticed. It was high water between one and two o'clock A.M. and P.M.; the rise of the tide was fourteen inches; and the flood came from the westward. Observations were obtained which determined the position of the landing-place to be in lat. 71° 23½'N., long. 156° 20'W., agreeing closely with the observations of Mr. Elson.[144] Then, bidding adieu to their good-humoured and admiring entertainers, the party set out on their return. They were arrested that evening by the ice; but next morning (August 5) it opened and allowed them to proceed. At a late hour they reached the camp of the Esquimaux in Dease's Inlet; and, after liberally recompensing them for the loan of their canoe, directed some of the men to follow to Boat Extreme where it would be left for them. Then continuing their route all night, at five A.M. on the 6th, the party rejoined the main body of the expedition.

We commenced our return the same afternoon and, being favoured by a light wind and an almost open sea, we sailed all night; and, next day (August 7), at noon, reached Cape Halkett. We then steered across Harrison's Bay; the wind increasing to a gale we shipped much water, but persevering, under treble-reefed sails, at three P.M. of the 8th, we landed safely at "Fawn River", within view of the point where our survey commenced. The position of this encampment was ascertained by good observations to be in lat. 70° 25'N., long. 148° 25'W. The gale having moderated, we re-embarked the following afternoon; and, running without intermission before a fresh breeze, we reached Demarcation Point to breakfast on the 11th. Several showers of snow fell during this run and it was piercingly cold. The Romanzoff and British mountains had assumed the early livery of another winter. The ice in Camden Bay was still very heavy; but it protected us from the dangerous swell to which we became exposed after passing Barter Island. Soon after leaving Demarcation Point, the ice became so closely wedged that we could no longer pursue our way through it. The following day (August 12) it opened a little, and the weather becoming fine, we put out and advanced for a few hours, when the mountainous swell and heavy ice obliged us to seek the shore, which we reached with some difficulty between Backhouse River and Mount Conybeare. There we were detained until the 15th. The icebergs which begirt the coast were of great size and of every imaginable shape; but from the summit of a hill, six miles inland, we had an unbounded prospect of the blue ocean stretching to the north. The pasture in the deep valleys among the mountains was luxuriant; herds of reindeer were browsing there, and we procured some venison. In the night of the 14th, the stars and aurora borealis were first visible. The following morning we resumed our route, and the weather continuing nearly calm, we reached the western mouth of the Mackenzie on the 17th, and there encamped.

---

144 Thomas Elson, the mate of Captain Frederick Beechey's *Blossom*, which had been blocked by ice at Icy Cape, had reached Point Barrow in the ship's barge in the summer of 1826. For further details see Beechey, *Narrative*.

The first Esquimaux seen during our return from Boat Extreme were at Beaufort Bay, but from thence to the vicinity of the Mackenzie we were continually falling in with small parties many of whom we had seen on the outward voyage. We maintained a friendly intercourse with them all, and they were very anxious to know whether we were soon to visit them again. The habitations on Tent Island were abandoned in consequence, we understood, of an alarm that the Loucheux meditated a descent to revenge the murder of their friends.

We have but few general remarks to add to the foregoing narrative. The tides all along the coast were semi-diurnal, the flood coming from the westward. The rise, however, was strongly affected by the winds and ice; and our opportunities of observing were but few. At Boat Extreme, the average rise was fifteen inches; high water from one and to two o'clock A.M. and P.M. The rise generally decreased to the eastward; and at Point Kay it was only eight or nine inches. That the main sea is clear and navigable by ships during the summer months, the long rolling swell we encountered on our return, and the view obtained from the mountains, furnished tolerable proof. We likewise saw some whales on our return. The prevalence of east and north-east winds during the early part of the summer is a remarkable fact. We were indeed favoured by a westerly wind for five days on our return; but this was almost the only exception. At a more advanced period of the season, however, the winds blew more from the west and north-west. It is now certain that from Kotzebue's Sound to Cape Parry there is not a harbour into which a ship can safely enter,[145] but it must be a very unpropitious season that would not admit of achieving that portion of the Arctic navigation; and another year ought certainly suffice for the remainder, whether the voyage were commenced from Barrow's, or from Behring's Strait. On this subject, however, we shall be better able to offer an opinion if successful in our next summer's operations.

The natural history of the coast from Return Reef to Point Barrow is meagre in the extreme. In the botanical kingdom, scarcely a flower or moss was obtained in addition to the collection made on other parts of the coast. In zoology, reindeer, arctic foxes, one or two lemmings, seals, white owls, snow buntings, grouse (*Lagopus salicite et rupestres*), and various well-known species of water-fowl, were the only objects met with; while in the mineralogical department there was not a rock *in situ* or boulder stone found along an extent of more than two hundred miles of coast. The variation of the compass was found to have increased from one to three degrees at the corresponding points where observations had been made by Sir John Franklin. At Boat Extreme, on the other hand, it was only twenty-one minutes greater that that stated by Mr. Elson at Point Barrow, where, by continuing the proportion, the quantities would coincide. The moon was never once visible during the whole outward and homeward voyage till our return to the western mouth of Mackenzie,

145 This is inaccurate: Pauline Cove, on the south side of Herschel, provided an excellent sheltered anchorage for American whaling ships for several decades around 1900, while the harbour at Tuktoyaktuk provided a safe haven for the ships supporting oil exploration in the Beaufort Sea in the 1970s and 1980s.

where a set of distances was obtained, and the longitudes of the other points reduced back from thence by means of a very valuable watch, generously lent to the expedition by Chief-Factor Smith.

The map of our discoveries will be prepared and transmitted to your Honours in the spring.[146]

Our ascent of the Mackenzie has not been characterised by any circumstance of particular interest. The weather continued calm and beautiful; and the journey was performed entirely by towing, in which manner we advanced at the rate of from thirty to forty miles a day.[147] The river has fallen very low, and the fisheries have, in consequence, been unproductive, causing a scarcity of provisions both at Fort Good Hope and among the natives. We saw a good many of the Loucheux, but the Hare Indians were all dispersed in the interior, searching for subsistence. From the coast up to Point Separation moose-deer[148] were numerous – being quite undisturbed; but from our first falling in with the Loucheux, no vestiges of either moose or reindeer have been seen. We reached this place yesterday with half of our summer's stock of provisions still forthcoming, and we are now awaiting with impatience the arrival of our outfit and dispatches.

Some Indians from Great Bear Lake have brought us intelligence of the party sent to establish our winter-quarters. They were stopped in Bear Lake River by ice during the whole month of July; lost one of their canoes; and it was not until about the 6th ult. that they passed Fort Franklin,[149] after which they had the prospect of an unimpeded passage across the lake. The continued easterly winds were the cause of this vexatious detention, during which the Dogribs kindly supported our people with the produce of their nets.

*Sept. 8.* We have this morning received Governor Simpson's letter, dated London, 11th of November last, and have to offer our lively acknowledgements to your Honours for the interest you have been pleased to express in the success of the expedition, and in the welfare of our selves and party.

We have duly received the journals of Captain Back's last expedition, and are glad to find that his new undertaking is in no way to interfere with our original instructions.

Your Honours may rest assured that our efforts in the cause of discovery and science next summer, to the eastward of the Coppermine River, will not be less zealous than they have this season been in another field, and we are sanguine in the hope that they will be crowned with equal success.

Our supplies for next season have come to hand. They were delivered in very indifferent order at Portage la Loche, and there was a considerable deficiency in the

---

146 See fig. 7.

147 The casual tone of this remark obscures what is in fact a remarkable feat of endurance on the part of the boat's crews!

148 Although the moose (*Alces alces*) is generally considered a species of the boreal forest, its range extends a considerable distance north of the treeline everywhere west of Hudson Bay, and reaches the Arctic Ocean from Cape Parry west, including the Mackenzie Delta and the North Slope of Alaska.

149 Now Déline. The winter quarters occupied by Franklin and Richardson near the west end of the Keith Arm of Great Bear Lake in 1825–26 (Franklin, *Narrative of a Second Expedition*).

weights of the pemican; but with the quantity saved of this year's stock there will be provisions enough for next summer's operations, and we have no further demands to make upon the Depot for goods.

We send two men express to Great Slave Lake with this dispatch, and to meet the spring packet, the non-arrival of which causes us some anxiety; and tomorrow we take our departure for winter-quarters.

We are, etc.,
Peter W. Dease.
Thomas Simpson.

• • •

## *8. Peter Dease's journal, continued*[150]

**6 September 1837.** This afternoon 2 Boats with an Outfit Cast up, 10 Men & 4 Indians, but have no other Dispatches than a letter for John Ritch with the Bill of Lading. Our letters will be sent with the Boat that brings the outfits for this Post & Fort Good Hope, as we were not Expected to return so Early. We must therefore Keep these Boats untill our Dispatches come.
**7 September 1837.** Not arrived yet.
**8 September 1837.** In course of the morning the last Boat under charge of J B Belleau arrived. Received our letters from the Southward, but none from the Governor[151] or Council. He had not arrived yet at Norway House 17 June – very late indeed. The Boats were sent off for Bear Lake with orders to wait for us about Fort Franklin and each boat to take 5 Bags Pemican & 1 of flour from the Cargoes left at the mouth of the River in order to Save the Small Boats[152] going up that River which is very Shallow.

• • •

## *9. Extract from letter from Thomas Simpson to George Simpson, 8 September 1837*[153]

Fort Norman, September 8th, 1837.

I am proud of the interest felt by their Honours [the directors of the Hudson's Bay Company] in our expedition: and hope that our present dispatches will show that your confidence in our zeal has not been misplaced.

I do not know what pecuniary reward is attached by the King's orders in Council to that portion of the Arctic navigation completed by us this summer; but I trust that you will ascertain this point, and not allow our interests to be overlooked.

You were pleased to tell me that my services as your Secretary at Red River had earned a Chief-tradership; now, however, I have the *exclusive* honour of unfurling the

150 MCM M214.
151 Governor George Simpson.
152 The sea boats, *Castor* and *Pollux*.
153 Alexander Simpson, *Life and Travels*, 255–6.

Company's flag on Point Barrow, and of thus uniting the Arctic to the Western Ocean, – which I humbly think entitles me to the second step.[154] Consider, I beseech you, the importance of the geographical problem solved; the able officers whom it baffled; the rewards conferred upon them for what they effected; and do not reject my just claims, although I am one of your own relatives. I have always confided implicitly in your kind sentiments towards me, and feel that they will be fully displayed on the present occasion. Such preferment, instead of cooling my zeal, would animate and inflame it, and at the same time give me the standing requisite to the creditable production of our travels to the world. Mr. Dease is a good, honourable man. I believe I have acquired his friendship, for in everything, even to the plan of our little Fort Confidence, he has adopted my advice, and has left the direction of the march entirely to me; the result proves that it has not languished under my directions; *indeed in such a service the surveyor must, of necessity, act as guide. While the whole onus of the duty thus rests on me, I cannot help feeling sore that you should have considered it necessary to entrust another with the command.*

We take our departure tomorrow for winter quarters; and hope to get our admirable little sea-boats taken some distance up Dease's River this fall, in order to lessen our portage labours in the spring.

• • •

## *10. Extract from letter from Thomas Simpson to Alexander Simpson, 8 September 1837*[155]

Fortune and its great Disposer have this season smiled upon my undertakings, and shed the first bright beams upon the dark prospect of a North American life. Yes, my dearest brother, congratulate me, for I and I *alone*, have the well-earned honour of uniting the Arctic to the great Western Ocean, and of unfurling the British flag on Point Barrow.

On the 23rd of July we supped on Franklin's Return Reef, and having a noble run across two large bays (at the bottom of the first of which rose a majestic branch of the rocky mountains washed by two great rivers), on the 27th we were brought to a final stand by the ice at Boat Extreme, in long. 154° 26'. Nearly one third of the distance still remained unexplored, and I resolved to complete it on foot, Mr. Dease most handsomely volunteering to remain with the boats and thus secure our retreat. I took five men with me, and in addition to our arms, instruments, and necessaries, we carried a small canvas canoe for crossing rivers. It was desperately cold, and the fog so dense as to be almost tangible. We were wet to the waist with wading through salt creeks and shallow inlets, and if our case while marching was bad, in the camp it was infinitely worse.

On the second evening by God's Providence we fell in with the first camp of the Western Esquimaux, and from them I procured an *oomiak*, or family canoe, with

---

154 In other words, Thomas Simpson felt that his performance on the expedition to Point Barrow merited promotion to Chief Factor. George Simpson's reaction was probably one of spluttering incredulity at this effrontery.

155 Alexander Simpson, *Life and Travels*, 257–60.

which we crossed some deep inlets and bays that we never could have walked around; and, being a highly seaworthy craft, we now carried on through thick and thin, and at 2 A.M. of the 4th ult. attained the object of my ardent wishes. We were soon surrounded by a crowd of natives, to whom Cape North is a place of great resort; but they were, in appearance, most friendly; and a fine old fellow took charge of a letter to the Russians, briefly noticing our proceedings. The women danced around us, and were indefatigable in their blandishments. They are w—s without an exception. The men are a set of lousy, good-humoured, thievish, pimping rogues. Without fire-arms I should be sorry to trust to their tender mercies, notwithstanding their smiling physiogs.

I rejoined Mr. Dease at 5 A.M. of the 6th, and we continued our return the same day. We were favoured by a clearer sea and strong westerly gales, and after a few detentions from the ice and storms, we regained the western mouth of the Mackenzie on the 17th of August, after a most successful cruise of thirty-nine days. We ran many a desperate risk in that cruise. In Camden Bay we were caught by the ice three miles from the land, had to throw out our cargoes upon the floating masses to save our sweet little boats, and after making several portages, the oars serving as the perilous bridges from fragment to fragment, we at last got boats and ladings secured on a large floe where we passed a horrible night. On another occasion, in a vast bay beyond Return Reef, we sailed twenty-five hours in a dense fog, thirteen of these out of sight of land, before we could effect a landing. We never encamped but for ice or contrary winds, it being constant daylight; and, when weary, slept by watches in the boats. The thermometer ranged between 30° and 40° of Fahr., and we wore our winter dresses, and Esquimaux waterproof boots. Yet with all these precautions few of the men escaped without severe sore throats.

I am charmed with the scenery of the Mackenzie; the weather, after we re-entered it, was beautiful, and ascending with the towing-line at the rate of from thirty to forty miles per day, we reached this place on the 4th instant. I have since been incessantly employed on public dispatches, which is my apology for this meagre outline; but you shall hear from me once or twice from winter quarters.

• • •

## *11. Letter from Peter Dease and Thomas Simpson to chief factors, chief traders, and gentlemen in charge of Posts for the Hon. Hudson's Bay Company, 8 September 1837* [156]

Fort Norman, 8th September 1837

(Via Sault Ste. Mary's & the Grand River)
On Service
Dear Sirs,

We have the pleasure of forwarding for their Honours, the Governor & Committee and for the Governor & Council, Dispatches connected with the Expedi-

156 HBCA D.5/4, f. 334.

tion to which we have been appointed, and have to request that you will afford every facility in forwarding them as Expeditiously as means may admit. We trust also that the same readiness will be observed in forwarding any dispatches intended for us, also any Boxes or Parcels to our address, especially Books and Astronomical Instruments.

We feel much satisfaction in stating that the party who accompanied us are all returned safe, and that the object of the Expedition has been successfully attained.

We have the Honour to remain,
Dear Sirs,
Your most obedient servants
Peter W. Dease
Thomas Simpson

• • •

## *12. Letter from Thomas Simpson to Donald Ross, Norway House, 8 September 1837*[157]

*Private*

Fort Norman, 8th Sept. 1837

My Dear Sir,

*We* have been *waiting* here for four days for our outfit and letters from the Southward, and this morning our wishes have been *partially* gratified, but where is the spring packet? If it came soon after the departure of the boats, why was it not forwarded in a Canoe to overtake them? Had we been in His Majesty's service, it would have been otherwise, yet none of His Majesty's officers were able to do what we have accomplished. Being now disburdened of "the winter of my discontent" I shall first proceed to acknowledge your most welcome epistle of 16th June, though I think that you have omitted to acknowledge several of mine, and long ones too, if I recollect aright. We are much indebted for your attention to our provision department, and though L'esperance and his troops eat nearly a third of our provisions we have saved enough out of this summer's stock to make up all deficiency. The twine, cassette etc. have come safe to hand. I feel rather anxius about our worthy Governor; he was well and hearty when I last heard from him (12th November), but what changes may not a day or even an hour bring forth! His sweet lady and daughter too were well & thriving. So Grave[158] turned back; well, I thought so; the Company have surely dispensed with *his* services; and poor Baker[159] whom I have heard reviled and despised, brought home his ship with her valuable return cargo in safety. I have horrible stories of the Yankees in the Cala, and strange stories are circulated respecting Capt. Carey's R.R. troops who seem to have carried famine into that land of milk and honey. The D—l take the natives, for my part I owe them hatred and not pity. I have picked up a

157 BCA ADD.MSS 625.

158 Captain John C. Grave of the Company's ship *Prince Rupert.*

159 Captain Henry Baker of *Prince of Wales.*

*drop* here and there of European news. Remember, friend, how I used to *rain* such intelligence upon you, when I was in the focus to & from which it radiated. Now do thou likewise. But I must back this injunction by telling you the results of this summer's cruise on the icy seas. They well merit the name, and I am now suprised how we have returnd to tell the tale; but our star has kept the ascendant, and its benign influence carried us safely through perils of no ordinary kind. Our principle was risk all for glory, and The Highest was our particular. We reached the Ocean by the most westerly mouth of the McKenzie (which Sir John Franklin vainly sought) on the 9th July, and thus escaped a repitition of the farce at Pillage Point; 19 of the rogues came off to us, we gave each a present, and when they wanted to bother us, I let fly a ball over their heads which soon made them bob and sheer off. We met many parties after that, but our arms always awed them till indulged, and then they became daring and excessively troublesome. They stole several things but we made them restore "at the point of the bayonet" The men are a set of lousy, good humoured knaves and the women are *all* thorough paced wh—s. We had many a rough adventure among the ice; one in particular in Camden Bay, when we got *caught* three miles from the shore; by throwing out the cargo upon the floating masses we rescued our sweet little boats from destruction, and then making bridges with the oars we carried from piece to piece, and in fine collected the whole upon a large floe where we got cooled to some purpose, for next morning when the ice opened and allowed us to escape, almost every man found himself with a severe sore throat. The water proof boots which we purchased from the natives were a blessing to us, and on many occasions preserved the health of the party. On the 23rd of July we supped at Franklin's return reef. The coast to the westward is very low and shallow, and there are some large bays, rivers and mountain ranges which were duly christened after their Krongs?. There is a scarcity of both fuel and inhabitants along these desolate shores. On the 27th at midnight we reached Cape Govr. Simpson, beyond which, during the four remaining days of the month, we only advanced as many miles. Young ice began to form, and to all appearance our voyage was at an end. But the enthusiasm of a highlander is not easily crushed, and yr. h. st.[160] undertook to complete the business on foot with five men. We marched through opaque fogs and many ice cold creeks till the afternoon of the second day when, just as we came to an inlet stretching out of sight to the southward, my eyes were blessed with the view of a small camp of the Western Esquimaux. We marched up to them and after disspelling their fears, bargained for one of their "oomiaks" or women's boats, which carried us nobly through thick and thin to the [illegible word] and long sought for point Barrow. At 2 AM of 4th Augt. I had the honor of unfurling the Company's flag there, and of taking possession of all our discoveries in His Majesty's name. We rejoined Mr. Dease at Boat Extreme on the 6th and on the 17th regained the McKenzie. Point Barrow is 50 miles direct and half as many more actual distance from Boat Extreme. There were two considerable camps of the Esqx. on that Point, and they were the most friendly of any we met with, though from the smallness of my party, they had us most in their power. I addressed a letter

160 Your humble servant.

to the Russians through them, and was sorry I was not bound out by Beehring's Straits myself in my new fangled but excellent craft. I named one of the bays I discovered after my dear friend Donald Ross Esquire. Accept my thanks for your and Mrs. R's kind wishes, and remember me most kindly in your family circle.

[Hole in paper] delighted with the scenery of the McKenzie, the mountains etc., and of course saw the sun at midnight more than once. Our building party were detained by ice in Bear Lake river[161] for a whole month, so that we are likely to find "Fort Confidence" still in embryo. Thither we hasten as soon as we have sealed our despatches, and from thence you will (D.V.) hear from me again, before we start for the Coppermine, on our return from which I hope to exclaim with the ancient poet "Sistimus hic tandem, nobis ubi deficit orbis".[162]

Meanwhile farewell, and believe me ever,
My Dear Sir,
Yours faithfully and sincerely,
Thomas Simpson.

• • •

## *13. Peter Dease's journal, continued*[163]

**September 9 1837.** Passed the day writing our Dispatches for their Honours the Governor & Committee[164] and the Governor & Council.[165] Appointed Peter Taylor with F. Wentzel to take them up to Ft Resolution. Had some Rain & light Snow this morning, but cleared again afternoon.

• • •

## *14. Fort Norman journal, 6–10 September 1837*[166]

**Wednesday 6th.** Two Boats arrived with the Expedition outfit. as Mr. Dease is near himself the Boats only go off to-morrow …
**Thursday 7th.** … The Expedition men repairing their Boats …
**Friday 8th.** Mr. Dease sent off his two Boats to Bear's Lake. the two Small Boats remain here till Sunday. their is two Indians of this Post employed in their Boats …
**Saturday 9th.** The gentlemen still busy writing their letters …
**Sunday 10th.** Mr. Dease and Mr. Thomas Simpson take their departure for their winter quarter Bear Lake … Cold weather and rain and Snow. Puce Cupe went with them as guide and to assist them in the portages that is both long and many.

• • •

161 Great Bear River.
162 "We are finally establishing here, where the world fails us."
163 MCM M214.
164 Governor Sir John Henry Pelly and the Committee in London.
165 Governor George Simpson and the Council of the Northern Department at Norway House.
166 HBCA B.152/a/16, f. 8d–9.

## *15. Peter Dease's journal, continued* [167]

**10 September 1837.** At 7 A.M. this morning,[168] after seeing our Dispatches off, we left Ft. Norman with a Strong head wind.[169] Got to the River[170] at 3 P.M. Got the boats loaded and Embarked at 5, and got up about 3 miles. The water very Shallow and bad tracking in most places.

**11 September 1837.** The Castor was injured to day and had to unload & Pitch a part of the plank that was Split.

**12 September 1837.** Saw Some women below the Rapid whose husbands are hunting. They gave us a Bustard, but complain of Starving. Got safe up the Rapid and landed at 7 P.M for the night.

**13 September 1837.** Saw one of the Indians who Embarked in the Boats[171] at Ft Nn. to come up the River, but being on the opposite Side the River we did not take him up.

**14 September 1837.** At 10 A.M. got to the lake[172] & Crossed the Bay to Fort Franklin.[173] Strong SE wind. Found the 2 Boats waiting for us at Ft. Franklin. Have a nett down out of which they had some fish. As the wind is Strong, gave them 2 long netts to put down & passed the day.

**15 September 1837.** About 11 A.M. the wind moderated. The netts gave us 40 fishes; [they] were taken up & we embarked at Midday, and with the oars only we advanced about 20 miles. The Evening fine and Calm. One of the Large Boats makes us lose much time as it is a very dull going boat.

**16 September 1837.** The weather Calm all day. Had to use the oars. Our progress was about 40 miles.

**17 September 1837.** About 9 A.M. today the wind rose from the S.E. rather ahead, and in Crossing the mouth of Narre Ella Bay, the Boat (Castor),[174] pitching heavily in the Swells, Split the Garvel Streak that had already been injured. So much so that had not one of the Large Boats fortunately been near, the Boat must have Sunk & the property been Sacrificed to save the lives of the crew. The property was taken out & the Boat towed ashore East Side of the Bay. Wind Stormy remainder of the day, which was required to replace the damage & dry the property that was wet.

**18 September 1837.** The wind had moderated so as to enable us to embark at 6 A.M. We proceeded to near the Point of the Scented Grass Mountains.[175] The wind rising, the Swells were too large to advance. We therefore landed. Some Snow fell afternoon, weather Cold & Cloudy.

---

167 MCM M214.

168 The day began with hard frost and a snowstorm (Simpson, *Narrative*, 191).

169 The headwind was so strong that they had to track the boats downstream (Simpson, *Narrative*, 193).

170 Great Bear River.

171 The boats which had preceded them.

172 Great Bear Lake.

173 Which they found in a ruinous state (Simpson, *Narrative*, 194).

174 This was Dease's boat.

175 Now known as the Scented Grass Hills, forming the peninsula between Smith Arm and Keith Arm.

**19 September 1837.** A Continuation of the wind Kept us all day ashore.[176] Some of the Indians went out above the hills to hunt but had no success although they say they fired 12 Shots at Rein Deer very near them. Had some light falls of snow.
**20 September 1837.** This morning the wind turned a little to Eastward but the Swells yet too high to allow us to proceed. The Indians were out again with no better luck than yesterday. Had some light Showers of snow.
**21 September 1837.** The weather as bad as yesterday. Indians out again. One of them Came back about Midday to tell us they had Killed 3 Rein Deers on the Mountain. 8 Men were Sent with them, and brought in the meat at Sunsett. The wind has moderated, and gives us hopes of getting off Tomorrow.
**22 September 1837.** The wind freshened during the night but at 10 A.M. we loaded & pushed off. The Swells very high. Wind ahead. Proceeded only about 4 miles to near the Extremity of the Point of Scented Grass Mountain & put ashore, as we must either form the Circuit of Smith's Bay, which is very Deep, or Cross the Lake to Cape McDonnell, about 25 miles across.
**23 September 1837.** The Wind Shifted round to S E. The Swells having abated since last night, we pushed off at 6 A.M. and made for Cape McDonnell.[177] After pulling to windward about an hour, hoisted Sail. Got Sight of the Cape between 11 and 12. Wind increasing. Desjardins, in one of the Large Boats, mistaking the Course, pulled for an Island south & W of the Cape & passed the night, the Boat being leaky. The Sea Boats out of Sight. This caused a loss of 4 hours Sail wind. This morning I took a passage in one of the Large Boats.
**24 September 1837.** Started at 4 A.M. under Sail. At 10 were in Sight of Narraghasse.[178] Proceeded untill 7 P.M., the Small Boats ahead. Put up among Islands E of the high Rockey ones called Narraghasse.
**25 September 1837.** Started at Same hour as yesterday.[179] At 12 got in the Channel of a large Island;[180] from E End in about 3 miles to Dease's River. We got to the place where our party are at 4 PM.[181] John Ritch and his men in health. Norquay the fisherman had cut his foot a few days ago and is off Duty about the Buildings,

---

176 Finding their canvas tents little protection from the cold, Dease and Simpson had a caribou-skin teepee erected and found it much more comfortable (Simpson, *Narrative*, 196).

177 With high waves and a strong wind, the spray freezing as it landed on boats, rigging, and sails, this was a risky venture (Simpson, *Narrative*, 197).

178 The Narakay Islands, a group of high, rocky islands on the southeast side of Dease Arm.

179 "Joined the small Boats this morning" (Dease's original note).

180 Now Ritch's Island.

181 This, the expedition's winter quarters, named Fort Confidence, lay about five kilometres west of the mouth of the Dease River (Simpson, *Narrative*, 199). Simpson's description of the site is as follows: "The station judiciously chosen for the establishment was a wooded point, on the northern side of a deep and narrow strait, formed by a large island. It commands a fine view of the lake to the east and west, and the rocks form a natural landing-place for the boats at the very door" (Simpson, *Narrative*, 203).

which are not very forward, as the Party landed here only 16th Ulto.[182] The Store is finished & the frame of the Dwelling House up & roofed. Chimneys not yet begun. The Natives as well as our Hunters have all been attacked with a Sickness that reduced them much and many of them are yet suffering from its Effects. In consequence of the Hunters' families having suffered, little or no provisions have been collected and no fall fishery yet commenced. But 3 netts are down and fish not abundant. Began writing to send back the Boat for Fort Simpson & 2 of the men for Ft Good Hope who are here with the others. 2 men, Gullion & Lamothe appointed for Ft Norman, are to pass the Winter here at Mr. Mowat's request.

• • •

### *16. Letter from Peter Dease and Thomas Simpson to the governor, chief factors, and chief traders of the Northern Department, 25 September 1837*[183]

Fort Confidence, September 25, 1837.

Gentlemen,

After a long and stormy passage across Great Bear Lake, during which we experienced the rigours of winter, we arrived here in safety this evening, in company with the servants and Indians employed by Chief-Trader Macpherson to transport our outfit and provisions for the ensuing summer. These people will be dispatched back to Fort Simpson tomorrow, and we embrace the opportunity to give you a brief sketch of the summer transactions at this place, and of our prospects of maintenance for the winter.

The site judiciously chosen by John Ritch for this establishment is about three miles from the mouth of Dease's River, in a strait formed by a large island with the northern shore of the Lake. He reached here with his party on the 16th ult., and since then has completed a store and erected the shell of a dwelling-house.

We are concerned to say that the Chipewyan hunters brought from Athabasca have quite disappointed our expectations. After indolently wasting the best part of the hunting season, they are now laid up with influenza, which has been going the

182 On 17 August, according to Simpson. Ritch and his party had found that the Great Bear River had only recently broken up, and was flowing between ice cliffs twelve metres high in places. Tracking the boats upstream under these conditions had been extremely dangerous, since the ice was rotten in places. It took the party a week (until 10 July) to reach the top of the rapids. Thereafter they found that the lake ice had just started to break up, and was driving down the river in terrifying masses. The boats barely escaped annihilation and one of the Indians' canoes was lost. It took the party two weeks to cover the fifty kilometres from the head of the rapids to Great Bear Lake. They thus reached Fort Franklin only on 6 August and took a further ten days to cross Great Bear Lake. No open water was visible from Scented Grass Mountain, but after three days they managed to push west along shore leads to the Acanyo Islands in Smith Arm. An inviting open lead tempted them northwards across Smith Arm, but when it petered out they had to struggle through heavy ice for twenty-four hours to reach the north shore. From there they followed the shore east to the mouth of the Dease River. Thus a trip of four hundred kilometres had taken forty-five days (Simpson, *Narrative*, 200–2).

183 Alexander Simpson, *Life and Travels*, 263–5.

round of all the Indians in this quarter. The Dogribs, however, promise to bring some provisions. In the interval, the nets at the door have supported Ritch and his men, and we shall immediately make proper disposition for our winter fisheries – the most distant of which is about fifteen miles from hence.

It is most fortunate that we got two small boats built at Fort Chipewyan last spring instead of a large one; and that we have brought these fine craft safely hither, for there is not in any part of Great Bear Lake, on this side of Fort Franklin, timber fit for the construction of a boat of any desciption.

We have only to add that our utmost efforts shall be exerted to achieve the service prescribed us next summer; and we trust that our success will be equal to that with which Providence has crowned our labours of this season.

The winter may be said to have commenced here, the ground being covered with snow, and the frost already severe.

We shall again do ourselves the pleasure of addressing you by the Mackenzie's River March express. Meanwhile,

We have the honour to be etc., etc.,

Peter W. Dease.
Thomas Simpson

• • •

## *17. Letter from Thomas Simpson to Donald Ross, Norway House, 25 September 1837*[184]

*Private*

Fort Confidence, 25 Septr. '37

My Dear Sir,

We reached our embryo establishment this evening, and are now snugly tented alongside of our future dwelling house. John Ritch has done a great deal for the short time he has been here, and I anticipate an agreeable winter within the Arctic Circle, with an early jaunt to the Coppermine, to explore the best track for our long portage. Hope speaks great things in my ear, which I trust will be fully realized. Our prospects in the grub line are so so; little or no dried provisions, but the promise of a tolerable fishery. This is a well chosen spot and commands a fine view of two long vistas of the chrystal lake, formed by the angular sides of a large opposite island. But a truce descriptions, for Mr. McPherson's people return hence tomorrow, and I must have done. In the expectation of hearing from you ere long,

I remain, My Dear Sir,
Every yours faithfully,
Thomas Simpson

Donald Ross Esq.

• • •

184 BCA ADD.MSS 625.

# CHAPTER SIX
# The First Wintering at Fort Confidence

## Introduction

Dease, Simpson, and party moved from their tents into the dwelling house on 24 October. Throughout the winter the majority of Dease's journal entries are concerned with the efforts to obtain food, namely through hunting caribou (and to a lesser degree muskoxen), either by the expedition's own hunters or by Dogrib and Hare Indians who congregated at Fort Confidence and who moved to and from hunting camps on an irregular basis. A great deal of effort was also expended on fishing, first with nets, and, after freeze-up, to a larger extent by ice-fishing using lines. The catch, initially abundant, decreased alarmingly over the winter. Here, too, the fish delivered to Fort Confidence was derived both from expedition personnel and from local Indians fishing at locations along the shores of Great Bear Lake and in nearby lakes and rivers. Dease carefully recorded the numbers of fish and the weight of meat delivered to Fort Confidence. For the entire period of the winter (26 September 1837 until 5 June 1838), he recorded that the hunters delivered a staggering total of 25,954 pounds (over 11½ tonnes) of caribou meat, representing some 260 animals, as well as 379 caribou tongues, 3636 pounds of muskox meat, 74 geese, and 3 swans. During the same period the fisheries produced 5153 fish, the weight of which was not specified. In neither case do these totals include the fish and meat consumed by the hunters and fishermen themselves. These latter quantities must have been substantial, since repeatedly the expedition leaders found themselves engaged in relief measures to rescue groups of local Indians from starvation.

The other activity which absorbed a great deal of time and effort, although not so carefully recorded by Dease in his journal, was that of cutting firewood. Already by 28 November the temperature had dropped to -42°F at night and Simpson recorded that the average temperature for the second half of December was -33½°F and for the entire month of January -30°.[1] A decade later, during his wintering at

1 Simpson, *Narrative*, 216.

Fort Confidence in 1847-48, Dr John Richardson recorded that the average temperature for the month of December (based on fifteen to seventeen observations per day) was -38.69°F.[2] Richardson found that his "predecessors had cut down most of the timber within a mile of the house."[3]

On 6 December a group of Indians arrived at Fort Confidence in a state of fear and panic; they had encountered the tracks and camp of a stranger and were afraid of being attacked. Simpson went to investigate and deduced that the tracks were those of a member of a party bringing mail to Fort Confidence who had been unable to find the station. Then on 29 December the mystery was solved when an Indian named Le Sourd arrived, escorted by some of the local Indians. Along with Peter Taylor, he had been entrusted with bringing the expedition's mail from Fort Simpson. Taylor had fallen sick and progressively weakened, and to aggravate the situation they had become lost and ended up in McTavish Arm, instead of Dease Arm. It was Le Sourd who had caused the panic among the Indians, while wandering around looking for the station. Taylor had eventually died and, after burying him, Le Sourd had run into the group of local Indians who brought him to Fort Confidence.

On 27 March Simpson set off to reconnoitre a route to the Coppermine River and to choose a site for a staging depot on the Coppermine or one of its tributaries. He returned on 6 April, having selected a site for the depot on the Kendall near where it flows from the Dismal Lakes, and some fifteen miles (twenty-four kilometres) from its confluence with the Coppermine. On 9 April he set off again with six sledges laden with supplies, bound for the depot on the Kendall River. Two men (George Flett and Hector Morrison) were to stay there to guard the depot. Simpson returned on 15 April. Two weeks later he set off to explore "Simpson's Bay," now Hornby Bay, at the head of McTavish Arm, which he had seen while searching for Le Sourd; he returned to Fort Confidence on 13 May. By then final preparations for the summer's exploring trip were well in hand.

## DOCUMENTS

### *1. Peter Dease's journal, continued*[4]

**26 September 1837.** The Boat for Ft Simpson was drawn up and pitched today. Writing letters, informing our Friends of our Safe arrival here & the State of affairs. Several Hare & Dog Rib Indians about the place, among whom are Some yet Sick.[5] The men were allowed rest to day, and appointed to different Duties about the Buildings tomorrow. Got 2 more netts put down to day. But 50 fishes were taken. Very few taken in the day time, white fish and Trout.[6] Few other Species of fish here.

---

2 Richardson, *Arctic Searching Expedition,* 2:251, table 10.

3 Richardson, *Arctic Searching Expedition*, 2:65.

4 MCM M214.

5 This illness was similar to influenza (FNB B, p. 29).

6 White fish (*Coregonus clupeaformis*); lake trout (*Salvelinus namaycush*).

**27 September 1837.**[7] To day Nahlhal, one of our Hunters, with his Wife & Child came in all very unwell. Tells us that Toozeyaze & Eegeen's Wife are very low, almost despaired of, and suffering from Want of Provisions at same time. 4 men working at Erecting a Chimney. 3 Sawing logs for floorings. 4 Cutting Wood for a Dwelling House for the men & 3 making netts. Laurent with an Indian attending the netts & John McKay, Servant. The produce of the netts was but 35 fishes.[8]

**28 September 1837.** Sent George Sinclair with a little Provisions to the Hunters and 2 doses of Salts.[9] Gave some also to the Hunter who is here and his Wife. Several of the other Indians also applied for medicine and we rendered them what assistance we can. I am sorry to say we are not very amply provided in that way, as I never thought we would have occasion for such medicine. The wind was Easterly, the Weather fine. The netts gave us 72 fishes.

**29 September 1837.** A light Snow fell last night. The day fine. All hands at work. Afternoon G. Sinclair came back and reports that Toozeyaze and the woman are very ill. Eengeen yesterday Killed a Doe and 2 young rein Deer, & wishes some one to be sent for the Doe for us here. Main Pocque or Cuthand (a Dog Rib Indn.) got the loan of a Gun and Some Amunition to go & hunt for us. Sett off this morning.[10]

**30 September 1837.** This morning Mr Thos. Simpson accompanied by Geo. Sinclair and the Boy Frans. Boucher with 3 Indians gone to the Hunters for the purpose of Encouraging them & Send home the meat of the deer as he intends remaining a day or two there. The Hall Chimney up; Another half way in one of the Rooms. Sawyers at Work. The day fine and mild. No Ice formed yet.

**1 October 1837.** The night was Cold. A thin Ice covered part of the Channel between this & the Island.[11] Injured the Bark Canoe to get to some of the netts. Got 89 fishes today. An Indian brought us 5 trouts. Cuthand came back without success; says he wounded a Deer but Could not get it.

**2 October 1837.** Snowed most part of the night. Wind SSE. Our Hunter appears very ill this morning; his Wife also. 2 of the Indians that left this with Mr. Simpson came back with the meat of the Deer. Say that Eengeen has Killed 4 more. They are to return tomorrow. The netts took 80 fishes to day.

**3 October 1837.** The Indians were Sent back to the Hunters' tent. The men at the Buildings getting on pretty well. Mr Simpson Came back this afternoon with Geo Sinclair & the Boy. Brought a Deer and left one at the Tent for the Indians to bring. Cuthand off again with 3 young men. Lent 2 netts to an Old Man to fish for us on South Side the Island.

**4 October 1837.** A Strong W Wind. The netts gave but 35 fishes to day. The men As usual Employed. The weather rather Cold.

---

7 "The Boat [for Fort Simpson] took departure for return" (Dease's original note).

8 Hence the men were issued meat for their supper (FNB B, p. 29).

9 Epsom salts, in the hope that it would cure their sickness.

10 The nets produced eighty-two fish (FNB B, p. 30).

11 Fishing Island, now named Ritch's Island.

**5 October 1837.** The Channel partly frozen over. 87 fishes brought by the Old Man we lent netts to. Had to go with the Boat to get our netts from under the Ice. But 4 were taken up in which 25 fishes were taken. 3 remain down yet.

**6 October 1837.** The wind which turned Easterly this morning Changed round by Southward to West in the afternoon, and had some Rain & Snow. 34 fishes were taken by the Netts we have down. 4 more were put to Westward of the House.[12]

**7 October 1837.** Rain with a Strong W Wind continued most of the night. Mr Simpson with 3 men & 3 Indians gone to the Indian Camp where they are fishing & will leave a man if any prospect of a fishery being productive. The Old Man came with 88 fishes and left us to go back. The weather Cloudy & Mild; some Snow fell. Wind W and NW throughout the day.

**8 October 1837.** Mr. Simpson came back with 189 fishes Collected from different Indians and our netts. Weather fine.

**9 October 1837.** The weather getting Colder. Our 6 netts took 66 fishes. The men at usual work. Evening Calm & Clear.

**10 October 1837.** This morning the Channel frozen over for about 2 miles Except a Small Space in the middle opposite the House here. Could not overhaul the netts. Strong E wind all day; Evening Cloudy. A very beautiful parascelne [paraselene] about the moon.[13]

**11 October 1837.** Strong wind during the night. Could only get at one of the netts which had 11 fishes in it. Egoothi came to advertise us that Eengeen has Killed 7 Deers of which 5 are for the House. I am glad to hear the Sick are getting better, but it is Supposed Eengeen also feels the Symptoms of the Sickness although not yet laid up by it. They are all without Snowshoes & have hard labour hunting now in the deep Snow.

**12 October 1837.** Sent John McKay with 2 Sleds & Dogs with the young man, and the Boy[14] will come back with him to bring a load. Sent 1 pair Snowshoes & 2 prs. frames of Birch (a Scarce article here) to the Hunters to enable them to hunt. Overhauled 3 netts which gave us 20 fishes. A Young man came from the fishery, the distance about 8 miles round by the Channel. Says the Channel is frozen over. Laurent sent him for Leather to tie the Sinks and floats for the netts. Also wants Ice chisels and a Hatchet, all which Articles were Sent by him. The wind Changed from E to NE and North. Weather becoming Colder, but the opening in the Channel here Extending considerably. We have no more fish; had to give out Rations of Dry meat. Our netts took but 11 fishes in the afternoon.

**13 October 1837.** This afternoon the news came back with the meat, and bring the unwelcome intelligence of Eengeen & one of his Children being laid up with the Sickness. Toozeyaze getting better but weak. Cuthand with 3 young lads came with part of the meat of 3 Animals he has Killed, which, with what was brought from

---

12 The men were plastering and building the men's house (FNB B, p. 32).

13 A mock moon, or a bright spot on a lunar halo, although Dease's wording suggests that he was referring simply to the halo.

14 François Boucher.

the Hunters amounts to 670 lb. – Another lad came from towards McTavish's Bay[15] from his Brother Nozanne. Says they have 3 Deers Killed, and they are numerous, but they have no amunition. Sickness prevented their proceeding to Ft. Norman. Got up our nett with 5 fishes only. An Indian came from the Fishery & returned.[16]

**14 October 1837.** Sent back the Boy to the Hunters' Tent. The Indians got some Amunition to enable them to hunt. The netts were Examined & 2 taken up. But 31 fishes were got from them. The Strait is nearly frozen over again.

**15 October 1837.** This Evening the Boy Came back from the Hunters' Tent. Says, being Sick they Sent him back here as they have no hope of Killing any animals soon. Nahlhal & his wife improving a little. The Indian that Assists Laurent at the fishery came for lines and hooks to fish with. Says that there is about 3 loads of fish collected & that the Old Indian who got the 2 netts has given 250 fish to Laurent for us. The Strait frozen over Entirely.

**16 October 1837.** Went with Felix & the Boy to the fishery where I find that including what was given by the Indians, Laurent has 700 fishes Collected of which we brought home 300. Egoothi came in the afternoon with the Welcome news of having Killed 6 Deers, 5 of them for us. Weather mild. 2 netts put down again. 26 fishes taken out of the others. An opening again in the Channel. The young man, Egoothi, says the Deer are numerous.

**17 October 1837.** Felix and the Boy were Sent out for the meat. Mr. Simpson goes to pass a few days at the Hunters' Tent. John McKey was for a load of fish and brought 103. Laurent, he says, took 93 this morning. Turned 4 pairs Snowshoe frames.

**18 October 1837.** Felix and the Boy Came back with 500 lb. meat. The Ice much broken in the Channel here. Weather mild.

**19 October 1837.** Sent back the Boy with his Dogs & Sledge; the other Sledge Broken. Laurent came from the Fishery; has, he says, been under the necessity of taking up most of his netts. Such a Strong current is formed when there is any wind that they are nearly carried away by it and hook among the Stones at bottom and must be torn to get them up. Got 2 pair Snowshoes barred[17] and one Knitted.[18] The last Chimney of the men's House was got up above the Square[19] today. Gullion assisting Ritch at working flooring & Doors &c.[20]

**20 October 1837.** Sent Laurent back. Felix repairing his Sledge to go to the Hunters' tomorrow. A Strong N & NW wind throughout the day.

**21 October 1837.** An Indian lad was sent with Felix to the Hunters. The wind from the Same Quarter but not so Strong as yesterday. Our netts cannot be got out yet. They are partly covered with Ice & can not be got without injuring them very much. Got 2 put down further to Westward[21] where the Ice is firmly frozen over.

---

15 McTavish Arm, the northeastern arm of Great Bear Lake.

16 With cords and leather for attaching floats and weights to the nets (FNB B, p. 34).

17 I.e. the wooden cross-pieces fixed in place.

18 I.e. the webbing woven.

19 I.e. above the stone fireplace.

20 Hector Morrison was making nets (FNB B, p. 36).

21 By Norquay.

**22 October 1837.** Weather fine and Clear; wind from the Same Quarter. Frans. Framand Complains of being unwell. Got 8 fishes.
**23 October 1837.** About Midday, Mr. Simpson came back with Felix and the Boy. Brought 4 Deers, Wt. 400 [pounds], of which 2 were killed by Mr Simpson. Nahlhal begins to go about. Killed one & Egoothi one. Deer are Plenty but the Hunters Weak. Got 15 fishes out of 2 netts.[22]
**24 October 1837.** Felix was sent for a load of fish. Brought 135. The 4 netts we have down took but 27. Got 2 of the other netts up and 6 fishes out of them. Sent Geo. Sinclair to remain with the Hunters, and the Boy also. They are to bring loads as Soon as they have Killed any number of animals. This afternoon we abandoned our Smoaky lodge[23] and removed to the House.[24] Ritch making Tables for the Rooms &c.
**25 October 1837.** Felix brought 144 fishes from the Fishery. Only 11 taken here out of 4 netts. A young man, Bro. to Cuthand, came to tell us that they have 2 Deers for us, which they wish us to send for. 2 Indians from the Fishery Came to beg some assistance from us, having lost most of their netts, they say carried off by the Strong Current. 2 other Indians came from a lake that is about 20 miles to the north of this; brought 42 Dry White fishes. They tell us they have 190 more collected for us. Got 2 pairs Snowshoes Knitted ready for use.
**26 October 1837.** Felix was Sent for another load of fish; brought 120. The fishery is diminishing daily, they say. The Indians are beginning to remove to another place where they Expect to take more fish. Our only chance is to follow, as we Know no particular Spot good as a Fishery. The 4 netts here gave but 15. Lent a nett to the Indians of the N Lake and a trench[25] as a trial. If the Fishery is good, we will send some one to try. Weather Hazy & Colder than it has been.
**27 October 1837.** Sent Felix to assist Laurent to remove & Sett his netts. I left this with the 2 Inds. from N Lake & got to their Tents. Find they have what they Said.[26]
**28 October 1837.** Passed the day at the Lake. Saw 2 herds of Deers. The Indians took 30 fishes out of 3 short netts & put down the one lent them.
**29 October 1837.** Left the Indians & Came back to the House. Felix brought 110 fishes. Says 120 remain there, Where they have been. They put down 2 netts yester-

---

22 The chimneys of the men's house were finished and mudded (FNB B, p. 37).

23 Their tent.

24 Simpson described the winter quarters as follows: "The houses were constructed on a very small scale to suit our means and the severity of the elements. They consisted of a log building, forty feet long, and sixteen broad, containing a chamber at either end for Mr. Dease and myself, separated by a hall, sixteen feet square, which answered the threefold purpose of our eating-room, kitchen, and an apartment of all work for the Indians. There was, indeed, the frame of a kitchen erected behind, but we were unable to complete it till the following year, when an observatory was also built. Our men's house was thirty feet long and eighteen broad, and, with the store, formed three sides of a little quadrangle fronting the south. The whole was habitable in a month after our arrival; but, from the smallness of the timber, and the difficulty of procuring enough of the frozen earth to cover the light roofs, our dwelling was miserably cold, the wind and snow having in many places free ingress. The men's quarters were rather more comfortable" (Simpson, *Narrative*, 208–9).

25 An ice chisel.

26 I.e. 190 dried whitefish for Fort Confidence.

day out of which 50 fishes were taken this morning. Our netts have taken very few and Catch at the Bottom which breaks them much.

**30 October 1837.** Felix was sent for the remainder of the fish and found but 104 instead of 120. Norquay was ordered to take up the netts here in order to go and try at the N Lake. The wolves run down & killed a Deer not more than a mile from this. 2 men were Sent for it & brought the meat; not much of it destroyed.

**31 October 1837.** Sent LaMothe to remain and assist Laurent at the Fishery. Felix goes with him to bring a load of fish if enough to load him. McKay making Sledges assisted by Geo Flett. Morin Carting logs for the Sawyers. Ritch & Gullion at Carpenter's work and Framand Squaring posts for building a Kitchen.[27]

**1 November 1837, Wednesday.** Fine mild weather – 8 lines were put down by Morin.[28] Evening Geo. Sinclair Came to tell us that the Hunters have Killed 6 Deer besides 2 Killed by himself, but at too great a Distance to go for now. At Same time the dogs that they had with them have run off after Deer & Strayed these 5 days. 2 of them Cast up here yesterday, but 4 are missing yet. The man has been now 3 days in Search of them without Success; a very Serious loss to us Should they not come back.

**2 November 1837.** Felix & Morin sent for the meat to the Hunters. 2 Small trouts were taken at the Lines. Framand & Geo. Flett are Cutting logs to build the Kitchen.[29] The weather fair & mild; Wind ESE. All hands Employed.

**3 November 1837.** This morning 3 Trouts were taken at the Lines. 4 more put down. The men came back with the Meat, Wt. 700 lb. Two Deers killed yesterday were destroyed by Wolves & Wolverines. One of the Dogs has got back to the Tent. There are yet three missing.

**4 November 1837.** The work going on as usual. No trout taken to day. We saw 6 Rein Deers crossing the Bay today. The Wind Changed to Westerly; Weather Colder than it has been yet. Mended a nett to be sent Norquay on Monday.[30]

**5 November 1837.** This morning 2 young men from the fishery came with the unwelcome news that Laurent's netts take very few fish, hardly a sufficiency to subsist himself & Companion. This is the place where the Natives were saying such a good Winter fishery might be made and in the end disappoints us. No reliance can be placed on their word.[31] These lads come from Cuthand to get ammunition to go after a Herd

---

27 The sawyers were Begg and McDonald. Norquay and Hector were sent to fish at the northern lake. Ritch and Gullion finished a cupboard in the hall (FNB B, p. 40).

28 Assisted by Dease (FNB B, p. 40). That he should himself be engaging in such work was probably a result of the fact that the Catholic members of the party were given the day off on this church holiday, and that Dease probably felt he should assist, given the urgency of the food situation.

29 Of the other men, McKay was building or repairing a sledge and Begg and McDonald were sawing lumber (FNB B, p. 41).

30 Felix and Morin set another eleven lines, making a total of nineteen (FNB B, p. 41).

31 In his field notes, however, Dease partially excuses the Indians by the remark: "they not being in the habit of remaining steadily at any station long enough to judge if there is a regular winter Fishery" (FNB B, p. 42).

of Musk Buffalo[32] that are at no great distance from their camp. Sent James McKay to Sleep at their Tents in order to be ready Early tomorrow to Join the hunt.

**6 November 1837.** Three Trouts were taken at the Lines to day. Afternoon a young man came from the Camp with 6 Deer Ribbs, 3 Tongues & 4 fishes. He tells us that Eddaw-delly, one of them, has put up 2 Deer for us of 3 that he Killed, & wants Amunition as Deers are plenty. At Sunsett George Sinclair & the Boy Came, bringing the Body of Old Rabbits Eyes (a Corpse). The poor old man has been Sick some time and Expired yesterday.[33] The Hunters will be here Soon. The meat of 3 Deers was brought at same time, Killed by Eengheen & Nalhahl. Felix & Morin were Sent this morning to the Lake where Norquay is, for what fish they may have to Send us.[34]

**7 November 1837.** This morning Mr Simpson with McKay, Sinclair & the Boy, left this to go & Endeavour to kill a few Deers. A Coffin was made & the Body buried. Morin & Felix brought 214 fishes given by the Indians.

**8 November 1837.** This Afternoon Eengheen came to tell us that the party are Encamped near the River.[35] They wish their property to be sent for, as they left Every thing at the Camp where the Old man Died. The Lines took but 1 trout. I had a nett put down yesterday and one to day at the Point of the Island[36] to Westward. Only 1 White fish taken. Blowing fresh; Weather mild. Evening arrived Nozanne and 3 other young men with 40 Ribbs, 26 Tongues & 126 lb. fresh meat. They tell us they have the legs of 3 Deers at their Tent which is 2 days walk from here, & that the loss of the Powder Sent by his Brother in the Fall is the reason they have not more. Deers are numerous. The Hunter[37] returned to their Tent.

**9 November 1837.** Felix was sent for the Hunters property. 3 of the 4 Indians that came yesterday were Sent out to hunt, but one of them Came back. The Hunters & Families came. Nalhahl has been out along the Hills & killed a Deer which was sent for – 115 lb. Missed several shots at others. The lines took 3 Trouts; the 2 netts nothing. The 2 Indians from the Lake where Norquay is, Came with a note from him stating that his netts take few fishes. He has but 60 collected. The Indians brought a pair Snowshoe frames and 2 Rein Deerskin Coats [for] which I sent them the Skins to have dressed & made up.

**10 November 1837.** The Hunters were out to day, but had no Success. The other came also without having killed any thing. The lines took but one Trout.

---

32 Muskoxen (*Ovibos moschatus*) were widespread throughout the mainland tundra from the Mackenzie Delta to Hudson Bay and throughout most of the arctic islands, except Baffin Island. They also penetrated short distances south of the treeline, as in this case. They were almost exterminated from the mainland by commercial hunting between 1860 and 1915. The species was given complete protection by the Canadian government in 1917, since which time it has made a spectacular comeback and has reoccupied most of its historic range (Barr, *Back from the Brink*).

33 The dead man was Nalhal's stepfather. Dease suspected that "this unfortunate death may, I fear, derange our Hunters for some time" (FNB B, p. 43).

34 At Fort Confidence, meantime, the men were clearing the ground for the kitchen (FNB B, p. 43).

35 Dease River.

36 Fishing Island, now named Ritch's Island.

37 Eengeen (FNB B, p. 46).

**11 November 1837.** This morning Morin was sent with Nozanne & party to remain a few days with them & put up any animals they may kill. He will bring the meat they have & go and see if any good place for fishing is in that Bay. Egoothi went with the Inds. to N Lake to see if there are many Deer in that direction. 4 trouts taken in netts and at Lines.
**12 November 1837.** A Cold NE wind. Afternoon Mr Simpson & party came back with 150 lb. meat including part of a Deer that Sinclair Killed with what the Indians had put up for us, and 100 fishes from the Fishery.[38] The netts have taken more for some days past than they had previously. Egoothi came back; has no Encouraging news from that Quarter.
**13 November 1837.** The Hunters were Settled to leave this tomorrow. Sinclair & McKay are also to go and try their chance for a few days towards McTavish's Bay. They are to pass by the Fishery & leave a Sledge and Dogs for LaMothe to bring a load of fish. The Boy (Boucher) to go with the Hunters.[39]
**14 November 1837.** Felix was sent to Norquay's fishery. The Hunters off.[40] The men here plastering the Building intended for a Kitchen, but which will not be finished this Season. It will therefore Serve in mean time for a fish & lumber Store until means admit of completing it.
**15 November 1837.** Norquay & Hector Morrison came back with Felix. Brought back 39 fishes. LaMothe came with 81 from Laurent's. A young man from Charleau's Camp came to tell us that he with his Brother (Called Big Head) & Charleau have 13 Deer for us at their tents 2 days from here. Felix, Norquay & McDonald ordered to go with them tomorrow. 3 Trouts were taken at the Lines.
**16 November 1837.** Lamothe was sent Back to the fishery. The men[41] & Indians off and 3 Trouts taken at the Lines.
**17 November 1837.** A Door was put up to the building intended for Kitchen.[42] Weather boisterous; Wind & Snow.
**18 November 1837.** Laurent came from the Fishery. Says that but few fishes are taken now & that he is at a loss where to try any more. Lamothe Strayed on his return and only got yesterday morning back. Fortunately the weather was not very Cold or he might have perished. 3 Small trouts taken.[43]
**19 November 1837.** This Evening 2 young men arrived from a Lake up Haldane's River where they say plenty fish may be taken. They have some collected and a few Skins Meat.[44] The Principal man of the Party is Called the Big White man.[45] Nothing taken to day at Lines.
**20 November 1837.** Laurent returned to his fishery. He is to make another trial near

38 Laurent's fishery.
39 Felix went to check the lines and found they had caught nothing (FNB B, p. 46).
40 Along with Sinclair and McKay (FNB B, p. 46).
41 Felix, Norquay, and McDonald.
42 Some of the men were also cutting firewood (FNB B, p. 47).
43 McKay and Sinclair returned after an unsuccessful hunt (FNB B, p. 47).
44 I.e. meat worth a few Made Beaver, the latter being the standard unit of exchange in the fur trade.
45 Alias Gros Blanc (FNB B, p. 48).

the Old Fall fishery, where the Channel between the Island[46] and main land widens. The men[47] came back with 646 lb. meat and have left about a Sled load yet at the Tents. Norquay, C. Begg & Hector Morrison ordered to go and try the Fishery at Haldane's Lake, and to collect what fish or meat the Indians may have or Kill for us. Gullion is to go & remain at Laurent's fishery. Felix to return[48] for the meat remaining at the Tent. Geo. Sinclair to remain with them to Secure the animals they may Kill. McKay to sett lines here. Geo. Flett & François Framand to cut and cart fuel. Ritch at Sundry Jobs in the House. Wm. McDonald to go for 3 Deers Killed by Dawaddelly towards Cape McDonnell.

**21 November 1837.** The men off in the Different Directions stated yesterday. Morin came back this afternoon with a young man, his load of meat 340 lb. They have Killed nothing worth mentioning since they left this. The travelling from here to the Bay[49] very bad, principally over Barren grounds that afford no Shelter from Wind or weather. Wood very Scarce to make fire along the Way.

**22 November 1837.** Nothing new occurred. The young man passed the day here to rest.

**23 November 1837.** The young man left this to return. Felix came[50] with but 70 lb. meat, Wolves or Wolverines having destroyed part of the meat.

**24 November 1837.** McDonald came back with 2 Indians; brought 250 lb. meat & 7 Tongues. Saw young Boucher yesterday who told him that the Hunters have killed 4 Deers. 3 of which were near the place he saw him. The Fishery[51] has failed entirely; the Indians leaving it.

**25 November 1837.** Sent Felix with McDonald for the 4 Deers mentioned yesterday. Morin sent to assist Laurent to take up his netts & try again towards the Old Fishery. An Indian is to Shew them the Place. Some Snow fell these 2 days past.

**26 November 1837.** Nothing transpired.

**27 November 1837.** Morin came home; has assisted Laurent to put down 3 netts & to remove their baggage to the Island where they are to put up their lodge. He tells us the Indians Stole about 60 fishes out of Laurent's Cache before leaving the Place.

**28 November 1837.** Felix & McDonald came back; brought but 3 Deers. Could not find the 4th. Laurent with Lamothe came from the Fishery. The former has had his feet frozen badly & must remain now untill healed. Sent Morin with Frans. Framand to put down the netts and try in different Places. The Thermo. was at -42° this morning. Laurent will go back with them to mind the netts.

**29 November 1837.** Weather mild. Nothing taken at the Lines.

**30 November 1837.** No Success at our Lines. LaMothe Cutting Fuel in place of Framand.[52]

---

46 Fishing Island, now Ritch's Island.

47 Felix and party (FNB B, p. 48).

48 With Charleau (FNB B, p. 48).

49 McTavish Arm.

50 From Charleau's (FNB B, p. 49).

51 Laurent's fishery.

52 The sun did not rise above the southern horizon on this date (Simpson, *Narrative*, 215).

**1 December 1837.** A fine day. No arrivals from any Quarter. Nothing from the Lines.
**2 December 1837.** Had some Snow with a Strong Wind. The lines give us nothing.
**3 December 1837.** LaMothe gone to the Fishery. The Weather fine for the Season.
**4 December 1837.** LaMothe came back and says that Morin has put down 4 netts at the old place and taken a few fishes. Has about 50 Collected; will be here soon. McKay removed some of his Lines to deep Water, 20 fms., but no Success attends them.
**5 December 1837.** The weather moderate. No arrivals from any Quarter. Our Express I Expected before now. Cannot imagine what keeps them.
**6 December 1837.** This Morning the Boy Came from the Hunters to tell us they have 12 deers put up. Egoothe is unwell and the others, since leaving this, had to make Snowshoes of a larger Size which Kept them for Some time from hunting. Animals numerous. Morin came with Laurent from the Fishery; brought 60 fishes of a very Small Size Compared with those taken there in the Fall. This Evening Nozanne, his 2 Brothers & Uncle with 2 Women & 3 Children came, under a great alarm. Have seen the track of a Strange Snowshoe near the Camp, such as they are not accustomed to See & Suppose it to be Enemies. But by their account it seems to me as if it is Some White Man with temporary Substitutes for Snowshoes. The track by their description is that of a Round flat Snowshoe not turned up as the Northern Indians do. They say positively it must be Strangers or Whites. They have a Gun and have used Cotton rags near an Encampment. They saw where some meat was left buried under the fireplace, Chopped or minced with fat. The person had Killed a Deer. One of these Indians Slept about 2 miles from the Stranger & in the morning found his fire burning yet, but would not follow his track to ascertain who it was. One of the party had been sent hither previously to advertise us to go for some Animals they had Killed, but upon seeing this, all fled and have overtaken the messenger. I could not persuade them they have nothing to fear from Enemies. With Some difficulty they were prevailed with to return with Mr. T. Simpson & 3 men, that we may know who it is or form some opinion from observation.[53]
**7 December 1837.** Mr. Simpson with Felix, Morin & G. McDonald left this with 3 of the Indians. The Boy with McKay left this for the hunters' tent to Collect as much as they Can & Encourage the hunters to Exert themselves. Laurent left this to return to the Fishery. In the Afternoon Sinclair with a lad Came from Charleau's Camp with news of their having 10 Deers for us there. We are much in want of Dogs & Sledges.
**8 December 1837.** Sinclair & Companion remained. The families of the others left this to Join them.
**9 December 1837.** Sent George Flett with the Wood Carters for a load to Charleau's tent with Geo. Sinclair. The lines were taken up today & gave us 3 trouts. Afternoon

---

53 Simpson and Dease surmised that it might even be a party from Captain George Back's expedition in HMS *Terror* (FNB B, p. 53; Simpson, *Narrative*, 209). Back had orders to take his ship to Repulse Bay, cross the isthmus to Committee Bay, and explore the coast west from there (Back, *Narrative of an Expedition in H.M.S. Terror*).

C. Begg with a Young Indian Came from Haldane's Lake where he Says the fishery is not near so abundant as they Stated it to be. They got 130 fishes and a little meat from the Indians, most of which has been used. They have been 4 days on the Way here. Saw the Cuthand & 3 others at Limestone Point about 15 miles W. of this.[54]

**10 December 1837.** Weather Cloudy & mild for the Season.

**11 December 1837.** Sent the Indians with John McKay to put down 8 lines where the last lines were. Settled him to return with Begg tomorrow & instructions to Norquay to follow the Indians when they leave that place as soon as the fishery fails, to come along the Lake, and bring us news as Soon as they Can.

**12 December 1837.** Begg left this Early. Afternoon Mr. Simpson & party with our Indian came;[55] brought 150 lb. meat. 2 Deers have been destroyed. Their Alarm Seems from what was found to have been caused by Some one in Search of this place, & Suppose it to be the bearers of the Express who do not Know the Lake and mistake our Bay for the other.[56] The man who came near the Indian's Camp had temporary Snowshoes & had Split from some birch bark for lighting fire more Expeditiously with Matches. As only the Marks of one person was seen, and the Indians declined going on the track where they had Seen Smoak before, nothing positive could be Known, but I fear much that the timidity of the Indians may be the cause of loss of life and perhaps all our dispatches.

**13 December 1837.** The young man remains to rest himself and Expects to see some of Charleau's party on the return of Geo. Flett. The lines took 1 trout.

**14 December 1837.** No arrivals from the Hunters. The men Carting Fuel. One trout taken at the Lines.

**15 December 1837.** The young man left this alone to return. Felix went for a load of fish & brought 97. George Flett came home with one of the party & the Lad who left this in the morning turned back with them. They have Killed 7 more Deers Since Sinclair left them & have now 16 for us. The load brought is 240 lb. The Boy[57] Came from the Hunters with Egoothe. They have now 22 Deers put up for us. Ordered Felix with the Boy for the Hunters & Morin with McDonald to Charleau's party tomorrow.

**16 December 1837.** The men as appointed left this. Lines give nothing.

**17 December 1837.** Laurent Came from the Fishery with 50 fishes taken this day out of 5 netts. One Trout taken at the Lines.

---

54 Four of Cuthand's group joined the band at Haldane Lake, but Begg thought that they would not be able to lay in any surplus for the fort; indeed the distance involved was probably too great to make such an endeavour worthwhile (FNB B, p. 55).

55 Simpson and his party had reached the area where the strange tracks had been seen (near McTavish Arm) on 9 December. They lit a fire on a conspicuous spot where it was visible for a long distance at night. On a low point by the lake shore they found the remains of an old fire, the camp of a single person, a cache of caribou meat, and the birch bark kindling mentioned by Dease. They deduced it was the messenger they were expecting from Fort Simpson with dispatches, a Cree named Le Sourd, but assumed he had lost his way. In case he returned to the campsite, they placed markers to guide him to the nearest Indian camp (Simpson, *Narrative*, 210–11).

56 Confuse McTavish Arm with Dease Arm.

57 François Boucher.

**18 December 1837.** Nothing worthy remark. The man returned to the fishery. Weather turning Cold. This Evening the Thermo. Stood at -40°.
**19 December 1837.** Hazy Weather not so Cold as yesterday. This Evening one of the young men of Nozanne's party Cast up in great distress apparently. Has again fled from the Camp under the impression that his 2 Companions who were out hunting have been killed by Enemies near their own lodge. When some noise was heard as of Strange people, a Call and Whistling, the Women became alarmed, and he being on his return from an Excursion in Search of Charleau's Camp, found the women in this State of Alarm, which Communicated to him, and without having taken off his Snowshoes, fled, followed by a Boy who he placed near a root in Concealment & came on Since yesterday after Sunsett. I know not what to make of this Second Story unless proceeding from the Same cause as the first. But the fright from the former alarm evidently has not thoroughly Subsided yet. They had 7 Deers killed when he left the Tent.
**20 December 1837.** The young man passed the day here. Turned 1 pair Snowshoe frames. Felix & the Boy Came back with 480 lb. Meat. The Hunters have killed 7 more Deers since last news. Morin & McDonald also came with 530 lb. meat. That party have also Killed 5 Deers since last accounts. There remain now 15 there and 20 at the Hunters' Tents. 40 fishes were brought from the Fishery by Felix & the Boy, the produce of 3 netts to day, better than I Expected. Geo. Flett is to go with Felix & the Boy to the Hunters' Caches, as they have seen vestiges of the Indians going in that direction. As they may Steal, Morin & McDonald will return to the Same place they were last.
**21 December 1837.** The men off as appointed yesterday. Afternoon Cuthand & his Brother Jack, came from Limestone Point with 106 lb. meat on a Sledge, which is all they have done since Octr. Their families have crossed over to the South Side the Island to Join the other Indians no doubt. 2 trouts taken today.
**22 December 1837.** Nothing new. Cuthand left this to Join his family. The weather Severely Cold.
**23 December 1837.** The Same Cold weather; the Thermo below -40°. We feel anxious about our Dispatches from Council. They Should be here before now.
**24 December 1837.** This morning Framand brought 30 fishes from the fishery. Felix with Flett and the Boy brought 760 lb. meat from the Hunters. 2 Deers have been destroyed in one of the Caches. Morin & McDonald Came with 540 lb. from Charleau's party; remains 7 in that Quarter, and 12 at the Hunters. They brought also 84 fishes. Severe Cold.
**25 December 1837.** All hands resting today & got wherewith to improve their usual fare by the addition of Some flour & fat.[58]
**26 December 1837.** Framand returned to the fishery. All the others Cutting and Carting Fuel. Wind westerly.
**27 December 1837.** Sent Felix, Morin & McDonald for meat to the Hunters. The Boy is to remain with them. Wind from the Same Quarter.

---

58 The temperature was -52° (FNB B, 60). To celebrate Christmas, a dance was "followed by a supper consisting of the best fare we could command" (Simpson, *Narrative*, 218).

**28 December 1837.** No arrivals from any Quarter. I fully Expected some news from Nozanne's party about this time, from the Alarm of the young man, which may be some trick of his among his friends that he has given to us under another colouring to prevent Suspicion of the other cause. However they must be all murdered, or all Safe, otherwise we would have heard something.
**29 December 1837.** This afternoon had the Arrival of Sourd, one of Ft. Simpson Servants, accompanied by 2 young Indians from the Camp that were about the fishery some time ago. Sourd is the Bearer of our Dispatches and has the Melancholy tidings for us of the loss of one of our best men, Peter Taylor, who was his Companion from Ft. Simpson, which place they left on the 10th Oct. with the dispatches in a Small Canoe. With it they got up part of Bear Lake River where they left it and travelled on foot.[59] At McVicar's Bay the deceased complained of Weakness and oppression of the Lungs. Part of their property was left & Sourd Carried the Remainder, leaving but his Gun to carry. In this State by Slow Journeys they got on, to near the Place that the Indians took the Alarm,[60] when finding himself too weak to walk, told Sourd he could travel no farther. A Campment was made where it was hoped he might recruit by rest & food as they did not want. They were on the South Side of the Bay within, I Suppose 12 or 15 Miles of where the Indians were, and it was from that Camp that Sourd made one of his Excursions looking about for the Establishment and Hunting, that he Slept near the Indians & was the Cause of the Alarm they took. He killed Several Deers returning to his friend, who he would not abandon while alive,[61] and by his account it is now 16 days Since he Expired. A most lamentable Case, as Mr Simpson was within a few miles of them when he found Sourd's Cache, but the Natives Could not be got to go where they had seen the Smoak on the opposite Side the Bay. The Cause of his Death, as far as we can Judge, was an old Consumptive Complaint,[62] perhaps aggravated by Cold and fatigue. He had been 15 days about that Camp before P.T. died, and he Says it is now 16 days Since he died, which must have been about the 12 or 13th. After having interred the Body as well as Circumstances admitted, he Came on, travelling & hunting his Way towards Cape McDonnell. Not Knowing the many Deep Bays of the Lake, he was on the Eve of giving up his Search and turning back when on the 25th he fell in with the Camp of Indians from whence the 2 lads Came with him. Our Dispatches from Ft. Norman were taken by Deceased, Peter Taylor, near to Ft Resolution where meeting Young Mr. McLeod on his Way to McKenzie's River with the letters, he sent them on by young Wentzel and turned back with Mr. McLeod and got to Ft. Simpson about 8th Oct., our letters being dated 9th. The very late and

59 They made good progress on smooth ice along the southern and eastern shores of the lake, although the centre was still open (Simpson, *Narrative*, 212).

60 On McTavish Arm, which they did not know existed, and mistook for Dease Arm, the location of Fort Confidence.

61 Le Sourd buried Taylor in a grave which he dug by thawing the ground with a fire (Simpson, *Narrative*, 212).

62 I.e. tuberculosis.

backward Season was the Cause of a very late arrival of Govr. Simpson at Red River and led him to think our Expedition to the Westward would not have been successful last Summer and advises perseverance by returning next Summer in that Event, which would be a difficult undertaking from our Present Station.[63]

**30 December 1837.** Nothing new. Nothing taken at the Lines some days past.

**31 December 1837.** Afternoon a young man & boy Came from Sinclair's party with news that they have added 17 to the 7 Deers they had remaining. Felix, McKay and Toozeyaze came from the Hunters; brought 360 lb. meat & 36 Skins Scraped. Morin & McDonald brought 560 lb. meat with 55 fishes from the Fishery. The 2 lads that came with Sourd were sent back with a note to the fishermen to come for New Year's day. Nozanne with a Brother arrived with some meat, Wt. 100 lb. This Brother coming to Join them, was the Cause of the last Alarm, being within hearing of their lodge but did not know it.

**1 January 1838, Monday.** The men from the Fishery came to pay the Compliments of the Season. They have got 150 fishes Collected at the Tent. In the afternoon Norquay, Begg & Hector. They have not collected any thing since they left this. Have Spent what the Indians gave them and have been poorly off for some days back. A Flag Staff was Erected this morning & the name of Fort Confidence bestowed on our present Establishment. An addition of Flour, Fat & Tongues added to the usual meat Rations & the day passed as pleasantly as could be Expected under our present Circumstances, the loss of one of our Companions being the only drawback we find to being quite Contented with our prospects & past Season.

**2 January 1838.** Nothing done to day. Several of the men ailing a little.

**3 January 1838.** Sent Laurent back with Gullion to Assist him at the netts & LaMothe as Wood cutter. Norquay & Hector Morrison go also to fish there and will have half the netts in charge. Morin went with Framand to the Fishery. They came back with 149 fishes. Nozanne and Brother left this. They will Join Charleau's party. McKay, Felix, Morin & McDonald to go for meat Tomorrow.

**4 January 1838.** The men Started Early, C. Begg with Framand Cutting Wood, Geo. Flett Carting. One Trout 31 lbs. taken at the Lines; 2 lines lost. The Thermo. at -36°. Sourd to remain with the Hunters & the Boy.

**5 January 1838.** Nothing new. Our Lines take so little not worth Keeping them down.

**6 January 1838.** This afternoon a young man came to tell us that Dawadelly has 2 Musk Buffalo for us & wishes them to be Sent for. Left the Camp yesterday.

**7 January 1838.** Severe Cold. Wind W. morning, turned to E. The men came back with their loads; brought 14.[64] Remain 9 there. Young man remains.

**8 January 1838.** The men remained Carting fuel for their houses, to allow their Dogs rest. Their paws getting Sore from the rough Snow.

---

63 By his closing comment, Dease means that the late breakup on Great Bear Lake would make it almost impossible to attempt the exploration of the coast westwards to Point Barrow from their base at Fort Confidence. Dease and Simpson had now received George Simpson's letter of 30 June (see p. 68).

64 Caribou.

**9 January 1838.** McKay, Morin & McDonald sent for the 9 Deers. Felix & Norquay to go for the Buffalo. All Started Early. Strong wind afternoon.
**10/11 January 1838.** Nothing of any moment occurred. Cold Weather.
**12 January 1838.** Afternoon Felix & Norquay came back with 290 lb. meat, much gnawed by Dogs. The Hunters have nothing new. Evening came the men. Say that much of the meat is Wanting. Did not see the Indians or Sinclair. Flett went to the fishery & brought 160; Felix & Norquay 80. Remain about as many more there. Have 13 netts down and are doing pretty well.
**13 January 1838.** The men Carting fuel. Weather Cold & Calm.
**14 January 1838.** The Same work as yesterday. The Cold much the same.
**15 January 1838.** Two Indians came from the Camp with 10 fishes, 12 lb. meat and 12 tongues. The men Sent off for meat. Strong wind & drift. George Flett was Sent for a load of fish.
**16 January 1838.** Flett brought 164 fishes. The netts produce more than they did.
**17 January 1838.** Morin & McDonald came home with but 350 lb. meat. Felix and McKay gone to the Lodge.

• • •

## *2. Letter from Thomas Simpson to James Hargrave, York Factory, 17 January 1838*[65]

*Private*

Fort Confidence, 17th Jany. 1838

My Dear Sir,

On the 29th Ulto. I had the satisfaction of receiving your much esteemed letter of 20th June. You do not acknowledge one of mine, written I think from Red River, but I hope you duly received it.

Many thanks for your kind wishes in reference to our present enterprise; I make no doubt you will be pleased to learn that the honor of completing our western discoveries was won by your old YF[66] friend, who with his own hand displayed the Company's banner on the eagerly sought for Cape Barrow, the keystone of our hopes and wishes. My party consisted of five men, and the distance travelled from "Boat Extreme", where I left my worthy colleague, was 60 miles. The first 30 (and painful ones they were) we performed on foot, more through mud and freezing salt water than on dry land, amidst fog too of almost tangible density. A great inlet stretching to the Southward rendered our farther progress hopeless, when most providentially a small camp – the first we had seen of the Far West Esquimaux – greeted our delighted eyes. We advanced rapidly towards them, but the dreadful apparation of White men, with guns on their shoulders, terrified the inmates to such a degree that throwing themselves into their canoes they abandoned every thing and pushed out

65 Glazebrook, *The Hargrave Correspondence*, 260–2.
66 York Factory.

of reach. By aid of my vocabulary, I however succeeded in bawling them ashore, and soon obtained one of their large baidars, in which myself and followers embarked.

Two large villages of the natives witnessed our ceremony of taking possession at Cape Barrow and joined their shouts with ours. It was there that Mr. Elson with the Blossom's armed barge, judged it unsafe to land in 1825; but I found the savages well disposed, perhaps on account of the well known craft in which we navigated, and her evident *emptiness*, probably too because we kept strictly on our guard, among such overwhelming numbers. We regained the mighty river of Mackenzie on the 17th of August, and on the 25th September, after being overtaken by winter in our passage of this great and stormy lake, we found our building comrades, who had only arrived here the day before we entered the Mckenzie. They had been stopped nearly 40 days by ice in the river & lake, for the summer all over the country was one of almost unprecedented severity. That fact I humbly think enhances the merit of our exertions, and though it on many occasions led to a dangerous intimacy with the polar ice, it entirely absolved us from that sharp scourge of northern travellers, the musquitoes.

This is a barren mountainous region, and the climate I need scarcely add is intensely cold. I however braved its worst, having until Christmas chiefly occupied myself on hunting excursions and exploring the country with the Indians. A successful initiation into the mysteries of reindeer hunting, and the discovery of another great arm of the lake,[67] terminating halfway between Dease's Bay and the Coppermine river, were among the fruits of these journies.

In March I propose leading the way for our boats & baggage to the Coppermine. I do not in the least apprehend that Capt. Back's researches will interfere with ours; indeed I suspect that he has much under-rated the land portage from Repulse Bay to the Gulph of Boothia. King's book[68] is the most venemous thing I have read for a long time. We have ample means of falsifying many of his assertions. Sir John's[69] is a dry, prosing concern. Having received no news papers, we are almost totally ignorant of public events; but, as regards our own circumscribed world, we are delighted with the splendid result of last year's trade, and trust that neither disease nor shipwreck will recur to cloud such fair prospects.

I suppose you are at this time enjoying yourself in our dear father land. I hope your jaunt may prove a complete renovation of your health, and send you back fortified with a fair rib,[70] a comfort which popular rumour arrogates both to yourself and to your self-denied friend Mr. Finlayson.

Sincerely reciprocating your kind wishes,
I remain. My Dear Sir,
Ever yours truly,
Thomas Simpson

• • •

67 McTavish Arm.

68 King, *Narrative*.

69 Sir John Ross's book, *Narrative*.

70 Accompanied by a wife.

## *3. Letter from Thomas Simpson to Donald Ross, Norway House, 18 January 1838*[71]

*Private*

Fort Confidence, 18th Jany. 1838

My Dear Sir,

On the 29th ult. I was made happy by receipt of your welcome favor of 26th July and regret that the monotonous character of an Arctic winter furnishes no materials for framing an equally interesting reply.

You are already in possession of my letters of last September, from Fort Norman and from this place. It is therefore needless to revert to our past successes. These, however, are much enhanced by the extraordinary inclemency of the summer, if I can so term a season when an unsetting sun struggled often ineffectually to raise the thermr. to the freezing point and when not a musquito dared show his blood sucking snout. N'importe, we did the deed, and the Governor's unsanguine letter, authorising us to return to the *western* coast (forgetting that Great Bear Lake is not passable till August!) makes me feel the greater pride that my tramp from Boat Extreme to Point Barrow should have effected an object otherwise impossible; and, had it even been possible, has saved the Honble Company an expense of two thousand pounds, besides placing in Prince Rupert's crest a laurel which the rest of the world strove in vain to pluck for more than two centuries. Had I been in His Majesty's service, I should have expected some brilliant reward, but the poor Fur trade has none such to bestow. I will have the honor and trouble of publishing our travels, no doubt, but the subject is so hackneyed and exhausted, and there are so few opportunities for *vivid* description among interminable ice and almost tangible fogs that little remains to be won in that line. *Professor Christie's*[72] style I certainly admire, but for that venomous scoundrel King, I hold in my hands the means of swamping the whole credit of his book, an affidavit of James McKay and George Sinclair that his story of their fatal encounter with the Esquimaux is utterly false and unfounded.[73] Nor are proofs wanting that he has slandered an honorable body of men, in regard to their treatment of the aborigines; but of these in their own time and place. I do not anticipate any interference from the gallant Capt. Back,[74] and after wading through Ross's

71 BCA ADD.MS 625.

72 Simpson is most probably alluding to the paper by S. Hunter Christie, "Discussion of the Magnetical Observations made by Captain Back, R.N. during his Late Arctic Expedition."

73 Thomas Simpson is alluding to a rumour (repeated by Dr Richard King) that at the Back expedition's farthest point at Point Ogle a party of Back's men, detached on an independent search, had killed some Inuit. Back learned of this rumour only on his return to Britain and made a conscientious attempt to establish the truth, with inconclusive results. The affidavit by Sinclair and McKay to which Thomas Simpson alludes, scotching the rumour, is not mentioned elsewhere. For more details of this rumour and Back's attempts to pin it down see Barr, ed., *Searching for Franklin: The Land Arctic Searching Expedition,* 65, n. 2.

74 In his expedition in HMS *Terror* to Repulse Bay.

wordy journal I am strongly inclined to suspect that the former officer has formed a somewhat short estimate of the land carriage from Repulse Bay to the Gulph of Boothia. It is not impossible that we may now try two summers on the eastern coast, to effect which we must live as Indians; that, however, is become quite natural to me, as until Christmas I accompanied them in many of their hunting excursions, can now kill my own venison; and had besides the honor of discovering another great arm of this vast lake,[75] surrounded by lofty mountains, and extending more than half way between this place and the Coppermine river. It is somewhere in it that my unfortunate servant Peter Taylor lies buried. What a pity we were ignorant of its existence last spring! Yet I have every hope of accomplishing our long and laborious portage,[76] and intend setting out in March to survey the best route for our boats & baggage.

Mr. Dease is a very worthy man, well acquainted with the management of Indians and of an inland establishment, but it is no vanity to say that every thing which requires either planning or execution devolves upon me.[77] He never would have dreamed of planning a third campaign, and though we live together on the happiest terms, he evidently feels that he is not in his right place.

With respect to political events, we are almost as much in the dark here as we were physically during the long absence of the sun; even of country affairs it is but a faint twilight that has reached us, but we are delighted with the splendor of last year's returns, which we trust will replenish the void of hard-earned gold caused by some many preceding disasters.

Warmly returning your own and Mrs. Ross's kind wishes,
Believe me to be, my Dear Sir,
Ever yours faithfully,
Thomas Simpson

Donald Ross Esquire

• • •

## *4. Peter Dease's journal, continued*[78]

**18 January 1838.** Felix, McKay & Geo. Sinclair came with 420 lb. meat. The Indians have got 9 Deers out to the Westward of their Tents, towards where our Chipewyan Hunters are, that they have seen 3 days ago. The Old man has Since arrived. Made 6 pair Snowshoe frames for us.

**19 January 1838.** The men resting. Sinclair Says Our Hunters, he heard, had put up

75 McTavish Arm.

76 To the Coppermine River.

77 Considering that Dease had successfully run the New Caledonia District for five years, this is nonsense.

78 MCM M214.

35 Deers, but they will not come by the same route. The old man putting Bars to Snowshoe frames.

**20 January 1838.** This Afternoon 2 lads from the Hunters came with a tally of 35 Deers. All the Indians drawing towards the same Quarter, which will drive away the Animals.

**21 January 1838.** All Hands for the voyage under orders for tomorrow. Sinclair & McKay to remain with the Hunters. Sourd & Boy to come back. LaMothe to come & Chop wood with Fs. Framand instead of C. Begg who will have to go with Sourd to take the Express to Ft Simpson.

**22 January 1838.** The men off as Ordered. Sent orders to Norquay to come with LaMothe from the Fishery.

• • •

## *5. Letter from Peter Dease and Thomas Simpson to the governor, chief factors, and chief traders of the Northern Department, 22 January 1838*[79]

Fort Confidence, Jan. 22nd, 1838.

Gentlemen,

We have now the honour of acknowledging Governor Simpson's letter, dated Norway House, June 30th, 1837; and received on the 29th of December.

At the same time it is our melancholy duty to report the death of one of the bearers – Peter Taylor –, occasioned by a pulmonary complaint of old standing; and accelerated, perhaps, by the fatigue of the journey. He lingered a long time after he became unable to advance further, during which he was carefully attended by his companion (a Maskegon Indian named "Le Sourd"), who shot several deer around their encampment, and on the poor fellow's decease interred his body. This distressing event occurred on a large arm of the Lake, stretching halfway between Dease's Bay and the Coppermine River, which was first discovered by Mr. Simpson on a hunting excursion some time previous. Le Sourd's track, while vainly searching in Simpson's Bay for the establishment, alarmed a small camp of Dogrib Indians stationed in the vicinity to such a degree that they fled hither in the utmost terror. Conjecturing the true cause, Mr. Simpson immediately started to bring relief to the strangers; but, after six days journey, during which he made the circuit of the northern part of that extensive bay, he returned without being able to trace their retreat. It was at this very date (12–13 December) that Taylor expired. Le Sourd then resolved to retrace his steps to Fort Norman; but at Cape Macdonell he fortunately fell in with another party of Indians, who conducted him hither.

We have entered on this long relation in order to preclude the possibility of misrepresentation; as in a late spiteful production we observe, among other calumnies, that the Company's officers are charged with wilfully exposing to starvation the people under their command – an accusation as false as it is malicious.

---

79 Alexander Simpson, *Life and Travels*, 265–9.

In consequence of Governor Simpson's departure for Norway House before the arrival of the Athabasca boats, no notice has been taken of our dispatch of the 31st of May. It contained an offer on the part of Chief-Factor Macleod to conduct a party of Chipewyans to meet us at the mouth of the Great Fish River this ensuing summer, with a list of supplies required for that service. A precaution of this nature was suggested in our original instructions, and was discussed in Mr. Dease's letter of the 3rd of January; but is now altogether nugatory. While we publicly offer Mr. Macleod the tribute of our thanks for his gallant proposal, we beg to assure the Governor and Council that, though this channel of retreat be cut off, our ardour in the noble cause in which we are engaged knows no abatement.

Notwithstanding the extraordinary severity of last summer, the blank on the *western* coast was minutely explored, as will be seen by the accompanying chart [see fig. 7]. That part of Governor Simpson's latest instructions, which recommends a second attempt in the same quarter, is therefore anticipated; indeed, from this point such an undertaking would with our means have been quite impracticable, as the ice on Great Bear Lake seldom breaks up till the month of August. But as the coast to the *eastward* may present a field for the labours of two seasons, we have availed ourselves of the full powers granted to us by applying to the gentlemen in charge of the Athabasca and Mackenzie's River districts, to forward hither, this ensuing Fall, an additional supply of pemican, leather, sledgewood, dogs, and some articles without which we cannot subsist. Our men have consented to prolong their engagements, if necessary, on condition of their wantages now enclosed to the gentlemen in charge of Norway House being forwarded to Great Slave Lake by the Athabasca brigade; for ere the lapse of a fourth year they will all be worn to the buff. For ourselves we make no demands whatever.

The region surrounding this desolate abode is mountainous, stormy and unsheltered. Both men and dogs are often severely frost-bitten while traversing it for food. We have tried various winter fisheries, both far and near; but one after another they have failed before the rigour of the cold. One, however, ten miles to the southward, still maintains a few hands; others are dispersed among the distant camps; and the few residents are supported by our Chipewyan hunters and the native Indians. We are happy to add that the influenza at last passed over without occasioning more than one or two deaths, and these among the aged; our own people are all in perfect health.

When we arrived here on the 25th of September, Dease's River was already fast. We were therefore unable to transport our sea-boats by water to the height of land, as we had contemplated. The long and toilsome portage will be commenced by stages in March; and Mr. Simpson will precede the party for the purpose of surveying the most practicable route to the Coppermine River. We propose taking our final departure from this place on the 1st of June. Another opportunity will occur of addressing you before that period.

We have the honour, etc., etc.,

Peter W. Dease.
Thomas Simpson

• • •

## *6. Letter from Peter Dease and Thomas Simpson to the gentleman in charge of Norway House, 23 January 1838*[80]

Fort Confidence, January 23rd, 1838

Dear Sir,

In consequence of a third voyage to the eastern coast in 1839, proposed by us, it became necessary, in order to obtain the consent of our people, to promise them a supply of necessaries from Norway House by the Athabasca brigade of this season. A list to that effect is enclosed, which you will be pleased to complete; and we have requested the gentleman in charge of the Athabasca district to transport the property to Great Slave Lake.

We are unfortunately unprovided with a nautical almanac for the year 1839, and unless we can obtain one in time, we shall not consider ourselves justified in venturing our party in an unknown country. We have therefore to request, that immediately upon receipt of this an express be dispatched to York Factory, and another to Red River, with the enclosed letters; and that the book, together with any instruments etc., for the expedition, be forwarded at any cost, so as to overtake the Athabasca boats; or, if that be impossible, at least to reach us here in the spring of 1839.

Our authority from the Governor and Council fully empowers us to make the above or any other demands; while we feel assured that your own good feelngs towards the expedition will ensure your particular attention to them.

We remain,

Peter W. Dease.
Thomas Simpson.

• • •

## *7. Peter Dease's journal, continued*[81]

**23 January 1838.** They brought 120 fishes from the Fishery. Arrived 2 young men from Gros Blanc near Haldane's River. Say they have 9 Deers for us, but we cannot go for them now. Told them to put up the animals untill we can Send for them.
**24 January 1838.** Weather very cold. The young men left this to return. Norquay returned to the fishery.
**25 January 1838.** Cold Clear weather.
**26 January 1838.** Felix came back this Afternoon with 325 lb. meat. Left the others to go to the last Cache. Laurent came from the fishery.
**27 January 1838.** McDonald, Morin, Sourd & Boy came today. The loads amount to 1035 lb. Weather very boisterous afternoon. It is Settled that Morin will go with Sourd to Ft. Simpson with Jack, an Indian required as a witness of the murder, that was Committed by Cadien & party[82] and another Indian that was with Nozanne.

---

80 Alexander Simpson, *Life and Travels*, 269–70.

81 MCM M214.

82 In 1835 Cadien, a voyageur from Red River, with two accomplices, murdered eleven Hare Indians. They were sent to Montreal for trial; Cadien was sentenced to banishment and the others acquitted (Simpson, *Narrative*, 221).

The 2 latter not yet arrived. The men say that Several Indians are to Come with loads of meat from the Cache to assist us to get it in Sooner.
**28 January 1838.** The weather very Cold and Boisterous. Snow drifting. The men, Felix, Begg, McDonald & the Boy to go for meat tomorrow. 154 fishes were brought by them yesterday.
**29 January 1838.** The men left this but the weather very Severe & Windy. Afternoon arrived 4 Indians, Babillard with 3 others, from the Rocks on the South Side of T. Simpson's Bay.[83] Brought 240 lb. Ribbs, 20 Tongues & 3 Rein Deer Robes. They have been 7 days on the way but did not travel Constantly. Two of these left Ft. Norman last Spring & the 2 others in the Fall. Fell sick & being out of amunition found it more Convenient to come here than go there. Sent Laurent to remain with Charleau & Nozanne's party to collect the animals they may kill.

• • •

## *8. Extract from letter from Thomas Simpson to George Simpson, 29 January 1838*[84]

Fort Confidence, January 29, 1838

Throughout the whole country, last summer was remarkable for its inclemency. We ourselves were detained a fortnight by the ice in Great Slave Lake. Ritch and his assistants were, after they separated from us, stopped for nearly forty days; and, had it not been for our indomitable perseverance, and my singular journey from Boat Extreme to Point Barrow, that interesting point would never have been reached; nor could the Company have gained that distinction by any renewed exertions of ours *this* season from our present position, as the ice of Great Bear Lake, which seldom opens before August, would have presented an insurmountable barrier to such an attempt.

Winter overtook us during our passage of this vast and boisterous lake in September, and we were threatened with being set fast; but by a bold push across from the western to the eastern side, guided by chart and compass, we safely reached our infant establishment on the 25th. Dease's River was already frozen, and our plan of transporting our little sea-boats up its waters to the height of land was therefore unavailing. The climate here is indeed intensely severe, and the mean temperature of the current month is lower perhaps than any upon record. The bare and stormy character of the country, and the long disappearance of the sun, contribute greatly to aggravate the rigour of the cold. I was, however, determined to inure myself to the worst; and, in order to eke out our scanty subsistence, spent the greater part of the time until Christmas in hunting excursions with various parties of Indians; during which I discovered another extensive arm of this inland sea, terminating nearly east from hence, fully half-way to the Coppermine River. It was somewhere near that bay that poor Taylor breathed his last. Had its existence been known to us last spring, it would have been our wintering-place; although, from its scanty clothing of stunted, stragglng trees, and its lofty encircling hills, it would have afforded but a very sorry residence.

83 Now named Hornby Bay, at the head of McTavish Arm.
84 Alexander Simpson, *Life and Travels*, 271–3.

It was a month after our arrival here before the houses were habitable; though very small, they still remain in a miserable, unfinished state; and, from a neglect of the fall fishery, our living was for a long time literally from hand to mouth. Mr. Dease is a very worthy man but he is dull and indolent. We live together, old wife and all, on the happiest terms; and I now understand that mysterious science – the management of an inland post – as intimately as if I had spent the last long nine years on the north side of Portage-la-loche.[85] Time flies quickly enough even in this desolate abode; though, not having Captain Back's good fortune in obtaining newspapers and periodicals,[86] the rest of the world is dead to us. My stay at the house is, however, drawing nigh to a close, as I intend starting in March to survey the route to the Coppermine, and to mark out the best ground for the boats and baggage to follow. The plan of the portage by stages is entirely my own, and I have every hope of effecting this laborious duty successfully. The provisions saved by the rapidity of our last summer's voyage will meet the heavy consumption on the portage without drawing on our seastock.[87]

• • •

## *9. Extract from letter from Thomas Simpson to Alexander Simpson, 29 January 1838*[88]

Fort Confidence, Jan. 29, 1838

My last to you was from Fort Norman in September, announcing our safe and early return from our first glorious campaign; I should say *mine*, for mine alone was the victory.

Our passage of this great and stormy lake in September was long and severe. Winter overtook us, but we safely reached our infant "Confidence" on the 25th of that month. The country was then covered with snow. Dease's River and all the smaller lakes were fast. The climate is, indeed, intensely cold, the land mountainous, and the scanty clothing of stunted and straggling spruce firs confined to the immediate borders of the lake. Both men and dogs got severely frost-bitten while bringing meat across the elevated and stormy region, where there is not a shrub for shelter. Until Christmas, I took my share of all these hardships, accompanying the Indians on their hunting excursions, and learning to kill my own reindeer. I also discovered another large arm of this inland sea, stretching half way between us and the Coppermine.

For that river I intend starting in March, to lay down the best route for the transport of our boats and baggage. My stay at the *house* is therefore very short – too short, indeed, for my literary employments; but, this first and hardest winter past, I shall probably have time enough.

85 Thomas Simpson is implying that he has learned in one winter what it had taken Dease a lifetime to master.

86 A reference to Back's wintering at Fort Reliance at the head of the East Arm of Great Slave Lake in 1833–35, when he enjoyed relatively regular communications, via the Company's network, with the outside world.

87 Their stock of pemmican and other provisions for the summer's exploration.

88 Alexander Simpson, *Life and Travels,* 273–7.

All our letters despond of our last summer's success, in consequence of the extreme inclemency of that season, and the Governor tells us to try the western coast again. Ha! ha! he forgot that all Great Bear Lake lies in the way, and does not clear out till August. My singular march from Boat Extreme to Cape Barrow put the coping-stone to that job. Would that the eastern one were already commenced!

We have a pretty long lease of night in this quarter, the sun being invisible at the house from the 30th of November till the 12th of January. Our winter fare consists of carriboo and musk-ox meat, whitefish, and an occasional huge trout. We were threatened with starvation at the outset; but, by dint of dispersing all hands, we got over that, and now enjoy abundance. Our buildings are small, as the climate and our means demanded: their position is on a strait opposite a large wooded island, three miles distant from Dease's River, the main feeder of the lake. I have not yet seen another spot where an establishment *could* have been erected.

When fatigued with writing, chart-drawing, and astronomy, I have a resource which you would hardly have expected here, in an excellent little library; which, besides scientific books, and a regiment of northern travels, contains Plutarch, Hume, Robertson, Gibbon, Shakespeare, Smollett, and dear Sir Walter. It is well that we came so provided; for our *friends* have not thought fit to send us any of the publications of the day.

Mr. Dease and I live together on the happiest footing; his old wife, a little grandchild, and a strapping wench, a daughter of his brother Charles, joining our mess. Dease is a worthy, indolent, illiterate soul, and moves just as I give the impulse.

With respect to the dangers to be encountered, make yourself quite easy on my account. They are great, indeed, compared with those of *civilized* life, but really present nothing appalling to people who have traversed the interior of this wild country. Our plans, thanks to my own foresight, are all admirably laid, and can scarcely fail to maintain us in plenty, and to ensure success. I am no wild theorist like Dr. King; all my proceedings are based on calculation and knowledge. On that foundation, and a humble reliance on a stronger Arm than man's, do I build my hopes.

I believe I have now run over everything that would interest you from this remotest of white men's dwellings; let us, therefore, turn to friends at home.

My last letter from our invaluable mother is of the 19th of February, 1837; at which time, in consequence, I suppose, of the Prince Rupert's cut-and-run from York Factory, she had not received a single letter from me dated in 1836; yet I wrote her from Red River once or twice during winter 1835–36. I wish there may be fair play in all this. She seems resigned to poor aunt Jane's death; but draws a truly affecting picture of the dying sister's being unable to take a last leave of her blind brother. O God! is such the end of this miserable life? They whose hopes are not fixed upon a better are, indeed, worse than the beasts that perish. Our dear mother is apparently comfortable and happy, enjoying the respect of all who know her; she well merits our filial regard, which I hope will never fail her. Both she and our worthy friends at Bellevue are now descending into the vale of years. How I long to embrace them all once more – to behold again our loved mountain land! Why were we born poor and friendless, when many a dolt inherits a fair estate? When or where we shall meet again, God only knows! May he preserve and prosper you, wherever your lot may be cast!

• • •

## *10. Letter from Peter Dease and Thomas Simpson to Murdoch McPherson, Fort Simpson, 29 January 1838*[89]

Fort Confidence, 29th Jany. '38

Dear Sir,

In consequence of Governor Simpson's instructions to us of 30th June last, we contemplate performing a third voyage along the eastern coast in 1839. With a view of enabling us to prolong the operations of the Expedition for that year, the Council have again authorised us to call on the Districts of Athabasca and MacKenzie's River for any facility or assistance in Men, Goods, Provisions, Indians, Craft, etc., etc. we may require.

Our first and most essential want is a supply of 30 Bags of Pemican and Grease which we have written the Gentleman in Charge of Fort Chipewyan to provide. We shall likewise require a good supply of Dressed Moose Leather, Birch Wood for dog Sleys, snow Shoe frames, amunition, Tobacco, and some other articles of which we send you a requisition. We are well aware of your personal anxiety to advance the interests of the Expedition, and therefore beg that you will compare notes with the Gentlemen in charge of Fort Chipewyan and Great Slave Lake[90] and endeavour by every possible means to complete the list now transmitted.

To render these supplies to their destination here this ensuing fall you will be pleased to employ a boat's crew of *servants* and Indians, our own summer establishment now reduced by the loss of a valuable man being too weak to attempt that duty.

With this Packet we send 'Le Sourd', Pierre Morin, an Indian Guide, and "Jack", the witness ordered out by the Governor in Cadien's case whom we have left it to yourself to settle with regarding his journey to Norway House. Our intimation on the subject being merely through a private letter from Mr. William Mowat,[91] have the goodness to inform us what steps have been taken towards bringing Cadien & his accomplices to justice, for surrounded as we are by the Slave Indians we have a personal as well as a general interest in the issue.

We are very desirous that "Le Sourd" should be sent back hither immediately with the return Express, in order to conduct your two men, La Motte & Gullion to Fort Norman in April, and at the same time to take charge of our last letters to the Honourable Committee and the Council. We are willing to make the man a gratuity for this laborious duty. The return party would therefore consist of "Le Sourd", Morin, the Indian Guide, and a lad of ours, young Wentzel, if he arrived from Slave Lake and is required to complete the Party. Otherwise you may retain him for the next Express. We expect only the Ship letters by this first Opportunity. But we have also another demand to make per return of the Bearers which we trust you will comply with, as indeed it is essential to the transport hither of the Sledge runner irons

89 HBCA B.200/b/10, pp. 20–2.

90 Fort Resolution.

91 See biography, p. 305.

which your black smith has made for us, and to our subsequent operations on the Copper mine Portage. It is as follows:

6 good dogs with harness complete
5 " Birch Sledges – our own being nearly all worn out
10 Large dressed Moose Skins
the runner Irons already mentioned
the nails etc. noted in the enclosed Memo

with any books or instruments that may have come to hand. We shall anxiously look out for the return of the People with these things about the 20th of March. Meanwhile with sincere esteem and regard,

We remain, Dear Sir, Your obedt. Hble. Servts.,

P.W. Dease
Thomas Simpson

Wantages of the Expedition per Boat, Fall 1838,
Independent of what is requested per Return of Bearers

| | | |
|---|---|---|
| 24 | Bags Pemmican | From Athabasca |
| 6 | " Grease | do. |
| 60 | Large dressed Moose Skins | Athaba. & McK.R. |
| 2 | " Leather Lodges | Athaba. |
| 6 | Good Sledge Dogs | Ft. Chipn. & Slave Lake |
| 10 | Wood for 10 dog Sledges full serd. & good | McKenzie's River |
| 20 | do. 20 axe helves | do. |
| 15 | do. 15 pairs stout Snow shoe frames | do. |
| 1 | Keg Clear Gum | do. |
| 1 | " Tar | Athabasca |
| a quantity of good Touch Wood | | McK's River |
| 1 | Keg Gunpowder | Athaba. & McK's River |
| 1½ | Bags Ball | do. |
| 2 | Rolls Sweet Tobacco | do. |
| 2 | Cents. Gunflints | do. |
| 6 | Hand Dags[92] | do. |
| 12 | Clasp Knives | do. |
| 4 | doz. Yew hdles do. | do. |
| 6 | Worsted Caps | do. |
| 12 | Flannel Shirts | do. |
| 12 | Cotton do. | do. |
| 1 | pair blue Strouds | do. |
| 1 | " White do. | do. |

• • •

92 Daggers.

## *11. Peter Dease's journal, continued*[93]

**30 January 1838.** The Indians passed the day here. Weather boisterous & Cold. Afternoon Jack, Big Head & 6 others Cast up. Have 6 Sleds loaded, partly their own & part meat Killed by the Hunters: 343 lb. meat, 32 Tongues & Some Skins. They are now all assembled in the Same Camp with our Hunters, which can be of little advantage to themselves or to us, As they will Scare away the Animals further and pass their time in playing their property[94] or Idle.

**31 January 1838.** The Babillard and his party left this to return to his Camp. The Indians who are to accompany the Express got Some Wants Supplied, And Every thing prepared to Enable them to leave this tomorrow. They will have full allowance of Provisions for 20 days, which ought to be enough, but have Amunition also in case of want to fire at Animals or any thing Else.

**1 February 1838.** The day turned fine & mild. The Express was sent off under the Care of P. Morin & Sourd, accompanied by Jack and Dzeeimbarre, two Indians.[95] The other Indians also left this. Framand & LaMothe Cutting fuel & Flett Carting it.

**2 February 1838.** McDonald & C. Begg came back with 500 lb. meat & 36 fishes from the fishing tent as they passed there. The weather Same as yesterday.

**3 February 1838.** Felix and the Boy came back with 578 lb. meat and 118 fishes from the fishery. The Weather still mild & fine for the Season.

**4 February 1838, Sunday.** Service as usual was read to all the men. Begg & McDonald left this to Sleep at the Fishery on their way back for meat. The Boy & Felix will Start tomorrow morning.

**5 February 1838.** The men off early. This afternoon 3 of our Hunters came from their Tents since morning. They have Killed 10 more Deers Since last accounts. The Express men passed at their Tents on the 3rd early. Two old men are coming with a little meat & a Sledge made for us by each of them.

**6 February 1838.** Two of the Hunters went for fish and brought 105. The 2 old men brought 120 lb. meat & some Skins. They were Settled with to return tomorrow.

**7 February 1838.** The Hunters & Indians left this Early. Afternoon Felix and the Boy Came with 530 lb. meat. Saw Geo Sinclair who says the Deer is Scarce now. They left Begg & McDonald this morning, to pass by the Fishing Tent, while they came a Shorter way.

**8 February 1838.** The loads of McDonald & Begg Amounted to 520 lb. meat. I am Sorry to hear that the fishery is not so productive as it has been some time back. The 2 last arrived men and the Boy Ordered to go for the 10 Deers last killed tomorrow.

---

93 MCM M214.

94 I.e. gambling away their property.

95 They were instructed to cross Great Bear Lake via McTavish and McVicar arms, then south via a series of rivers and small lakes, i.e. east of the Franklin Mountains, to Fort Simpson (Simpson, *Narrative*, 223). They were told to come back as soon as possible with dogs and sledges, with a number of moose skins and iron for sled runners for transporting the boats to the Coppermine.

**9 February 1838.** The men off Early. Sent to the Fishery for fish. 110 brought, which is all they had. The Weather Colder than for some days past.
**10 February 1838.** The men came back with the meat Sent for, but 550 lb. Wt. Did not see any of the Hunters or Indians. Suppose they have raised Camp to Cross the End of T. Simpson's Bay.
**11.12. February 1838.** Nothing new or interesting. Weather severely Cold.
**13 February 1838.** This Day Gros Blanc and a young man Came with 2 Sleds. Their loads Amt. to 186 lb. Ribb meat, 36 lb. Pounded meat, 5 lb. Grease and 16 Tongues. They have put up the Legs of 7 Deers and one Musk Ox at their Lodge, which we will send for as soon as we can.
**14 February 1838.** Sent to the Fishery; 112 fishes were brought.
**15 February 1838.** The Boy and Felix sent with Gros Blanc for the meat they have. The day was Windy with a light Snow but not very Cold.
**16.17. February 1838.** The men cutting & Carting fuel. 80 fishes brought from the fishery.
**18 February 1838.** The Boy with Felix came back with 350 lb. meat being all they had.
**19 February 1838.** Ritch & C. Begg were Sent to look out for Wood fit for boards to repair the Boats, also for Oars, but after being out most of the day Came back without Success.
**20 February 1838.** The men were out again for the Same purpose, and without any better Success. 104 fishes were brought from the Fishery.
**21 February 1838.** Ritch & Begg Cut wood today for 10 Oars in the Bay West of this.
**22 February 1838.** The Wood was brought home. 74 fishes brought from the fishery. About Midday Egoothi, Little Slave Hunter & Nozanne's Brother Came with news of their Having 53 Deers Killed and nearly all Collected at the Same Cache. At Same time tell us that they Killed 2 Musk Buffalo yesterday which are near hand. The men were sent for them & brought them, Wt. 540 lb. The Slave and Chipewyan Hunters are together. Babillard and the other Camp are beyond them. The last Camp, they Say, have Stolen 5 Deers from Different hunters.
**23 February 1838.** Sent the 4 men that have Dogs[96] for meat. John was sent with them that the young man, Egoothe, might Shew him where he and Nalhahl Say there are some fine trees for Sawing, not very far from the bottom of the Bay East of this. A Strong Wind and Drift of Snow. John came back almost perishing with Cold. The Indn. could not point out the Spot to him.
**24 February 1838.** Geo. Flett was sent for fish. Gullion came out with him. Brought only 40 fishes all they had. Fishery failing fast.
**25 February 1838.** Nothing new.
**26 February 1838.** Norquay brought 20 fishes from the Tent. Took but 7 this morning out of 3 netts before he came away. Weather Cloudy & Strong Wind.

96 Felix, Begg, McDonald, and the boy (FNB B, p. 74).

**27 February 1838.** The men came back. Nalhahl & 2 other Indians with them. The latter brought Birch Wood for Boat timbers & Snowshoe frames. 1135 lb. meat received. They have Killed about 20 more Deers Among them but most of them Kept for their own use. Including the latter Killed Deers, 62 are put up by their accounts.

**28 February 1838.** The men and Indians passed the day resting. Sent for fish and 43 were brought. Norquay and Geo. Flett ordered to go with the others for a load of meat, as we will not have too much time to get all home.

• • •

## *12. Letter from George Simpson to Peter Dease and Thomas Simpson, 28 February 1838*[97]

Messrs Dease & Simpson London, 28th Feby. 1838

Gentlemen,

I have to acknowledge your letter of 31st May addressed to the Governor & Committee, which was received here about the close of November, too late to be replied to with any chance of an answer getting to hand before your departure from Bear Lake in the prosecution of your voyage to the Eastward.

The Governor & Committee look with much interest and anxiety for the report of your survey to the westward of last summer, and they sincerely hope it has been successful & unattended with any accident or misfortune either to yourselves or your adventurous followers.

If from the unfavorable state of the season or any other circumstance you have not succeeded in the accomplishment of that part of the survey their Honors concur in the opinion, contained in my letter of 30th June from Norway House, that it will be better to repeat the attempt this ensuing summer, than to proceed immediately to the Eastward, thereby taking 3 summers to accomplish the whole survey instead of two as originally intended.

In your several recent communications you refer to a letter written by Mr. Dease of 3rd. Jany. wherein that Gentleman appears to have suggested some deviation from the plan of the voyage to the eastward contained in my original instructions of 2nd July 1836. That letter I have never seen, and none of the Gentlemen I met with at Norway House last summer seemed to recollect or to understand the precise object of such deviation. I was therefore, in my letter from Norway House of last summer, unable to give any opinion or instructions thereon, and can now by the directions of the Governor & Committee only say that the desire of the Board of Direction and the wish of every one interested in the Fur Trade is, that the objects of the Expedition should be accomplished, leaving it to you to determine the mode or manner in which that is to be done, and to enable you to accomplish those objects, the Honble. Company's resources in the country, in Men, Goods, provisions, etc., etc.

97 HBCA D.4/23, pp. 269–71.

are hereby and by the enclosed Circular, placed at your disposal, of which I feel assured you will make a discreet & proper use. I have to beg you will from time to time send full & explicit reports of your operations in duplicate to the Governor & Committee and the Northern Council as the deepest interest is felt in the success of your mission, not only by every branch of the service, but by the Government & the British Public.

Capt. Back who sailed in the Terror Bomb in 1835 with the view of continuing the survey of the coast from the Eastward, got no further than Nottingham Island, where he wintered,[98] and his ship sustained so much injury from Ice, that he was under the necessity of relinquishing the object of his expedition, and of returning to England where he arrived in the month of October last & I do not think it is the intention of Government to make any further attempt soon to prosecute that object.

The undermentioned articles applied for by Mr. Simpson are herewith forwarded, viz. a dipping needle, two thermometers, nautical almanacks, 3 dozen Perryean Pens, and with best wishes for your success,

I remain, Gentlemen,
Your mo. obed. Sert.
Geo. Simpson

To the Gentlemen in charge of
Districts & Posts of the Honble.
Hudson's Bay Company
Gentlemen,

I have the Governor & Committee's directions to instruct you to meet the demands of Messrs. Dease & Simpson or either of these Gentlemen, in Clerks, Men, Goods & Provisions, etc., etc. for the use of the Hudson's Bay Northern Discovery Expedition as it is intended that the objects of that expedition shall be carried into effect as far as the Company's resources in the country may admit, whatever inconvenience may arise therefrom to either branches of the service.

I am, Gentlemen, your mo. obed. etc.,
Geo. Simpson

• • •

## *13. Peter Dease's journal, continued*[99]

**1 March 1838.** Sent 6 men with their Sleds for meat. The men here at usual work, which is to Supply the House with Fuel and lay up some.
**4 March 1838.** Frans. Framand brought 23 fishes from the Fishery.
**5 March 1838.** The men came back from the Hunters with 1662 lb. Fresh Meat & 46 lb. Ribbs without the Bones. They say that there remains now but about 2 Sled

98 Not exactly; the ship spent the winter adrift in the pack, passing Nottingham Island in the process. See Back, *Narrative of an Expedition in H.M.S. Terror.*

99 MCM M214.

loads in the Large Cache, & 9 Deer in the others. 9 Indians of Babillard and EddawDelly's parties came also. They brought 462 lb. Ribbs, 70 lb. Pounded Meat, 3 lb. Grease & 72 Tongues. They Say that in Crossing the Bay the day before yesterday they saw the Snowshoe tracks of 2 Indians who are not of their party and must be some Strangers. They were recent therefore they fired several Shots in hopes they might hear them.

• • •

## *14. Letter from Murdoch McPherson to Peter Dease and Thomas Simpson, 5 March 1838* [100]

Fort Simpson, 5th March 1838

Gentlemen,

I have the pleasure to acknowledge your letter of 29th January, received here on the 25th ulto.

Your immediate demands upon this place are complied with by the present conveyance so far as the means here at present could admit. The most essential, and indeed the only deficiency is on the quantity of dressed leather and the number of dogs and sleys sent which amount only to half your demand of them, but it is all that this place can at present produce. The Iron Works etc. that you called for by the return of the men in September have been completed and forwarded from here by way of Martin Lake[101] in Jany. and I trust that you have received them before now. It being therefore unnecessary that "Le Sourd" should return at present I sent him on with the Packet to return with the Winter Express from YF wherever he may meet it, and as the only man I have here capable to perform that duty, to proceed across Land to Fort Confidence with the Letters etc. that may come forward for the Expedition by that Express, providing he will be here by the 1st April. After that period the season would not admit of his going from here to Bear Lake and back to Fort Norman before the disruption of the ice on McKenzie's River. In this case La Motte & Gullion will take your Dispatches to Fort Norman and the Guide who now accompanies Morin undertakes to guide them, being an Indian of that place. Young Wentzel has not yet returned from Slave Lake.

The Wantages of the Expedition "per Boat Fall '38" will I expect be fully provided, at least such of the articles as are separately required from McKenzie's River you may depend on, and with the exception of the dressed Leather, what you asked jointly from Athabasca & McKenzie's River will be provided from McKenzie's River Outfit, should Athabasca be deficient in Means to provide the share (at the time required) that I have referred to that District in my arrangement already transmitted to Mr. Chief Factor McLeod on that subject. For the Provisions, Leather and Dogs, I am sorry to have to say that the Expedition must entirely depend upon Athabasca, of which I have notified the Gentlemen in that District. A Boat's crew

100 HBCA B.200/b/10, pp. 28–30.

101 Now Lac la Martre.

will be employed from this place for the transport of those supplies to their destination at Bear Lake, conformably to your request. But to avoid as much as possible any loss of time in that Lake, which might lead to serious consequences, it would be very desirable that one of your summer men, or a trusty Indian from Fort Confidence, should be at Fort Norman on the 5th September to Guide the Boat to the Establishment.

On the subject of Cadien and his accomplies in the affair of Fort Norman the Company have taken the opinions of eminent Council in England and of the Law Officers of the Crown in Canada as to the proper course of proceeding against them, and it has ben determined to forward them to Canada for trial, where they were accordingly sent last summer. This you will please to intimate to such of the Hare Indians as may visit Fort Confidence, and also how deeply we all lament what has taken place.

Morin is detained here longer than I would wish, waiting for the Iron Works which our young and exceedingly slow Black Smith takes a long time to make. The work will, however, I hope give satisfaction.

With much esteem and regard, and my fervent Prayer to the Almighty to guide and protect you in your arduous, but glorious enterprize.

I am, Gentlemen,
Your Obedt. hmble. Servt.
M. McPherson

• • •

### *15. Peter Dease's journal, continued*[102]

**6 March 1838.** Men and Indians preparing to be off tomorrow again, Except Geo. Flett who has no Sledge, one being useless. Afternoon 2 Indians from Ft. Norman arrived with letters from that Post and Ft. Good Hope, by which we have the Satisfaction of hearing that our Friends at these places are well, and have the means of Subsistence to take them through the Winter. This is the 10th day Since they left the Head of the River[103] for Fort Franklin. One of them Complains of being Sick & has a bad Cough.

**7 March 1838.** Sent off all the Same men except Geo. Flett, And Orders to McKay to come with them and bring 2 or 3 young men with him to Serve as Guides to the Copper Mine River, as it will be necessary to go and see how the travelling in that direction may be and have the road marked before beginning our Transport. Keep our young man here for the same purpose, and the Sick letter Carrier also remains for a few days. Others off.

**8 March 1838.** Gave a dose of Physic to the Sick Indian. Weather Cold in the Afternoon, A young man from Gros Blanc came to tell us that they have 24 Deers for us in the Haldane's River. He left the tent the day before yesterday. Five Lines were

---

102 MCM M214.

103 Great Bear River.

put down on Tuesday as a trial for Trout, but not one has been taken yet although Examined daily.

**9 March 1838.** The weather Cold. Wind generally Easterly. Got 2 pairs of Snowshoes Knitted, an article been great breakage of Since the Fall. Ritch brought in a pair of frames made by himself.

**10 March 1838.** The wind Changeable: Northerly Morning & Westerly after noon. Weather has a Hazy appearance; very Cold. Employed the Indians here undoing the line of Broken Snowshoes[104] which may serve for others. Nothing from the Lines.

**11 March 1838.** Severe Cold wind from Westward.[105] The men came back with the meat that was out Amounting to but 767 lb., much less than I thought, as there Should have been about 15 Animals out yet. The Expenses[106] are great, for people & Dogs Carting. 4 Indians came with them. Brought 15 pairs Snowshoe frames. Some are ready knitted for use. The Indians removing further off, the Deer being Scarce about where they are now.

**12 March 1838.** The Dogs allowed the day to rest. Sent to the Fishery and got 48 fishes, being all there was. Cold w. Wind yet.

**13 March 1838.** Framand, LaMothe, McDonald, Boy, John McKay & Norquay were Sent for the 22 Deers at Gros Blanc's Lodge with the young men that Came from him to advertise us. The other Indians left this to return to the Tents accompanied by the young man who brought the letters; also the one that was Kept for going towards the Copper Mine River, the Weather being yet too severe to travel over the Barren Grounds. I have therefore given them orders to be here in 10 days again.

**14 March 1838.** The wind continues from Westward. Weather Cold. Nothing of any importance. Did not go to the Lines.

**15 March 1838.** The night Cold but the day fine. 4 Trouts were taken to day at the Lines.

**16 March 1838.** Nothing new. Wind Westerly morning & turned to the Eastwards Afternoon.

**17 March 1838.** This afternoon the men came back from the Indians' tent with all the meat they had, 1220 lb. Very small Animals considering there were 22, hardly averaging 60 lb. Each. The lines gave nothing to day.

**18 March 1838.** Nothing worthy notice occurred. Service read as usual.

**19 March 1838.** Norquay was sent to resume his Duties at the Fishery, and orders were given for them to put down lines for Trout, as Our Lines here gave 3 to day, Although the baits were old. The Boy was Sent for fish and brought home 82.

**20 March 1838.** Felix put down 5 more lines to day; the others baited afresh. The men Cutting & Carting fuel. Fine mild day.

---

104 I.e. recycling the *babiche* or thonging.

105 This was the coldest day of the winter. One thermometer (the one normally used) recorded -60°F; another -66°F. Not surprisingly, the men bringing the meat had frostbitten faces (Simpson, *Narrative*, 225–6).

106 I.e. consumption of meat.

**21 March 1838.** Two Trouts were taken to day. A Strong Westerly Wind with a drift of Snow.
**22 March 1838.** The wind from the Same Quarter but not so Strong. The Lines gave us but 1 Small Trout to day. Hector Morrison came from the Fishery last night, with the unwelcome tidings that the netts take but very few fishes, some days but 8 to 10 out of 6 or 7 netts.
**23 March 1838.** Nothing of any importance. Hector returned to the Fishery. Gave them some meat for that place as they do not take enough for their daily wants. 1 Trout taken at the Lines. No word of our Express or the Indians who were to have been here yesterday to guide Mr. T. Simpson to the Copper Mine River, previous to Commencing our Transport thither which we cannot defer much longer.
**24 March 1838.** The Same produce from our lines as yesterday. I paid a visit to the Fishery, where out of 6 netts they took but one Small Trout.
**25 March 1838, Sunday.** This afternoon our anxiety for the return of the Express was relieved by the arrival of P. Morin with the 2 Indians who left this with him, bringing letters with the Satisfactory intelligence of the Company's Ships[107] having Safely reached York Factory last Autumn, 24th Augt., without any accident & Sailed between 15 & 20 Sept on return. The business of the Interior Satisfactory in regard to the Affairs of the Company, but it is much to be lamented that a dreadfull disease (the Small Pox) has been introduced in the Saskatchewan by a Party of Stone Indians from there having gone towards the Missouri for the purpose of procuring or Stealing horses and got it among the Mandan Nation, and on their return, the infection Spread among them, which Carried off between two & three hundred men besides women & Children. It is much feared that it may Spread & depopulate most of the Country. From F. Simpson, C.F. McPherson mentions having Sent off the Ironworks made up for this Place by Indians 20th January, of which we have heard no tidings, unless it may be that the Carriers have mistaken their way and are at a loss to find out this Place, as the party of Indians who came here on the 5th Inst. Saw the tracks of two persons on Simpson's Bay[108] dragging a Sledge, apparently loaded & when they Saw this they deviated from the direct route to come here & were going apparently towards Cape McDonnell. I am of opinion that these are the Indians bringing the Iron from Ft. Simpson, but as it is doubtfull, defer sending to see. The Sourd who accompanied Morin home to Ft. Simpson has been Sent on with the Express & Should he be back there by the 1st April, will be Sent here with the Winter Express, which he is to bring from Slave Lake back to Ft. Simpson, as Soon as possible.
**26 March 1838.** Nothing worthy notice occurred. From the non arrival of any Indians to guide Mr Simpson to Cop. Mine River, preparations were made for his Departure with 2 men, James McKay & George Flett tomorrow to proceed to Cop. Mine River and ascertain the best Spot for making our Deposit of Provisions & then we

107 The news they brought included the report of the death of William IV and the accession of Queen Victoria (Simpson, *Narrative*, 227).
108 Now Hornby Bay.

will begin our Transport on their return. Our Lines took nothing to day but a very Small Trout.

**27 March 1838.** This morning Mr. Simpson with the 2 men mentioned, left this with 2 Sleds hauled by 4 Dogs & provisions for 11 days. The weather fine but an Easterly wind. In the Afternoon arrived the 4 Chipewyan Hunters with Nozanne. Brought 200 lb. Fresh meat & 76 lb. Ribb meat. George Sinclair & Laurent came with them. The Rein Deer, they say, are getting Scarce where they are, and the few they have Killed were required for their own Support. They saw the tracks of the 2 people formerly seen, which must be those who were Sent with the Iron Works, of whom we hear nothing.

**28 March 1838, Wednesday.** The Hunters passed the day here. From them I learn that one of the young men who was to have been here on the 22 fell sick on his return, which may account for their nonarrival. The Chipewyans themselves were talking of returning to their own country, but upon further Consideration and some advice to them they have for the present relinquished that Idea, and have directions to Come back & Endeavour to hunt to the N & Eastward towards the Copper Mine River by the time we may be transporting the property, which they promise to do. Laurent is to return with them. George Sinclair & the Boy are appointed on a voyage in Search of those whose tracks were Seen along the North Side of Simpson's & McTavish's Bays towards Cape McDonnell, as I much fear these unfortunate people may have perished through famine, Apparently not Knowing this Place.

**29 March 1838.** The Parties left this as mentioned. Hector came from the Fishery with 9 White fishes. Our Lines have no Success. The Weather Windy and Cold. McDonald & Charles Begg Employed Sawing Boards for the Boat Repairs & planks for making Trains[109] to drag them. Others Cutting & Carting fuel.

**30 March 1838.** The Weather most boisterous: wind & Snow; Weather Cold. Nothing of any moment occurred. Hector returned to the Fishery.

**31 March 1838.** The weather worse than yesterday. Nothing new.

**1 April 1838.** This afternoon 3 young men from Babillard's Camp with the two who brought the letters from Ft. Norman arrived with 3 Small Sledges. The 2 latter are on their Return to Ft. Norman; the others in Case their Services may be required about the Transport from here to the Cop. Mine River. The reason they State for not having Come at the appointed time is the Sickness of one of them, who is yet unwell, and a Brother of his Comes as a Substitute.

**2 April 1838.** The Indians passed the day here. The two letter carriers to take their departure for Ft. Norman tomorrow, and two of the 3 others are to go in Search of those who were to have brought the Iron. Have provisions given them for 4 days. 80 lb. Ribb pieces and 10 Rein Deer Tongues were brought by them. 2 trouts were taken at the Lines. Felix & Morin off for 8 Deers to Haldane River today.

**3 April 1838.** The Indians left this as mentioned. The other young man was Employed Knitting the foot lacing of a pair of Snowshoes. The weather fine & Clear but Cold for the Season.

---

109 Toboggans.

**4 April 1838.** The Lines gave 1 trout today. No word from the Fishery.
**5 April 1838.** This morning 2 Indians arrived from towards Martin Lake, Sent by their leader, to bring the Intelligence that the two Indians who were Sent with our Ironworks, for want of Provisions & not Knowing the Way, left their Sledge with the Irons along the Opposite Side of the Point and returned to the Camp. The 3rd day after, their Leader sent off these two to apprise us of the Circumstance & to go for the Irons, according to the Information of those who left them and they have hopes of finding them.[110] The 2 young men Sent to Search for them Came back, and the one who has remained here, being acquainted with the Country thereabouts was ordered to accompany these 2 new comers to go for them. Provisions given them to take their departure tomorrow morning. One trout at the Lines. Felix & Morin came back from the Gros Blanc's Tent with the meat they were Sent for, Wt. 590 lb. That Indian had Killed nothing since the messenger had been sent off, but was absent from his lodge when the men came away hunting. The weather fine but not warm. I Expected the return of Mr. Simpson today, but he must have been considerably delayed by the boisterous weather we have had.
**6 April 1838.** The Indians gone for the Ironworks. Afternoon Mr. Simpson came back & has been to the Copper Mine River and fixed upon a Place to deposit our Provisions & baggage on the banks of a Small River[111] that falls into that one about 15 miles from their Junction. This is the third day since he left the place intended for the Deposit.[112] The young man who remains was Sent to Collect some pitch. Brought a little but Says the weather is yet too Cold to get it properly. Norquay came yesterday from the Fishery of which he gives a very poor account. Nothing taken. He was Sent back to day with Orders to take up his netts and Lines & to leave it entirely as the netts get damaged considerably.
**7 April 1838.** The men employed Carting fuel. Two Chopping.
**8 April 1838, Sunday.** The Indians sent for the Ironworks Came this morning with them all Safe and Correct. Also the Letters from Mr. McPherson, which are dated 18th January. I am however much relieved now of Anxiety, as the loss of them would have been of great consequence to us, and I doubt much if the repairs required to our Seaboats Could have been Efficiently done without them. The men from the Fishery Came home, and We are now prepared to Send off the first part of our Transports tomorrow: 6 men wh [with] Sledges, 3 Indians to assist and Guide & 2 men to remain at the Depot. One of the Hunters (Chips.) with 2 of the Slaves came. Brought 1 Sledge with 120 lb. Fresh meat. They have Killed among them 30 Deers, and this happens unfortunately at a time when we cannot employ our people to go for the meat, as all our Sledges will be employed in the transport of our property which can be done only in 3 trips to that Place, taking from 7 to 9 days each.

110 "This is the 16th day since they left their tents" (Dease's original note).
111 The Kendall River, which flows from Dismal Lakes into the Coppermine.
112 Simpson and his companions ascended the Dease for three days till they reached treeline. There a blizzard pinned them down for a day. On the 31st, although the wind was still strong, they continued on their way. Striking a small stream, (the Kendall) they followed it down till they reached its confluence with the Coppermine on 3 April (Simpson, *Narrative*, 229–32).

**9 April 1838.** This Morning Mr. Thomas Simpson, who again accompanies the party, left this with Geo. Sinclair, J. McKay, P. Morin, F. Felix, Charles Begg, Willm. McDonald, Geo. Flett and H. Morrison. The 2 last mentioned appointed to remain in Charge of the property. The 3 young men who came on the 1st Inst. are also of the party and will be employed either to Hunt about there or to Come back, as may be Judged most Expedient. An Old Man with 2 boys came from the Camp with 135 lb. Rib Meat, 21 Tongues and 11 lb. Beat meat.[113] These Indians with those who brought the Irons and Hunters prepared to leave this in the morning. By the latter I Send word to them that they must Endeavour to come round on the transport road where Deer is very numerous, and will Save much freight & Expense of Provisions going and coming to & from the Place. Ritch has been working the Rudders, thafts &c. for the Boats.

**10 April 1838.** The Indians all left this. Afternoon, one of the young men from Gros Blanc's Camp came to tell us they have Killed 3 Musk Buffalo and that he left the Camp late this morning. Ordered such of the men as Can take Sleds to go with him tomorrow for the meat, as our Store is getting low: John McKey, F. Framand, LaMothe & the Boy, the latter with 2 Dogs on the Sledge. Norquay and Gullion Employed making netts.

**11 April 1838.** The men left this to go for the meat mentioned. Nothing Else of any moment.

**12 April 1838.** The aforementioned men Came back, accompanied by Gros Blanc, the 2 young men, 3 Women and 2 Children. Brought all the meat which Amounts to 620 lb. The Lines were taken up and 1 trout taken, as we have no fish to bait them.

**13 April 1838.** The Indians passed the day here. In the Evening, 5 young men of the Outer Hare Indian tribe came with a Sledge each, the Contents of which have not been Examined. Most of them are Suffering from Sore Eyes. Have been 7 days from their Camp which is about north of Ft. Franklin Inland.

**14 April 1838.** The Hare Indians' loads, Amounting to 214 lb. Rib Meat & 32 Tongues, they were paid for, and left us to return in the afternoon. Gros Blanc & party are to leave this in the morning.

**15 April 1838, Sunday.** The Indians left this to go hunting to the Northward where they Expect to find Musk Buffalo. Mr. Thomas and the party came back. They followed a different route from the first, which although considerably longer, being much farther northward, is far more level and less liable to cause any breakage of our Sleds, a great Consideration with us. Besides the advantage mentioned the present route is the one we will follow with the Boats, as from where they left the upper Part of the River[114] a land transport of only 5 miles leads to a Chain of Lakes[115] communicating by narrows. Thence commences the Small Stream[116] leading to the Copper Mine River. At about 15 miles from the latter is our Deposit, left in charge of

113 Pounded meat.

114 Dease River.

115 Dismal Lakes.

116 Kendall River.

George Flett and Hector Morrison with the 3 Indian lads to hunt about there for them as Deer is not Scarce. They say great numbers of Musk Buffalo along the way & Killed one. The Indians also Killed 2 Going, but out of the way of the Sleds to take up. 5 young men came from the Indians' Camp with meat.

**16 April 1838.** The Dogs and People were allowed to rest today & prepared to go with the next load which will take all the provisions required for our Summer Voyage along the Coast.

**17 April 1838.** Six men left this for the Deposit[117] taking 22 Pieces of Provisions for the loads, besides their own to go and Come. Afternoon the two hunters, Nalhal and Eengheen, with a Slave Lad Came, the former each a Sled with Dogs, the latter dragging one. The 5 young men [who] came on the 15th left this in the morning. Their loads amounted to 378 lb. Ribmeat, 68 Tongues, 90 lb. Beat Meat and 12 lb. Grease, the latter an article they bring very little of.

**18 April 1838.** The Hunters have Killed 34 Deers which they have put up in hopes we may be able to send for a load. Their loads amount to 500 lb. Fresh meat. The young man 64 lb. Rib meat & 4 Tongues. The Hunters prepared to return tomorrow. The young man remains to accompany the men after their return from the Cache or Deposit, for a load of meat, as we will get Short unless we can get in Some in time.

**19 April 1838.** The Hunters left us. Two men Continue nett making. Young Bouché different Jobs: Boat Work, Sleds for do.[118] along the Way, Boxes, &c., &c.

• • •

## *16. Letter from Peter Dease and Thomas Simpson to the governor, chief factors, and chief traders of the Northern Department, 20 April 1838*[119]

Fort Confidence, 20th April 1838

Gentlemen,

We now resume our report on the affairs of the expedition up to the present period.

Our last fishery entirely failed about two months ago, but our hunters continue to furnish us with a sufficient supply of reindeer and musk-ox meat. No fewer than six or seven guns have recently burst in the chase, but happily only one personal accident has ensued – the loss of a forefinger by one of our young Chipewyans. Both Indians and whites, we are happy to say, enjoy uninterrupted health.

The greatest degree of cold registered here this winter (60° below zero) occurred on the 11th of March; the *mean* temperature of the winter so far is lower by several degrees than any experienced by Franklin or Back, while the gales that here frequently accompany the *severest* cold constitute a fearful distinction. The easterly winds which rush down from the mountains of the barren grounds are the most intense;

---

117 I.e. the depot.

118 I.e. sleds for hauling the boats on.

119 Alexander Simpson, *Life and Travels*, 277–81; extract in HBCA D.5/5, f. 42.

the westerly more commonly bring snow. Not the remotest symptom of spring has yet appeared, and the thermometer this morning stands at 26° below zero.

Mr. Simpson has explored the barren grounds stretching from this place to the Coppermine River in three different lines. His first journey, performed with two men in the latter end of March, was solely for the purpose of discovery. On the second, he conducted our whole disposable force of men and dogs, laden with part of our provisions and baggage, for the ensuing summer, to a secure spot which he had marked in the first dwarf woods on the banks of a navigable stream, fifteen miles from its confluence with the Coppermine, and ninety-five from this place by the route adopted. That route leads through the plains of Dease's river, to a point where the latter turns off to the north-west, in lat. 67° 22'N., long. 117° 42'W., far beyond where it was seen by Dr. Richardson. A height of land of six miles conducts to a chain of dismal lakes, thirty miles in length, winding in a south-easterly direction through a narrow cleft in a low range of primitive rocks which produce not a tree or shrub and seem unfrequented by any living creature. During the preceding day's journey, musk-oxen were numerous, and several were shot. Mr. Simpson left the baggage in charge of two trusty men (provided with lodges, etc.), who are to remain at the station till we pass with the light boats in June. The remainder of our summer's provisions has been subsequently forwarded to the same place; and the boats will be transported up Dease's River, either on iron-shod sledges, when the snow melts off the ice, or by the first open water, acording to the advance of the season. Our object is to reach the northern sea not later than the 1st of July.

Chief-Trader Macpherson has favoured us at our request with a perusal of the minutes of the council for 1837. With extreme surprise we read, that our careful and meritorious boat-builder and summer-master, John Ritch, has been 'fined five pounds, for wanton destruction of the Company's property at Norway House last year.' The man solemnly declares himself unconscious of any such transgression. We have had good cause to notice his conduct during the trying emergencies of last season, in a very different manner; and in justice to him and to the expedition generally, we beg leave, with all due respect, to request a more satisfactory explanation of the grounds of this most unexpected charge, which has as it were accidentally come to our knowledge. Ritch has applied for his immediate discharge in consequence of this circumstance, but we are quite unable to dispense with his services, as, independently of the weakness of our numbers, we have no other person capable of the summer management of this place, on which you must be aware that the safety of the whole party essentially depends. The horrid affair of Cadien and his accomplices,[120] having been agitated among the Dogrib and Hare Indians by whom we are surrounded here, we were under the necessity of applying to Mr. Macpherson for information respecting the steps which you have been pleased to take in that dark transaction. These have been communicated to

---

120 "A murder of several Indian families. The principal, Cadien (a half-breed servant of the Company), was brought to Canada and tried for murder; of which, on the testimony of his accomplices in crime, he was found guilty. He was condemned to death; but his sentence was mitigated to transportation for life to New South Wales" (Alexander Simpson's footnote).

the natives, who appear satisfied with our explanations; we are thus relieved from considerable anxiety which we felt on account of Ritch and his fishermen, who will be left entirely at the mercy of these Indians during our absence on the coast.

The remaining ten men attached to the expedition will form our crews; and if again protected by Providence, we may hope to return successful to winter-quarters before the 1st of October.

We have the honour to be, Gentlemen,

P.W. Dease.
Thomas Simpson.

• • •

## *17. Letter from Lord Minto[121] to J.H. Pelly, 20 April 1838[122]*

Admiralty, 20th April 1838

Sir,

I beg that you will accept my best thanks for the obliging communication of your interesting Dispatch from Messrs Dease & Simpson, and that I may be allowed to congratulate you upon the success of an expedition reflecting so much credit upon the public spirit & liberality of your company, and upon the enterprize and intelligence of its officers.

I have the honor to be, Sir,
Yours faithfully
Minto

• • •

## *18. Peter Dease's journal, continued[123]*

**20 to 23 April 1838.** Nothing particular transpired. Nor have we any news from Ft. Simpson, although the departure of the men of Fort Norman has been postponed in hopes the Winter Express might yet come to hand before the 25th which we have fixed upon as the date of their departure for that Place, to be accompanied by the Boy Boucher, who is Sent for the purpose of Shewing the Way to the Boat over this Lake in the Fall, according to Mr. McPherson's[124] request that someone acquainted with this Lake might be Sent, that no loss of time may Cause delay in their getting here. LaMothe and Thos. Gullion with the Boy preparing for Departure In the Morning. The men with one of the 3 Lads left to hunt for the men, Came. Have deposited their loads Safe under Flett's care.

• • •

121 Gilbert Elliott, second Earl of Minto, was First Lord of the Admiralty from September 1835 until the end of Lord Melbourne's administration in 1841 (*Dictionary of National Biography*, 6:675–6).

122 HBCA A.8/2, p. 142.

123 MCM M214.

124 Murdoch McPherson.

## *19. Letter from Thomas Simpson to Donald Ross, Norway House, 23 April 1838*[125]

*Private*

Fort Confidence, 23rd April 1838

My Dear Sir,

The non arrival of the winter express at Fort Simpson at the usual period, on or before the 1st Inst., which alone can have prevented our zealous friend Mr. Mcpherson from forwarding it hither, is a most serious disappointment to every member of the Expedition. What extraordinary cause can have retarded its progress? It looks as if the existence of the Expedition in this miserable banishment, instead of quickening the conveyance of intelligence, as it most unquestionably ought to do, has the reverse effect. What care we for all the fine *words* of our friends; let them show us their attention in *actions*! Among other mortifications caused by the present miscarriage, is the not hearing from yourself; and as you did not write with the ship letters (which got here on the 25th ulto.) it almost appears to me as if you had dropped the correspondence. We are totally in the dark regarding every thing passing "in the world" and it is from a rumour current on this side of Portage La Loche that we have given up praying for old William the 4th (save his soul) and now imagine ourselves to be the subjects of a youthful Queen. Considering that we are in the way of naming countries and so forth, we think that their Honors might have condescended to announce to us so interesting an event. They do not deserve such servants as we are, as they do not know how to treat us. Our proceedings, I suppose, interest nobody; – such as they are you will see them in our official letter. The bringing the name of a faithful servant, like Ritch, into dishonor, immediately after that of a base deserter, was adding insult to injury, and the underhand manner of the fine is anything but creditable to the imposers. Have we been sent to the Arctic regions that our means & lives should be the sport of a tyrannical Council? Our only consolation in our wretched quarters at present is, that we have all ample allowance of blessed sun-light again, though it gives precious little heat as yet. For hard, dry musk ox & reindeer flesh too, we do not want, and notwithstanding the irritation of our spirits, our souls are firmly bent on reaping a fresh field of glory this summer, if Providence vouchsafe a continuance of His protection, without which we are frail, helpless worms.

Some one wrote us that you are to have a Deputy Govr. this year. I hear with pleasure that you sent home Jane under Miss Armstrong's charge; the girls who remain at R.R., without a governess, have, I should think, a poor chance of improvement, at least in feminine graces. It is to be hoped that the unfortunate Colony may not perish of famine. The smallpox that has been raging in the plains has perhaps thinned the Upper Settlement, which would be no great loss to humanity. We are so totally

---

125 BCA ADD.MS 625.

excluded from intellectual excitement here, that I cannot muster another idea. Remember me in the kindest manner to Mrs. Ross; and with unchangeable regard,

I remain, My Dear Sir,
Yours most faithfully
Thomas Simpson

Donald Ross Esquire.

• • •

## *20. Peter Dease's journal, continued* [126]

**24 April 1838.** The young man who remained here was Sent to Collect some pitch but only brought about 2 lb.[127] Being on the Eve of Separation and the last opportunity of Sending any news to our Friends for the Season,[128] Our Small Party were entertained with a Dance which was Kept up with much vivacity to a late hour. Every thing prepared for the Departure of the Boy, Frans. Boucher, Thos. Gullion and A. LaMothe for Ft. Norman, the two latter being McK. River District men appointed for that Post, and passed the Season here, from an apprehension of Scarcity of Provisions there for Winter. Six men with Sleds to be guided by the young man, who we Kept here for the purpose, go to the Hunters' Tents & with the Assistance of the Hunters and Indians are to bring all the meat they have put up. Other young man is to go & Join the Gros Blanc.

**25 April 1838.** The Express and other Departures this morning leaves 3 men, besides the Servant here. Afternoon Gros Blanc with the 2 young men, two wives and a young Girl, Came with 6 Sleds loaded with meat of 8 Musk Buffalo they Killed Some days ago. Weather Mild; a thaw.

**26 April 1838.** A strong S.E. wind, and thaw which reduced the Snow fast. The meat brought by the Indians yesterday Amounts to 760 lb. and 3 tongues of the Animals. One of the Women badly with Sore Eyes.

**27 April 1838.** The Indians left this to return. In the afternoon the two young men that were left with G. Flett came back. They gave us a note from him in which he mentions having sent them off, as they were doing very little there. Killed but 2 Rein Deer and helped to consume them, Although Animals are not Scarce.

• • •

126 MCM M214.

127 "The men cut and Shovelled off the Heap of Snow in front of the House" (Dease's original note).

128 Also to mark St George's Day. On this day the temperature rose to 0°C (32°F) at noon, for the first time since 17 October (Simpson, *Narrative*, 236).

## 21. *Letter from James Stephen to J.H. Pelly, 27 April 1838*[129]

Downing Street, 27th April 1838

Sir,

I am directed by Lord Glenelg[130] to acknowledge your letter of 17th Inst. enclosing a Copy of the Report from the Expedition fitted out by the Hudson's Bay Company, to explore the shores of North America to the westward of McKenzie's River together with a copy of the instructions issued to them.

Lord Glenelg desires me to thank you for this communication, and to convey to you his congratulations on the happy issue of an enterprize, no less honorable to the Hudson's Bay Company, than interesting to the scientific world in general. His Lordship trusts that the future exertions of the Expedition to explore the still undiscovered regions to the Eastward of point Turnagain, will be crowned with success, a result which He confidently anticipates from the judgement, energy and fortitude, already displayed by the Gentlemen employed on this service.

I have the honor to be, Sir,
Your most Obedient Humble Servant,
Jas. Stephen

• • •

## 22. *Minutes of Governor and Committee, 28 April 1838*[131]

Resolved, that in consideration of the zeal, talent and perseverance evinced by Mr. C.F. Dease and Mr. Thomas Simpson, in the conduct of the Discovery Expedition, and of the steadiness and good conduct of the Men employed there be paid from the Funds of the Hudson's Bay Company One Thousand Pounds & that it be divided as under, viz. Four Hundred Pounds to Mr. Dease, Four Hundred Pounds to Mr. Thomas Simpson, Two Hundred Pounds equally among the Ten men who accompanied the Expedition to the Coast.

• • •

## 23. *Peter Dease's journal, continued*[132]

**28 April 1838.** This afternoon Charleau & family arrived from Simpson's Bay. Brought 12 Ribs, 8 Tongues, 12 lb. Pd. Meat[133] & 2 lb. Grease. Met the men going for meat the day they left this. The Weather fine but not warm.

---

129 HBCA A.8/2, p. 143.

130 Charles Grant, Baron Glenelg, MP for Inverness and Fortrose, then for Inverness County, was colonial secretary in Lord Melbourne's second administration, starting in April 1835 (*Dictionary of National Biography,* 8:380–1).

131 HBCA A.1/61.

132 MCM M214.

133 Pounded meat.

**29 April 1838, Sunday.** This morning the men came back, Pacquet or Nalhal with him. Brought all the meat they had.
**30 April 1838.** This afternoon four men, C. Begg, McDonald, Framand & Cartier were Sent off to take the last loads with private baggage &c. to the Deposit where Geo. Flett is. Wrote a note to Flett intimating to him the time we intend to leave this, 15th June, & recommending all care of the property rendered there. Provisions for themselves to wait untill that time, are forwarded.
**1 May 1838, Tuesday.** Mr. Simpson prepared to go and Survey Simpson's Bay from where Dr. Richardson with Lt. Kendall[134] finished the Survey of McTavish's Bay & Connect it with the Extent of those Bays to Cape McDonald.[135] Afternoon a Courier Came to announce to us that a Party of Hare Indians, 27 in Number, are Coming with Provisions to trade, and demanding Tobacco to meet them, which Compliment was declined & told they might Come if they want Tobacco. Mr Simpson with McKay & Sinclair took their departure for the intended Journey. The Hunter's families with Some of the Slaves who hunted for us, with Babillard & family, Came and pitched their Tents near us intending to remain Some time here no doubt. They brought among them 80 lb. fresh meat, 332 lb. Ribs, 48 Tons.,[136] 70 lb. Pd. Meat & 5 lb. Grease.
**2 May 1838.** This Afternoon the Party of Hare Indians Arrived. A great bustle among the Natives of the different Tribes occurred after meeting. The day being too far advanced, did not Examine their loads. It is Musk Buffalo meat only they have brought. The Deer Scarce in that Quarter, they Say. They have been now 7 days on the way.
**3 May 1838.** The Provisions brought by these Indians were received, amounting to 1836 lb. Rib meat, 140 Tongues and 60 lb. Grease. They Passed the day to rest and Chat with those about here.
**4 May 1838.** Settled with the Party of Indians. 5 of them remain among the Indians here. The others took their departure to return. Some of the Hunters, getting Short of Provisions, apply for something to Eat. A Couple of them put down lines to Endeavour to get some Trout.
**5 May 1838.** Three young men including one of the Hare Indians left this to go and Join Nozanne about Simpson's Bay. 3 of the Hare Indians also asked permission to go and hunt About where Geo. Flett is in charge of the property, which was granted and they left this, being Supplied with a little Amunition & provisions & take a letter for Flett. Gros Blanc & party Came & gave in 130 lb. Buffalo meat.
**6 May 1838, Sunday.** This morning the men Sent with the property on the 30th Ulto. came back. Delivered a note from Flett, who received all in good order. The Snow they Say is disappearing in Some places, but the Weather is not Warm yet to the Northward. Deer Continues numerous about there, but no one hunting in that direction. The Indians say they Saw Bustards yesterday.

---

134 Dr John Richardson and Lieutenant Edward Kendall carried out their survey from Fort Franklin in April 1826 (Franklin, *Narrative of a Second Expedition*).

135 I.e. Cape McDonell.

136 Tongues.

**7 May 1838.** The Indians all Collected about us doing nothing but Amusing themselves. Eengheen put out 3 lines and brought us a trout to day. The men Employed: two Sawing, 2 Cutting logs, one Squaring oars and the others Carting the logs & fuel.

**8 May 1838.** This Afternoon the young man who left this with Mr. Simpson came back with the 3 young men that left this to go & join Nozanne, who they did not find where they Expected to see him. Mr Sn.[137] finding the travelling too bad to proceed across land to C.M. River,[138] Sent back the young man.

**9 May 1838.** Early this morning was much Surprised at the arrival of young Wentzel with 2 Indians, bringing the Dispatches for us,[139] Come by the Winter Express, all Safe according to List. They have been on the Way Since 5th April from Ft Simpson. The delay Caused partly by want of Provisions to come on from the Camp of Martin Lake Indians where they had to hunt for a Supply and partly, the Indians State, from the Laziness of young Wentzel along the Way. The news received by these letters of a mixed nature. The worst is the information of the Small Pox spreading with fatal Effect among the Plain tribes and along the Coast of the Pacific to the Northward as well as about Columbia River. The Company's Ships got Safe with their Cargoes to Y. Factory and left it in Good time to return to England with the Furs.

**10 May 1838.** Three Indians, Charleau with 2 others, left this to go towards Cape McDonnell to hunt. All hands Employed as usual. Ritch preparing the Boats for repairs as they require also additional binding to Strengthen them. Got Some pitch gathered by 2 young Indians who brought but a Small quantity.

**11 May 1838.** The letter Carriers are detained untill Mr. Simpson's return as they Expect to get to Ft. Simpson before the Departure of McK.'s River Boats for the Portage[140] and will afford another opportunity of writing our Friends. The weather rather Cold for the Season. The Snow dissolving but Slowly.

**12 May 1838.** Nothing of any Consequence occurred. The natives amusing themselves various ways, but none of them Exert themselves to go and Kill any thing for their Subsistence although they are in want of Provisions.

**13 May 1838, Sunday.** This morning Mr. Simpson came back from his voyage, after having Surveyed the whole Coast of the Bay from where Doctor Richardson & Lt. Kendall returned in 1826. There now remains but the Extent of Smith's Bay[141] & that of Narre ella[142] unexplored thoroughly in this Lake.

• • •

137 Simpson.

138 Coppermine River.

139 They also brought copies of the New York *Albion* and the London *Times* (Simpson, *Narrative*, 241).

140 Portage La Loche or Methye Portage.

141 Smith Arm.

142 Deerpass Bay.

## *24. Letter from Thomas Simpson to Donald Ross, Norway House, 13 May 1838*[143]

*Private*

Fort Confidence, 13th May 1838

My Dear Sir,

On my return this morning from my last snowshoe jaunt of the season (300 miles), your highly valued letter of 8th Decr. last was put into my hands. It contributed greatly to restore my good humour, which the long delayed arrival of the express had thoroughly overcome. I can, however, only acknowledge your attention, as the departure of the Indian carriers for Fort Simpson tomorrow, compels me to hurry through my correspondence at a hand gallop. Your kind expressions in reference to the welfare of the Expedition, and my own individually, are received with the same feeling that dictated them. Our operations, so far, have proceeded satisfactorily enough. After drinking tea at the Coppermine river, I led the way for our provisions & baggage, which are all safely rendered to the banks of a small but navigable stream (15 miles from its junction with the great river, and 95 miles from this place by the chosen route), where they remain under the charge of your old acquaintances George Flett & Hector Morrison.

After achieving this laborious service, I made the complete circuit of the great arm of the lake to the eastward, discovered by myself last fall, and partially explored, looking for poor Taylor, in December. It is from that journey that I am this morning returned. The most serious difficulty now before us, is the transport of our light boats to Flett's station. We, however, have little doubt of acomplishing that too, by one means or another; and then for ice, storms and fogs once more! Any thing is better that the comparative inaction of a 9 months' winter.

I am much obliged for your country news: what a pity that the destructive diseases abroad in the plains should intrude to mar the fair prospect of augmentation in the Fur Trade! That infernal Colony too, seems to be going to old Harry with rapid strides. If those in whom we are interested were removed, I should sorrow very little to hear that the expected deluge whelmed all the rest this spring. McAllum's prospectus, if acted up to, will certainly not disparage the boarding schools, to which, as a most valuable institution I wish complete and enduring success.

I am rather astonished that that respectable youth [Huber] should be so suddenly love-struck, and that by so plain a goddess; yet it was cruel in Uncle D. to interfere, as *advice* in such cases is always thrown away. You are, I suspect, quite mistaken in imagining D. himself "to have recorded a vow of chastity", as not many weeks hence you may see him with a fair tendril clinging to his side, transplanted from across the Atlantic wave! And that morose genius D. Logie will perhaps come out similarly appointed. Tell me how my Bear Lake prophecies shall have been fulfilled.

In [hole in paper] great political changes have doubtless taken place; nor is it there

---

143 BCA ADD.MS 625.

alone that crowned heads have resigned their ancient honors, and crumbled into dust, like ordinary mortals. The world is indeed a stage, and all who move upon it "merely players". May our parts be those of honor and of virtue and then we may look with confidence toward a higher sphere.

I am gratified to learn the welfare of Mrs. Ross and family; and with kindest regards to you all,

Believe me ever,
My Dear Sir,
Most sincerely Yours
Thomas Simpson

D. Ross Esquire

• • •

## *25. Peter Dease's journal, continued*[144]

**14 May 1838.** Ritch Working at the Boat. Sawing & Fuel. Sent for the Small Canoe that was left at the Fishery last Fall. Preparing to Send the Martin Lake Indians off Tomorrow.
**15 May 1838.** The Indian that brought letters from Ft. Norman in the Winter Cast up after having returned there for his family. Says he met the men on their Way there on the 3rd near Ft. Franklin & they would have reached that on the 5th if not on the 4th Inst. The Martin Lake Indians received the letters and were Supplied with Some Provisions & Amunition to return and left this. Some of the Indians also left this to go and Encamp about the Entrance of the River[145] Expecting to take fish there and perhaps may Kill a Deer occasionally, as they are not Scarce in that direction at a Short distance.[146]
**16 May 1838.** Nothing of any importance occurred. Bustards flying about.
**17 May 1838.** Sent George Sinclair with Laurent to the River to put down netts as the Indians take trout there with Hooks & Lines and Could we procure fish to Bait Lines here we might take trout I think.[147]
**18 May 1838.** George Sinclair Killed a Bustard yesterday & Nalhahl with Eengheen brought in 6 this afternoon. One of the Indians, it is reported, has killed a Musk Buffalo, and another a Rein Deer. Felix put down 6 Lines to day. Others Sawing and working about the Boats. One will be finished tomorrow.
**19 May 1838.** Laurent brought us no news of how they succeed at the River. The Indians tell us 6 fishes were taken by them. One of the young men brought 45 lb. Fresh meat of the Deer he Killed. Felix took 2 Trouts at the Lines; one of them weighed 43 lb.
**20 May 1838, Sunday.** The Hunters brought 2 Bustards, and 15 fishes were brought from the River. All the Indians Except our Hunters are there.

---

144 MCM M214.
145 Dease River.
146 The first goose was seen flying over Fort Confidence (Simpson, *Narrative*, 241).
147 Canada, Hutchin's, and snow geese were seen flying over (Simpson, *Narrative*, 241).

**21 May 1838.** The Lines took 1 Small Trout. The Hunters brought us 6 Bustards. I omitted to mention that on the 18th two Outer Hare Indians, a man and a Boy arrived in a pitiful Condition from Starvation. They left their friends & families in a very reduced State somewhere about the Head of Smith's Bay. Three of the Party have died in Course of the Winter of Sickness. Sent for Cuthand & Brother to go and take Some Provisions to them. The latter declined going. Only Cuthand left this with Some Ribbs & Pounded meat for them on the 19th, His Father being of the Party with 4 others & their Wives & Children. It has been Strongly recommended to the bearer to take Every precaution to Assist these poor Creatures & Save them from perishing of hunger which he promised to do.
**22 May 1838.** Late yesterday Evening A young man brought us 42 lb. fresh meat of a Rein Deer he Killed near the men's Station at the River as it was crossing over. The distance at which he Killed it 310 Paces. 7 fishes were brought from the Fishery and 11 Bustards.
**23 May 1838.** Rained hard. Rain & Snow most part of the day. 3 men Covering the Roof of the House with mud, it being open in many places. Some Indians Came from the River & Returned; received 15 Bustd.
**24 May 1838.** The Weather Cold & Cloudy. Some Snow fell. The Lines were taken up as they produce nothing.
**25 May 1838.** Nozanne's Brother came from the River. Gave in 23 lb. Fresh meat, and to the men at the River left 35 lb. Tells us that three of them who were out hunting have Killed 4 R. Deers & Wounded the fifth, which they are to Send for tomorrow. Sent off one of the Boats to the River, the other not being ready yet.
**26 May 1838.** Today Main Pocque or Cuthand Came back accompanied by a man and a Boy of the Party he took relief to. Left the remainder of the Party 4 days ago on their way here, thinks they will yet require assistance to Enable them to reach this. They left this to Join the Camp at the River after getting a meal of meat.
**27 May 1838, Sunday.** The party of Indians Came – A most miserable Sett – Consisting of 20 Persons, who with the Exception of three are old, worn out men & women with young Children. The Hunters from the River brought 3 Swans & 11 Bustards. Big Head & Brother brought 120 lb. of Deers meat of the Animals they Killed.
**28 May 1838.** All the last arrived Indians left this to go and Join those who are at the River.[148] Babillard's Brother with 2 other young men brought 99 lb. Fresh meat of Some Deers Killed. The Boat prepared to be sent to the River tomorrow. Mr. Simpson went there to Question Sinclair about the practicability of Sending the Boats on the Ice, as many different opinions are given as to the breaking up of the Ice in this River, also its depth of Water for taking up the Boats that way. It is Sinclair's opinion that there will be enough Water, and better to wait untill it is Clear of Ice than attempt taking them now. Sent McKay to go with him to Examine as far up above the Rapids as they can & to be back tomorrow evening. The Hunters went out fowling & Eengheen Killed 2 Deers.

148 Dease River.

**29 May 1838.** Sent 3 men for the 2 Deers, which were brought, 210 lb. Five young men came from the Northern Lake. Brought us 2 Geese & 40 Spliced trouts of a Small Species. 4 fishes brought by the men from the River. Charleau & an old man came from near Cape McDonnell with 20 Bustards. A very Strong ESE Wind all day.
**30 May 1838.** The Indians left this last night Except Charleau who took his departure this morning. McKay & Sinclair came back from the River, & think there will be enough water in any part of it for our Boats. The latter returned to the River. More Carting & Covering the House with Mud.
**31 May 1838.** An Indian from the River brought us 42 lb. fresh meat of a Deer that was Killed yesterday. The Geese seem to have passed on to-wards the Coast. Very few seen now. The meat was put out to dry the Ribbs as the Store is too damp.

• • •

## *27. Letter from J.H. Pelly and the Committee to Peter Dease and Thomas Simpson, 31 May 1838*[149]

London, 31st May 1838

Gentlemen,

We were much gratified by the receipt of your interesting Despatch of 5th September which came to hand on the 11th ultimo.

The report contained in that communication of the proceedings of the expedition under your command during the past summer whereby you have succeeded in connecting the surveys of Sir John Franklin and Captain Beechy and in completing the Discovery of that portion of the Coast of the Arctic Sea situated to the westward of McKenzie's River is most highly satisfactory.

The zeal, ability and judgement manifested in this arduous service reflect upon you much honor; and entitle you to our warmest commendations, and in order to mark our approbation of the conduct of the expedition we have voted from the funds of the stockholders of the Company the sum of one thousand Pounds as a Gratuity to be divided as follows, vizt. Four hundred pounds to each of you and two hundred pounds to be distributed among the Ten men who accompanied you to the Coast.

Copies of your Despatches of 5th September were forwarded to the Secretary of State for the Colonies, the First Lord of the Admiralty and the Royal Geographical Society who notice the important services rendered by the expedition in very complimentary terms as you will observe by the annexed copies of their communications on the subject.

We regret very much our inability to promote Mr. Simpson on this interesting occasion, but we shall have much pleasure in appointing him to the First Chief Tradership that may become vacant in the service.

We look forward with intense anxiety and interest to the results of your operations to the Eastward from Copper mine River this summer, and earnestly hope they may be crowned with the like success that has attended your recent labors.

---

149 HBCA A.6/24, ff. 290–3.

From your ardour, activity and judgement and the excellent conduct of your people, we are quite sure that no human effort will be wanting to continue the survey from Franklin's Point Turnagain eastward to Back's Great Fish River or Ross' Boothia Felix, and we shall hope to learn that you return by the Great Fish River to Slave Lake, there to pass the winter. But should you not be able to return by Great Fish River, and that it may become necessary for you to retrace your steps to the Copper mine River and Great Bear Lake, and that no hope remains of completing the survey the following year by that route, then and in that case we are desirous you should proceed to the eastern End of Slave Lake with your people in the course of the winter, there construct Boats, and at the opening of the Navigation Summer 1839 descend Great Fish River to trace the Coast westwards from Point Richardson's of Captain Back's survey, so as to connect the survey which you may have made this year easterly from Point Turnagain of Franklin, and then return to the mouth of Great Fish River.

If the season afterwards admits of it you will extend your survey Eastwards from the mouth of Great Fish River so as to trace the coast to the straits of the "Fury and Hecla", into any other opening into Hudson's Straits or to Barrow's Straits as the case may be, and thereby complete the different surveys of the northern shores of the American Continent.

We sincerely hope that you and your brave followers may be blessed with health and strength to go through with this arduous service which has been so prosperously begun, but your persevering to this extent must of course depend upon the strength and health of yourselves & party continuing to enable you to endure the fatigue.

Should all your endeavours to accomplish this great object by the route of the Copper mine River & Great Fish River fail, it is probable we may endeavour to have it effected from Churchill, by boat as far as it is practicable, the party being prepared to winter with the Esquimaux and to proceed from one camp to another of that Tribe along the Coast to the Northward and Westward until the Coast be traced from Hudson's Bay to the mouth of Great Fish River.

We are aware of the difficulty, hardship, and danger of the Service, but feel it cannot be in better hands and earnestly trust the honor of accomplishing this great work is reserved for you.

We have only further to add that we look to you with entire confidence for the accomplishment of this bold and hazardous Service and that to enable you to carry it into effect you are hereby authorized to make such requisition for men, provisions, craft, etc. to be forthcoming at such point or Points as you may consider necessary or desirable, and that all Chief Factors, Chief Traders, Clerks and servants are hereby instructed to comply with such requisitions, which we are satisfied they will cheerfully & promptly do, as one and all must feel the most lively interest in an enterprize which so deeply concerns the honor and character of the Hudson's Bay Company.

Governor Simpson has adressed you so frequently and fully on all matters connected with the expedition, and taken such effectual measures to meet and even to anticipate all your wants, that we have only further to say that you possess our warmest and best wishes and that we are, Gentlemen,

Your affectionate Friends,
J.H.P. Govr.
B.H. Dpt. Govr.
A.C.
H.H.B.
R.D.
A.C.

One Copy of the foregoing to be addressed to Great Bear Lake by McKenzie's River and the other to Great Slave Lake, each accompanied by the following Circular.

• • •

## *28. Extract from letter from J.H. Pelly and the Committee to chief factors and chief traders of the Northern and Southern Departments, 1 June 1838* [150]

London, 1st June 1838

Gentlemen,

... On the 14th of April we were much gratified by the receipt of Messrs. Dease & Simpson's report of the proceedings of the Discovery Expedition under their command during the past summer having date the 5th September when they were about to proceed from Fort Vancouver [*sic* – Norman] to their winter quarters at the eastern end of Great Bear Lake. The zeal, activity and judgement manifested by those Gentlemen in the execution of the important Service committed to their charge afford us high satisfaction, and the success that has attended their labors in connecting the surveys of Sir John Franklin and Captain Beechy of the Northern shores of the Continent to the westward of McKenzie's River reflect much honor on the expedition which cannot fail of being highly gratifying to every Gentleman interested in the Fur Trade.

As a token of our approbation of their exertions we have voted from the funds of the stockholders of the Company the sum of One thousand Pounds as a Gratuity to the expedition to be divided as follows vizt. Four hundred pounds to each of the leaders and Two hundred pounds to be distributed among the 10 men who accompanied them to the Coast.

Herewith we forward a letter to those Gentlemen in duplicate unsealed, one copy of which is intended to be retained at the Factory, another to be sent to Norway House & Red River and a third to be transmitted for the information of the Commissioned Gentlemen in the Saskatchewan, English River, Athabasca, & McKenzie's River Departments to the end that they and the other Gentlemen throughout the Northern District may be made acquainted with our views and wishes connected with the further operations of the expedition, in which the Government, the British

150 HBCA A.6/24, ff. 293–6.

public and ourselves feel the most lively interest, and in order that every assistance, support and facility may be afforded to enable us if possible to accomplish the great work of completing the different surveys of the Northern Shores of the American Continent, so successfully begun. A Chronometer will be sent in the Packet Box for the use of the expedition, which you will forward to Messrs. Dease & Simpson by the bearers of their Despatches.

This Discovery Expedition, although projected & entered upon solely with a view to the acquisition of scientific knowledge & information and unconnected with any views toward advantage from Trade, is now, we are glad to find, likely to be productive of benefit in the latter respect, as from the extent of Colvile's River we think there can be no doubt it is the outlet of many lakes in the interior Country, where Fish are sufficiently abundant for the maintenance of Establishments which might with advantage be formed with a view to collecting furs which must be numerous in that new Country.

It is desirable therefore that this Country should be examined with the least possible delay and to that end, we wish Mr. McPherson to equip a small party under the command of Mr. Bell or any other experienced leader to ascend any of the Rivers that fall from the westward into the McKenzie near Fort Good Hope and direct their course across the Country so as to intersect Colvile River and some of the other large Rivers discovered by the Expedition, with a view to the examination of the Country, to the opening of a communication with the inhabitants, and to the discovery of the best access to it from the McKenzie River.[151] This we think may be done in the course of next summer by Mr. Bell in whose activity Governor Simpson has much confidence. Mr. Mcpherson, leaving some trusty person in charge of Fort Good Hope during his absence, and if a favorable report be obtained of it we are desirous the Country should be established summer 1840. It is not necessary for us to enlarge on the subject as Governor Simpson will address Mr. McPherson thereon. Yet we cannot help remarking that our progress in exploring and occupying the great extent of new country accessible from McKenzie's River is exceedingly tardy and that it would be gratifying to us to learn that a greater degree of energy was perceptible in extending the trade of McKenzie's River than has for some years past manifested itself in that quarter ...

We remain, Gentlemen,
Your affectionate Friends,
J.H.P. Govr., B.H., Dept. Govr., A.C., W.H.B., R.D., A.C.

• • •

151 The hopes of the Committee were, of course, founded on a false premise. The Colville drains only the area north of the Brooks range, while the area west of Fort Good Hope is drained by the Yukon and such major tributaries as the Porcupine. Nonetheless, this false notion of the topography and drainage to the west of the Mackenzie would, in time, lead to the exploration of the Porcupine basin and to the establishment of Fort Yukon on the Yukon River near the confluence with the Porcupine.

## *29. Peter Dease's journal, continued*[152]

**1 June 1838, Friday.** Mr. Simpson went to the River and States it as the opinion of the natives that the River is free of Ice; to ascertain which Morin & Norquay were Sent with a Canoe there, that the latter may Go with Sinclair a days Journey up to see whether it is actually free of Ice, that we may take advantage of the high Water to ascend it. They will be back on Monday. The Party of Hare Indians that came Starving have again made their appearance yesterday Evening in order to go to the North Lake & fish & passed on today.

**2 June 1838, Saturday.** A fine day – The Chipewyan Hunters and families left this to go and hunt along the River with Some of the Slaves and will be on the way when we pass.

**3 June 1838, Sunday.** Nothing new. The men have been preparing their property for departure. Laurent Came from the River with a young man and returned. Each took a bundle of baggage thither.

**4 June 1838, Monday.** This morning 2 Hare Indians brought 26 Spliced trout and White fish taken at the River behind here, or rather a Small rivulet running from the hills. They left this to return. One of that tribe who has not been here yet, came & traded a Small Beaver Robe, 1 Beaver Skin & 2 Rats.[153] The Wind N & NE. Weather Cloudy & Cold. Sent some Amn. to 2 Indians. George Sinclair & Norquay Came back; have been up the river above the Rapid & Saw no Ice in it, but Snow on the Hills. Preparations are therefore making to take our departure Sooner than was otherwise intended. An opening of the Ice is formed in the Strait here and Extending fast.

**5 June 1838, Tuesday.** Sent the Provisions and other articles that may be wanted to the River property for departure. The Indians are all off to hunt along the River. We may Expect to see them on our way up.

• • •

152 MCM M214.
153 Muskrats (*Ondatra zibethicus*).

# Chapter Seven
# Eastwards, 1838

## Introduction

Leaving John Ritch, assisted by three men, to take care of Fort Confidence and to start a fall fishery for provisions for the following winter, Dease, Simpson, and party left the fort on 6 June 1838. They camped that night at the mouth of the Dease River, and next day they started up that river with the two sea boats, *Castor* and *Pollux*. Six local Indians were hired to help portage the boats from the upper Dease River to Dismal Lakes. The boats, gear, and baggage were portaged over the five miles of the portage route to Dismal Lakes over the period 13–17 June. Dismal Lakes were still icebound and hence the boats were mounted on sledges. Since there was a strong wind blowing on the 18th, sails were hoisted on the boats; this manoeuvre was so successful that the men had to trot to keep up. They reached the staging depot on the Kendall River on the afternoon of the 19th to find Flett and Morrison in good spirits; they had even accumulated some 400 pounds of dried caribou meat. Most of the Indians were sent back to Fort Confidence but two (nicknamed Larocque and Macacoose) were engaged to accompany the expeditions as hunters.

The party started down the Kendall River on the 20th. The Coppermine had only started to break up, but after a day of waiting, the ice had sufficiently broken up and the water level had risen enough to allow them to start downriver, although they had to wait a day for the ice to clear completely. On 1 July they portaged down past Bloody Fall, but the sea ice showed no signs of breaking up and hence they camped between the fall and the river mouth. Nearby they found an Inuit camp; despite their best attempts at indicating that their intentions were peaceful and friendly, the Inuit fled over the sea ice to some off-lying islands, abandoning their tents, belongings, and dogs.

It was not until 16 July that the sea ice broke up enough along shore to allow the party to start its coastwise journey; the intervening period was spent in fishing, hunting, and some local exploration. Progress eastwards along the coast was achieved only by taking advantage of shore leads which opened and closed in a frustrating manner. Repeatedly the men were forced to camp for hours or days until the ice moved out sufficiently to allow them to proceed. One positive aspect of this, however, was that the hunters were able to provide a steady supply of caribou meat, thus allowing them to economize on pemmican. No Inuit were encountered, but they found several caches of sledges, stone pots, snow shovels, and so on – winter equipment stored for the summer while its owners were inland hunting.

The party reached Cape Barrow (the western entrance cape of Bathurst Inlet) on 29 July. From there they headed east via Marcet Island and Wollaston Point; beyond this, ice delayed them repeatedly as they worked their way through the Barry Islands, and it was not until 4 August that they landed at Point Everitt on the east side of Bathurst Inlet. From here they worked their way north across the mouth of Melville Sound, rounding Cape Flinders on the 9th, but being stopped by ice later that day only two miles from Franklin's farthest point at Point Turnagain. There then followed ten days of frustration and disappointment; the ice stubbornly remained packed against the shore and the weather was vile, with strong winds and showers of rain and snow. Worse still, there was almost no driftwood, and fuel could be spared only for a cooking fire morning and evening.

Finally, on 20 August, Simpson set off, accompanied by seven men (including the two Indian hunters) to explore the coast as far as possible on foot, on the understanding that if ice conditions improved Dease would send one of the boats to overtake the overland party and continue the exploration eastwards more expeditiously. In the event this did not happen. Simpson and his group returned to where Dease and his companions were patiently waiting at Point Turnagain on the evening of the 29th. They had traced the west and north coasts of Kent Peninsula as far as Cape Alexander; eastwards from here, to Simpson's frustration, the sea appeared to be totally ice-free.

For the remainder of that day and the following day a belt of ice prevented them from starting back westwards, and even on the morning of the 31st they had difficulty in pushing through a belt of broken ice to reach the open water offshore. Thereafter, however, they were able to hoist the sails, and they made splendid progress on a direct course across the mouth of Bathurst Inlet to Cape Barrow. On the entire run back to the Coppermine they encountered no ice and they reached that river on the evening of 3 September: they had covered in four days what had taken them twenty-five days on the outwards journey.

Having cached the boats' masts, oars, and yards at the coast, and some pemmican and the canvas canoes at Bloody Fall, the men started tracking the boats up the Coppermine to a suitable location near the Kendall confluence. Here the boats and much of the equipment were securely cached for the winter and the party started overland for Fort Confidence. They arrived there safely on the afternoon of Friday, 14 September.

Especially by comparison with the previous season, the expedition's achievements in the summer of 1838 must have been disappointing. Owing to persistent sea ice, the expedition had been unable to penetrate any farther east by sea than Franklin had done. And to give him his due, had it not been for Simpson's drive and initiative, which resulted in his pushing eastwards for a further 85–90 miles (136–44 kilometres) on foot, the season's work, despite the time and effort involved, would have had to be rated a complete failure.

## DOCUMENTS

### *1. Peter Dease's journal, continued* [1]

**6 June 1838, Wednesday.** All our arrangements being made, and every thing required for the voyage rendered at the mouth of the River. John Ritch is appointed to take Charge of the Establishment, having J. Norquay, Frans. Framand & the Lad Wentzel under him, to fish, prepare the Buildings for another Season in case they may be required, & Collect what provisions the Indians may bring in course of the Summer, and Establish a fall Fishery when the Season comes, Should we not be back in time. Instructions are left with him, how to act in the Event of our not being here when the Supplies Come, or of our not returning. The day fine. We left the House at 8 P.M. with the dogs & Sleds, taking our bedding &c. to the River on the Ice, accompanied by the 2 men, that they may bring the Dogs back to the House, and Encamped at the boat Station. The River seems to be quite free of Ice and its Channel working its way in the Bay, opening its width through the Ice gradually. Here we find the Anemone[2] in full blossom and Some Catkins[3] Spreading & beginning to bud, while nothing indicates the commencement of vegetation yet at the House, although there is but so Short a distance intervening. The Boats are in good order and appear as yet to be tight, and well bound, having had additional bindings to the thafts, with intermediate timbers placed in them to Strengthen them, and the Rudders much Stronger than those we had last Summer. Being once more under way, we may Say, All the party sat round a good fire in cheerful hopes, under Providence, of a happy result to our labours.

**7 June 1838, Thursday.** The men were sent back to the house with the Dogs, and after taking breakfast we Embarked. The Crews as follows: the Castor, James McKay Steersman, François Felix Bowman, Charles Begg & John McKey Midn., P.W Dease Offr. The Pollux, George Sinclair Steern. Laurent Cartier Bown., Pierre Morin & Wm. McDonald Midn., Thos. Simpson Offr. After proceeding a Short distance found the River Shallow with Rapidous Points but Sufficient depth for our Boats, the Banks wooded with the Fir Tree[4] of rather Stinted growth, very twisted & Knotty. Small willows & black birch forming the underwood, very thick in most Places. Our advance was principally with the tow line & poles. The valley of the River Con-

1 MCM M214.

2 Probably the pasque flower or prairie crocus (*Anemone patens*).

3 Willow catkins.

4 Black spruce (*Picea glauca*).

tracts as we approach a Rapid Running between high Rocks rising to from three to five hundred feet on each Side. We put up for the night at 8 P.M. near the commencement of it.

**8 June 1838, Friday.** After leaving our Encampment a Short distance came to a Rock which has been detached from the Cliff near it and fallen in the Stream like a Pillar. Named by our people the Old Man of Hoy[5] from its resemblance to Some Rock bearing that name in the Island of Hoy in Scotland.[6] The Rapid is too Strong & obstructed by many rocks, caused a deep Cascade in the main Channel which obliged us to Carry all our luggage. The Boats were towed up light. A Strong Current continued for some Distance above it, where the Stream is again Smooth, which enabled the men to ply the oars. Banks of Ice in some parts, on one of which a Mark was Seen leading to 2 Rein Deers Killed by the Indians & put up for us. They were taken in. At 1 P.M. a young Indian came down to us, Saying the Hunters are at no great distance & have Killed several Musk Buffalo & Rein Deer & wishing we would send for some. As we are desirous of getting assistance to carry our Boats over the height of land intervening between the upper Parts of this River & the Dismal Lakes, Mr. Simpson and I accompanied the messenger to the Tents, a distance of about 3 miles, and Engaged 6 young men to accompany us as far as our Depôt that is on the Banks of Kendalls River.[7] They brought some meat, received from Different Indians, down to the Boats which we reached about 10 P.M.[8]

**9 June 1838, Saturday.** With three additional men to each Boat we advanced rapidly. Saw Several Deers. Afternoon the Channel became more Shallow with a Stoney bottom. Three of the Indians landed to fire at Deers that were seen at a Short distance from us. Two of them in the hurry put a Ball over a load of Shot the guns had in them, and in attempting to fire burst the two guns at once. Fortunately they received but little injury themselves. In some of the Shallow parts afternoon the Castor was injured, and rendered leaky. Put up at 8 P.M. at the Foot of a rapid running between two points of Granite Rock, and the Boat unloaded & drawn up.

**10 June 1838, Sunday.** After passing the Rapid a number of Shallow points occur for about 3 miles. The Stream again widens; the banks & bottom are Sand. At 10 A.M. a Musk Buffalo was Seen. Landed the Indians, who killed it. After taking break-

---

5 A sandstone sea stack 450 feet (137 metres) high, on the west coast of the island of Hoy in the Orkneys.

6 A large colony of cliff swallows (*Petrochelidon pyrrhonota*) and some hawks were nesting on the cliffs (Simpson, *Narrative*, 246).

7 Named by Dr John Richardson after his companion Lieutenant Edward Kendall during their overland journey from the mouth of the Coppermine to Great Bear Lake. The party crossed the river at the outlet of the Dismal Lakes on 14 August 1826 (Richardson, "Dr. Richardson's Narrative of the Proceedings of the Eastern Detachment of the Expedition" in Franklin, *Narrative of a Second Expedition*, 273).

8 Several Canada geese were shot and their nests were found among the rocks. Pintail ducks (*Anas acuta*) and "black diving ducks" – probably one of the scoters, namely the white-winged scoter (*Melanitta deglandi*), surf scoter (*M. perspicillata*), or common scoter (*Idemia nigra*) – were seen on the river, while willow ptarmigan (*Lagopus lagopus*) were quite numerous in the stunted forest (Simpson, *Narrative*, 248).

fast, proceeded. The river widens in some places and runs in a very serpentine course. At 4 P.M. a Channel from SE Joined near a conspicuous high Sand Hill coverd with Fir trees.[9] About 2 miles farther the river again receives another Channel. We continued in the North Eastern one, which became narrower as we proceeded & Shallow in many parts untill 8 P.M., when we came to a Succession of Shallow Stoney ridges with too little water to float the boats. Put up for the night in order to Examine our Course in the morning.

**11 June 1838, Monday.** The Weather became very boisterous during the night. A Strong wind with a Snow Storm Kept us at our Camp for the day.[10] Saw many Deers but none were Killed. On Walking along the banks of the Channel find that by Carrying our Boats and property about 2 miles, we may again advance some distance up the River.

**12 June 1838, Tuesday.** The Boats & property carried above the Shallows & loaded. Mr Simpson and I walked untill, approaching the Hills to the NE, we find the Stream again Shallow from Sand banks running from Point to Point. We waited for the Boats. They Joined us at 9 P.M. & put up for the night, this being the nearest part we can make to the Dismal Lakes which are about 5 miles from us.[11]

**13 June 1838, Wednesday.** Crossed to the S. side of the River & got the Property taken to a fine Sandy Hill near a Small Lake covered by Ice. The Boats were taken by the men & Indians half way to the Lakes in course of the day & returned to us at 7 P.M. A Wolf prowling about us, very anxiously. Upon looking round the Sand Hill we are Encamped on found her Den, containing a litter of young ones, which we got out after Some work. There are 4 young about 10 days old. the dam was Killed by one of the Indians and the pups taken to the Camp, where food was given them. They are a pretty Silvery grey Colour & I have no doubt may be brought up and domesticated.[12]

**14 June 1838, Thursday.** The baggage, rigging, &c. was Carried in 2 trips to the Place the Boats were brought yesterday.

**15 June 1838, Friday.** The weather proved so boisterous that no move was made.

**16 June 1838.** The Boats were taken on, with one trip half way to the Lake.[13] A Strong N wind all day.

**17 June 1838.** All the baggage rendered at the Lake. Some deers were Seen on the hills. Three Indians went out & Killed one, which they brought to the Tents. Geo. Sinclair was out to the Hills to Northward of us, and brought a piece of ore that

9 Almost certainly the isolated hill known as Hanbury Kopje. The river coming in from the southeast is in fact the Dease River, flowing out of Lac Rouvière. The stream which the party was now following was Sandy Creek.

10 This camp was at treeline, but enough dead, stunted spruces were found to provide fuel for the campfire (Simpson, *Narrative*, 250).

11 They had ascended the Dease about 70 miles (112 kilometres) (Simpson, *Narrative*, 251). The present portage route striking off from here northeastwards to the west end of Dismal Lakes is about 6 miles (10 kilometres) in length.

12 Simpson hoped they might be trained to pull a sled (Simpson, *Narrative*, 251).

13 Dismal Lakes.

has the appearance of Iron mineral with plumbago in it; affords a Red Shining paint to the Indians. Some of the Indians complain of fatigue & wish to return to their friends but were prevailed upon to continue.

**18 June 1838, Monday.** Got our Sleds put together & the Boats lashed on them; the luggage in them. Left our Camp at 5 A.M. The Ice quite Solid with Snow upon it in Some places; in others very rough walking, being Sharp in points under the feet. After proceeding for about 5½ miles took breakfast, which was cooked by collecting Small Willows. No wood growing along the Lake, the hills on Each Side being Rocks & barren heath, with much Snow yet in the hollows between them. A Strong wind prevailed most of the day, which in a bend of the Lake was favourable, and the Sails were hoisted on the Boats, affording the men great relief & Kept them at a very brisk pace for Some time.[14] Our day's travel was about half the distance from the last night's Encampment to Geo. Fletts lodge, say 15 miles, and put up for the night at 6 P.M. after having navigated about a mile in a narrow part that was free of Ice.

**19 June 1838, Tuesday.** The night was fine & Cool. Left our Encampment at 3.30 and got to the narrows between the two Lakes where the Boats were launched & took us about a mile free of Ice, when they were again dragged on the Ice a few miles to another narrow place which received the Water of Kendals River from SW.[15] Here again our Boats were launched for about a mile more. The lake again Expanding for about 3 miles took us all Safe to our Dêpot lodge[16] at 3.30 P.M. George Flett and Hector Morrison with the three Indians quite hearty. The latter being active hunters have Killed 24 deers since their arrival here on the 21st Ulto., which has enabled them to Collect about 400 lb. Dry meat. The property Safe and in Good order. The 7 Indians who came with us, with one of the three who were here, are to return. The two others, named by the men LaRocque and Macacoose, I have engaged to accompany us along the Coast as Sixth hands in the Boats.[17] Besides their activity as hunters may be of advantage to us. Kendals River, Flett tells us, has been free some time now and the Weather by his account has been finer here than we have had Since leaving the Establishment. Every thing prepared for departure tomorrow, as it is the opinion of our Guides that there is Sufficient depth of Water in the River to take down our Boats to the Copper Mine River with full Cargoes. The Irons[18] of our Sledges were taken off in order to send them back to the House by the Indians who return from here, and notes given them upon J. Ritch for the payment of their Service,[19] with provisions to take them back to their Camp.

---

14 Simpson (*Narrative*, 252) reported that they attained a speed of two knots, and speculated that it would become the basis of a legend among the Indians.

15 The river coming in from the southwest is now known as Teshierpi River; the name Kendall River is reserved for the river draining the Dismal Lakes to the Coppermine.

16 Tent at the depot where the Kendall River flows out of Dismal Lakes.

17 They were persuaded to have their hair cut, and the rest of the men clubbed together to outfit them in voyageurs' dress (Simpson, *Narrative*, 254).

18 Iron runners.

19 At a rate of one Made Beaver (the standard unit of exchange in the fur trade) per day, plus gifts such as tobacco and shoes (Simpson, *Narrative*, 254).

**20 June 1838, Wednesday.** At 1 P.M. After having written to Ritch and C.T. McPherson what arrangements we have made in the event of not returning in time to see the Boat in the Fall that is to bring our Supplies from Ft. Chipewyan and McKenzie's River, the Indians were Sent off, by whom also were Sent the young Wolves, brought here from Wolf Den Lake by Mr. Simpson, and now sent to the house to be reared and domesticated. Our Boats loaded with their whole Cargoes[20] and we proceeded on our Voyage down the Rapidous Stream of Kendal's River, which required Caution to avoid injuring the Boats, being narrow with many Shallow Stoney places forming rapids. The lower part is less so with Sandy Banks as we drew near the Copper Mine River, about half a mile from which a rapid occurs, the Stream running between two high Rocky Cliffs with violence. From the foot of this Rapid the Copper Mine River presented to our view a Solid surface of Ice, with exception of a very narrow Channel along the W Shore where pieces of Ice detaching from above were floating down a very Strong Current, which rendered the navigation extremely hazardous. We therefore put ashore for the night at the Point forming the Junction of the two Streams, which by observations taken by Mr. Simpson is in 67° [gap] N Lat. and [gap] W Long.
**21 June 1838, Thursday.** The water both in the Small River & Cop. Mine rose Since we landed considerably, near 12 Ins. perpendicular. In the morning Sinclair & Geo. Flett were Sent down to examine how the River is below some distance. The two Indians also went out hunting, and came back with the meat of a Deer they Killed. Sinclair & Flett also Killed one which they left along the Banks of the River; have been about 12 miles along the River, where they say the Ice has been broken up, and the passage is free, but a great deal heaped up along the banks on each side, the Current very Strong. The width of this part and a Couple of Islands not far below this, they think is the Cause of its not having broken up here. The Channel has widened considerably along Shore since yesterday.
**22 June 1838, Friday.** The water continued rising & the channel along Shore enlarged, which enabled us to Embark at 10 A.M. Much Ice floating down. We landed to take in the Deer Killed by Sinclair and proceeded. Saw many deers. Two more were Killed & taken in. Some Musk Buffalo were also Seen on the Hills. At 6 P.M. a contraction of the River running between high rocks forms a Strong rapid, which the quantity of Ice drifting renders dangerous with full loads. Part of the Cargoes were therefore Carried about ¾ of a mile and the Boats run down. The Swells of the Rapid were such that the Castor took in a considerable quantity of water. Put up below the Rapid for the night, near the Commencement of Some high Sand banks, and a gravelly or Stoney Island just below us. Some of the provisions got wet, but were put out, as the wrappers merely can be Soaked from the Short time the Water was in the Boat. The Wood has become more Scarce along the River as we approached this Rapid and we had to go some distance to Collect enough to cook our victuals.

---

20 The cargoes included twenty-eight bags of pemmican, six hundredweight of flour, and about three hundred pounds of dried caribou meat (Simpson, *Narrative*, 254).

**23 June 1838, Saturday.** The quantity of Ice floating down increased, rendering it unsafe to go down among it. Sinclair went some distance along the River and reports it as being Shallow and Rapidous. Saw a narrow pass which forms a Strong rapid with an Island in the midst, dividing the Stream into 2 Channels. An Observation was taken which places this Spot in 67° 23' 14"N. Lat. 116° 6' 51"W. Long: 51° 50 Var.E., apparently the place at which Sr. John Franklin encamped on the 12th July 1821.[21] Some Deers were seen but none Killed. Afternoon the Water fell considerably, which is supposed to be from the River having been dammed up by Ice, and has now forced its passage.

**24 June 1838, Sunday.** The floating Ice continued and we remained at our Camp. One of the Indians Killed and brought home a Deer, which are in poor Condition at this Season. Water rose again about 4 Inches.

**25 June 1838.** After breakfast, the river appearing pretty free of Ice, pushed off. Run the Rapid Sinclair had Seen and found the Stream generally Stronger than above the Rapid we first passed. Put ashore in order to Examine a long Chain of Rapids running over Shelving ridges of Rocks about 6 miles below Island Rapid, which we got down Safely, and again on approaching Escape Rapid[22] of Sr. John Franklin the Steersman went to Examine it, but as we afterwards found the height of the Cliffs opposite the Strongest part is very deceptive, as a dangerous rock near the right hand bank was concealed from their view, and only by the admirable presence of mind of Sinclair in running his Boat between it and the Cliffs through a very narrow passage was the Pollux saved from being dashed to pieces. My Boat in avoiding that danger, had to encounter very heavy Swells, at same time pass under a Small Stream which was throwing its waters from the top of the Cliffs like a thick Shower bath, pouring a great deal of water into the Boat before we could get Clear of its Stream. No other damage was Sustained, fortunately, and after baling out the Water we proceeded some miles when we got to a bend of the River which is much obstructed by Shallows & gravelly ridges. Below this Shoal, Rapidous Place high mud banks appear and an Expansion of the River.[23] At a distance the high Banks and mounds almost present to the imagination huge Icebergs, being formed of a Whitish Clay.[24] Below the first of these the passage was totally Stopped up by Ice and we had to retrace our Way Some distance to find a favourable Spot to Land at. Although we could not haul up our Boats they were put in a Safe Spot & well secured. On ascending the Hills we had a view of the Coast, the Bloody fall being about two miles from us. Some Deers were Seen and fired at but none Killed.

---

21 Franklin and party were on their way down the Coppermine by canoe from their winter quarters at Fort Enterprise to the Arctic Ocean on his first expedition to explore the arctic coast (Franklin, *Narrative of a Journey*, 345).

22 Named by Franklin on 15 July 1821 from both his canoes having narrowly escaped being swamped (Franklin, *Narrative of a Journey*).

23 The widening of the river just above Bloody Fall.

24 This whitish clay is a conspicuous feature of the lower Coppermine valley; it represents glacio-lacustrine sediments from Glacial Lake Coppermine which was dammed by glacier ice lying across the lower valley around ten thousand years B.P. (St-Onge, "Glacial Lake Coppermine, North-Central District of Mackenzie, Northwest Territories").

**26 June 1838.** One of the Indians (LaRocque) Killed & brought in a Deer. John McKey wounded one, but could not get a Chance to Kill it after a long pursuit. The Ice Cleared down to the Fall, but as that place is destitute of Wood and we have here plenty fuel of the drift wood brought down the River & left upon the Point we are now Encamped upon, Sheltered also from Cold Winds by the Hills, we prefer remaining at our present Station untill we can proceed direct to the Coast. A few Bustards, were seen flying about.
**27 June 1838, Wednesday.** Two men were sent to see how the Ice appears along the Coast. I went to the Bloody Fall, where is Still to be found a memento of the horrid deed that gave use to its name, and mentioned in Franklin's narrative,[25] but no appearance of the Boats left by Doctor Richardson in 1826,[26] nor of any Esquimaux having lately been on the w Side the River here. Upon the Hills the marks of a Camp appear where they Stretched Deer Skins, apparently last Summer or autumn, consisting of two or three families. Some of the men State having seen footmarks of them of this Season. The two men Sent to the Coast report that the River is free of Ice, but that it is fast to the Shore as far as they Could see on all Sides of the Coast. They brought an arctic Salmon[27] Killed by Gulls on the Shoals, quite fresh.
**28 June 1838, Thursday.** The forenoon was fine but the Weather became Cloudy with a thick fog afternoon. The two Indians were out and each Killed a Deer; the meat brought in. Some of the men who have been out to the Westward say they fell upon a river of considerable breadth and depth about 6 miles hence, running towards the Coast,[28] but did not follow its Stream any distance. They saw tracks of the Esquimaux of this Season near its banks.
**29 June 1838, Friday.** A fine Day. Some Deers were Seen across the River. A party crossed over and one was Killed by LaRocque & brought. The water is falling considerably in the River. Many Shoals now appear in the Shallow rapid that were covered when we landed here.
**30 June 1838, Saturday.** The Weather Cloudy & Cold. Nothing worthy notice occurred.
**1 July 1838, Sunday.** As the water is falling fast in the River, and it appears to be Shallow below the Falls, we considered it would be better to go down to the coast,

---

25 The allusion is to the massacre of an encampment of Inuit by the Chipewyans accompanying Samuel Hearne in 1771 (Hearne, *A Journey*). The "memento" to which Dease refers consisted of skeletal remains, also noted by Franklin and party in 1821 (Franklin, *Narrative of a Journey*). Dease and Simpson found two skulls (Simpson, *Narrative*, 262).

26 The *Dolphin* and *Union*, left here by Dr Richardson after his coastwise voyage from the Mackenzie Delta on 10 August 1826; from here he and his party walked overland to the mouth of the Dease River, where a party with canoes met them and took them back to Fort Franklin (Franklin, *Narrative of a Second Expedition*, 269). But Dease and Simpson did find ropes, tarpaulins, etc., remnants of the tents which Richardson had left standing (Simpson, *Narrative*, 262).

27 Arctic char (*Salvelinus alpinus*).

28 The Richardson River, named by Franklin after his companion Dr John Richardson in July 1821 (Franklin, *Narrative of a Journey*, 361).

and sett netts untill we may have an opportunity of commencing our Coasting voyage thence. We therefore proceeded downwards and had the Cargoes & property carried across, which to avoid the Steep Hills lengthened the distance to near half a mile. This labour occupied a considerable time. Leaving the Portage, the River is Shallow and rapidous for about ½ mile, where my boat Struck several times & became so leaky as to require almost constant bailing out the Water. We therefore landed below a Small Island and Steep Sand Banks on the East Side and unloaded & Drew up the Boat. About 9 Inches of the garvel Streak[29] was Split, whether in launching it over the portage or in the Rapids, I cannot Say. LaRocque after landing went above the Hills and returned after about 2 hours absence with a Deer and a Grey or Laughing Goose[30] he had Killed.

**2 July 1838, Monday.** Sinclair was out to day and reports having Seen two or three Esquimaux tents along the Coast, Eastward of the River but did not go near, not wishing to alarm them. The distance by his account about 5 miles from the River. 2 Geese and a Gull were killed to day.

**3 July 1838, Tuesday.** Accompanied by McKay, Sinclair & G. Flett I walked down to endeavour to have an interview with the natives. When we had got opposite the head of a Small Channel forming the Easterly outlet of the River above an Island, we saw a party in the act of encamping upon a rising ground forming a second bank from the River, as we were discovered by them at Same time. A Single individual advanced a Short way towards us while the others, who appeared to be 2 Women and a youth, fled towards the Island & Crossed the Channel, which is not more than knee deep, to the north end of the Island. The man, seeing their retreat apparently secure, withdrew, making many menacing gestures, and pushed off in a Kayack to follow the fugitives. By this time we had reached the Spot where their property lay, and 4 dogs tied. A lodge which they were preparing to erect, bags &c., &c. lay in Scattered Confusion. On seeing the Indian afloat in his Kayack I desired the men to remain where they were and walked down some distance alone, unarmed & making Signals of friendship, Calling him ashore, which after a little hesitation, putting some Questions which I could not answer, he turned back and came ashore, to induce him to which I remained about 60 yards from the waters edge. With many words & gestures, he came up to me & presented the point of a Spear after having broken the Shaft, the latter end turned towards me, and then breaking off the head of an arrow presented it in the Same manner. At my desire he sat down for a few minutes, during which time I endeavoured to make him understand that we were friendly to the Esquimaux,[31] and having given him a couple of buttons as a peace offering, he was allowed to depart, with a recommendation to bring back his family that their property might not be lost. After having got round the Island and Joined his family, we saw them all on the Ice, running out for Islands about 8 miles out to NE of

29 Garboard strake.

30 White-fronted goose (*Anser albifrons*).

31 The Inuk asked whether Dease and his companions had their families with them, which presumably he would have interpreted as a sign of peaceful intentions (Simpson, *Narrative*, 263).

us[32] in great alarm apparently, carrying the Kayack with them. On returning to the Camp, the Tents were Struck and we proceeded down to the Spot that the poor Creatures' property might not be destroyed by the Dogs.

**4 July 1838, Wednesday.** The property of the Esquimaux was Collected, and covered over with their lodge. Among the numerous articles contained in one of these Bags is a Small tin Cannister of a travelling Case, two gun Springs, some Iron drills made of Screws, and a Top of a large Tinsel tassel,[33] several Small pieces of Copper, some Shoes, meat half dried but half putrid. A large Winter Sledge was also left at the River edge, with a large bag Containing meat and various articles, utensils & Two Stone dishes or Kettles of an Oblong Square Shape. The runners of the Sledge measure 13 feet in length, 2 ½ in thickness, 13 in depth, with 14 transverse bars, tied on with Seal Skin line and Shod partly with Whalebone and partly with the horns of Deer, fixed on with wooden rivets. Part of the Jawbone of a Whale is among the rest, which measures 14 feet in length but is not entire in length, and has been Split & worked for convenience of carriage. The dogs remain about, one a very old one, remains by the property. Three others going about; one of them a Slut[34] has a young pup of which we have taken charge. This evening the dam, having got entangled by a cord hanging from its neck was brought to the Tent and appears docile enough. A nett was put down on the opposite Side of the River in an Eddy.

**5 July 1838, Thursday.** This morning a violent Wind came on with rain, and Continued all day. Had to unload and draw up the Boats. Among the Chips formed by the Esquimaux on the hill, some Pieces of the Mahogany Boats left by Dr Richardson in 1826 at the Bloody Fall, were found, but nothing seen to indicate what use it had been applied to. The wind was too Strong to Examine the nett put down yesterday.

**6 July 1838, Friday.** Weather moderated & Cleared this morning. In our nett 2 Arctic Salmon, 2 trouts, 6 Tullibees,[35] 2 Methys[36] and a Small Flounder were taken, which affords a pleasing variety to our Food. The two Indians were out and Killed 3 Laughing Geese. Wind N & Cold.

**7 July 1838, Saturday.** Had a thick mist forenoon. Weather Cleared by 2 P.M. The Ice appears from the northerly wind to be pressing in the Bay and upon the Island at the mouth of the River. George Sinclair has gone out and did not come back this evening. Some flowers being now in full Blossom were Collected.

**8 July 1838, Sunday.** At 4 A.M. Sinclair came back. Says he has been about 12 miles distance along the Coast, and thinks he saw at Some distance from him, on his return after leaving the Coast, 3 or 4 Tents of Esquimaux, but was not near enough to distinguish

32 The Couper Islands, named by Franklin after a friend of Dr John Richardson in July 1821 (Franklin, *Narrative of a Journey*, 361).

33 These items had been salvaged from the articles left by Richardson in 1826; this was confirmed by the presence of some painted pieces of wood, derived from his boats (Simpson, *Narrative*, 264).

34 Bitch.

35 Cisco or lake herring (*Coregonus artedii*).

36 Burbot (*Lota lota*).

Plainly. This was about 3 miles from the Coast near a Small Lake. 12 fishes were taken to day: trout, Salmon, Tullibee & 1 White fish. The Ice is yet fast on the Sands as far as we can see to Eastward.

**9 July 1838, Monday.** The netts gave us 12 fishes again to day. The Indians were out & Killed but one Goose. Wind from N to NE with rain. Afternoon Sinclair out hunting again.

**10 July 1838, Tuesday.** Came back only this morning. Says he saw the Place where the Esquimaux Tents were on the E side of a River which he could not wade across,[37] but they have left it, and apparently gone to the interior following along the River. The weather was windy with rain part of the day. 11 fishes were taken.

**11 July 1838, Wednesday.** A fine day; wind variable. Mr. Simpson and I went with a Party in a Boat to Examine the Bay to Westward which is Clear of Ice for about 6 miles along Shore. A point there runs out to ENE upon which the Ice being fixed, we could proceed no farther. To the W of that Point a narrow Bay runs in, which as far as I could Judge, appears to receive a River, but we did not go to the bottom of the Bay, fearing that the Ice might Close in between us and the River. Some Esquimaux tracks of not a very Old date I saw in the farthest bay, going Westward, the footmarks of a party of 9 Persons. Some deers were seen but none Killed. One Grey Crane[38] and a Duck was all the game Killed. In the evening after our return a violent wind rose from Northward. 21 fishes were taken out of the two netts. One only was put to replace them.

**12 July, 1838, Thursday.** The wind continued Strong all day. The Ice continued Close in Shore. Weather cloudy & Cold. 7 fishes taken to day, among which is a Small flounder.

**13 July 1838, Friday.** The wind variable, but most from Northward. The netts gave us 8 fishes. The Indians were out & brought 2 Geese. Macacoose says he saw two Esquimaux at no great distance from him, who were apparently out hunting and were near a Small Lake within gunshot from him. Having seperated from his Companion, he was afraid & retreated, but at same time made signs to them, waving his Cap, which they did not answer. Some flowers were Collected to day.[39]

**14 July 1838, Saturday.** No opening of the Ice yet. The nett gave us 9 fishes. Wind from North; rather Cloudy throughout the day. Vast numbers of musquitoes are about us these few days past, and are most numerous when the wind turns from E by the Southward to W. The Dogs left by their owners have been taken in Charge by the men, and seem reconciled to their Change. They are low, thicksett Animals, and although they were wild when loose, are not wicked when tied, never attempting to bite when handled. [N]or do they ever try to free themselves although Simply tied with a leather Cord of some length.

**15 July 1838, Sunday.** The Indians went out, and Came back late with a Buck Deer. Say they followed a River some distance that runs SW, and which has Wood along

37 Probably the Napaaktoktok River.

38 Sandhill crane (*Grus canadensis*).

39 The sun almost touched the northern horizon at midnight (Simpson, *Narrative*, 267).

its banks where they saw it. Appears to be the same River that Sinclair saw, whose estuary on the Coast is to the E of this about 10 or 12 miles.[40] The day was fine and Wind variable. The Ice seems to be dissolving fast among the Islands & along the Bay, a Channel opening from Eastward point to the Islands NE of us. We have to get fuel from the N Side of the Island where a considerable quantity of Drift wood has been driven by the Wind. 6 fishes were taken to day.

**16 July 1838, Monday.** The Ice has broken up in the Bay sufficiently to afford us hopes of Crossing it tomorrow. With that view at 8 P.M. the Boats were loaded and we left our tedious encampment. Took up the nett as we passed and landed on the N point of the Island, where a few Plants were Collected. From the vestiges found this appears to be a place often visited by the Esquimaux. Many muscle and Common Clam Shells upon the beach with sea wrack,[41] but the Water quite fresh all round the Island. The tide rose very little. Its highest rise in the River where we were did not exceed 6 inches. At Midnight having a fine Clear Sky with an open horizon north, we had a beautiful observation of the setting Sun, which did not entirely disappear, ⅛ of its Disk remaining visible. As the Ice formed the extent of our visible horizon it would lead us to conclude that the land to the northward of our position on the Island must be very low, as the distance according to our map is not such as to prevent it being Seen otherwise.

**17 July, 1838, Tuesday.** Taking advantage of an opening among the Broken pieces of floating Ice we embarked at 1 A.M. and working our way through much crush Ice we gained the E point where a Space of about a mile broad allowed us a clear pull for some time.[42] On approaching Berens Group of Islands,[43] the Ice again neared the Shore, and we landed on one of them to breakfast. Some vestiges of Esquimaux were seen upon it, but of old date. On the N Side the Ice packed on the Islands. We therefore continued along the Main Land and passed several Islands, the channels inside becoming more narrow untill 11 A.M. a Small River[44] was passed, and the Islands running out to the NE. The wind rose from SW and with a fine Breeze hoisted our Sails, but the Ice being yet on the points we had to Strike Sails & push through for about half a mile, and again an open space. Passed Sir Graham Moore's group[45] and continued untill 5 P.M. when the Ice became too close to proceed. We landed in time to Shelter ourselves from a heavy Shower of Rain, preceded by a few Distant peals of Thunder. Clouds gathering, the weather very Sultry and musquitoes Swarming about us, very annoying. Several Rein Deers were seen in course of the day and in two places Musk Buffaloe grazing. A few Shots were fired at a Deer without Effect. In one place one of these Animals was seen out on the Ice where it was

40 Napaaktoktok River.

41 Seaweed.

42 The Indians were quite excited by this, their first experience of travelling by sea; two seal pups intrigued them in particular (Simpson, *Narrative,* 270–1).

43 Named by Franklin on 21 July 1821 after Henry Berens, governor of the Hudson's Bay Company (Franklin, *Narrative of a Journey*, 364).

44 Probably the Asiak River.

45 Named by Fanklin after Sir Graham Moore, on 21 July 1821 (Franklin, *Narrative of a Journey*, 364).

much broken. Having either seen or heard us, [it] was roused and running for land among broken pieces, frequently was Seen to fall into the water and again get out. As no pursuit was made, I have no doubt it got Safe ashore. Our landing place is near Point Lawford of Sr. John Franklin.[46]

**18 July 1838, Wednesday.** The Ice too close packed to allow us getting away, Sinclair went out, but saw no Clearance. A light Wind arose from WSW. Sinclair says that a River[47] falls in about two miles to Eastward of us, and at its Estuary he saw marks of an Esquimaux camp with a Pole put up as a Signal that they have gone up along the River, which falls into the ocean by two Channels, one of which was too deep for him to wade across. Therefore he could not ascertain the number of Tents there may have been. A Strong current was in both channels, which are not distant apart, formed by an Island. The Indians were out also and brought in the most part of two Deers they Killed, one a buck in pretty good condition, the other, a Small one, is Lean. They saw several tracks of Esquimaux of a recent date. The tide seems to be very irregular. The highest rise was 7 inches at 3 P.M.

Our usual tormentors very numerous. The country in the interior appears to become more Rocky and barren as we advance. No Wood to be Seen growing, and but few creeping willows. Patches of Snow appear in many places where it has been collected by drift.

**19 July 1838, Thursday.** Had rain all this morning; the Wind from Westward. The Ice began to Clear a little about Midday, but rain continued for Some time. At 7 P.M. the Ice was removed from the land ahead of us but driven Eastward with the Wind. We took advantage of the opportunity and after getting Clear, hoisted our Sails for about 10 miles, making for an Island, where the inner Channel was Closed with Ice. We advanced about two miles outside a flow of Ice which afforded no passage to the Island. Our Efforts to push through small openings were vain, as the Ice was following and Closing upon us. Our only alternative was to pull back against a Stiff breeze and among floating masses to push our Way as we could to get to Shore in the Bay, which we at last Effected after a laborious Struggle. Got to a landing Place at Midnight, where the Boats were unloaded and drawn up. A thick fog had arisen after it ceased raining. Wind turned to SW. A Small stream flows in this Bay which is about 5 miles from the Western to Eastern Point. The latter has 2 Islands of Considerable height off of it. The Channel of inside one about ¼ mile broad.

**20 July 1838, Friday.** The wind Continued part of the day. From the Summit of the Rocks rising about 300 ft. a little distance from Shore, the Channel of the Island was Seen to be free of Ice. An opening along the Shore of the Bay allowed us to push off at 7 P.M. Opposite the Island we saw vestiges of an Esquimaux Camp, and landing to examine it, found it had consisted of five or six families, by the fireplaces, 6 Winter Sledges, several of their Stone Kettles, dishes, Snowshovels and other arti-

46 In fact Franklin named a group of offshore islands the Lawford's Islands, after Vice-Admiral Lawford (Franklin, *Narrative of a Journey*, 365).

47 Probably the Kugaryuak River.

cles deposited under their Sledges, with Stones to Keep them down.[48] This Camp seems to have been left for the Interior about a fortnight back, as near as we could judge.

We proceeded for three hours following along Shore, the Ice being in no part more than a mile out, and in many places leaving but Space to push through, and Shoving other pieces off Shore with our Poles, by uniting both Crews in that labour. We were again Stopped after rounding a rocky point westward of which a Small River runs in a narrow bite. After landing, following this Stream for about a mile from the Coast, I found it rushing through a chasm of high Rocks composed of Red porphoritic Granite & forming a beautiful fall of about 100 feet descent in different Cascades. The weather Cleared up & became fine and warm. A distant Cape to the NE in view. The Islands more distant out at Sea than hitherto. Had a distant Roll of Thunder not very loud.

**21 July 1838, Saturday.** A Space along Shore having Cleared we embarked at Midday and with Similar labour to that of yesterday in many places, got at the end of 7 hours to the Cape seen yesterday, which forms a high Projecting rocky Point, where from all appearances the Ice has not moved in the least yet this Season.[49] A narrow gorge near the point afforded us a good landing place, and here we found on both sides of this narrow pass (being about 200 yds. across the Point) deposits of Esquimaux, placed upon ledges of high Rocks, out of Reach of any animals, having used their Sledges as ladders to take up their property, that on the West Side being about 40 ft from the Base of the Rocks & covered over with Skins &tc. Kept down by weights of Stone. The one on the E Side not quite so high but Secured in a Similar manner. Some quantity of Seal Blubber is among the Store, as the oil running down the Rocks in considerable quantity, has by the Heat of the Sun been thickened upon the Rocks to the consistence of a Gummy Substance. The Camp, which seems to have consisted of 4 Small Tents, was on the East Side: pieces of Deer Horns, Stone vessels, Musk Buffalo horns lying about, and Some of their Sledges covering their Shovels, Stone dishes and Kettles. It is apparently Some time since they left this Spot, but in which direction we cannot ascertain. The wind from W of South most of the day. The Ice adhering to the Point very thick.

**22 July, 1838, Sunday.** The wind from SW brought on a dense fog at 5 A.M. which continued throughout the day. Geo. Sinclair went about 9 miles along the Coast but could see no distance ahead. Says he Killed a Seal at the Entrance of a Creek but it sunk to the bottom. At 5 P.M. the wind Chopped round to NE; fog continuing. Had a distant roll of thunder. Some of the men whose curiosity was excited to Know the contents of the Deposit,[50] were much Surprised to find in it a good Cotton Shirt and some Beads (apparently some of the articles left at Copper Mine River in 1826 by Doctor Richardson), many articles of their own Cloathing, Shoes, Skins & Some

48 Here, too, there were wood chips from Richardson's boats (Simpson, *Narrative*, 272).

49 The temperature reached 71°F in the shade and 88°F in the sun, the hottest day of the summer (Simpson, *Narrative*, 273).

50 The Inuit's cache.

Blubber in Sealskin Bags. All was carefully replaced and covered again, as it required I think a Strong Stomach to resist the Smell of the Blubber in the State it then was.

**23 July 1838, Monday.** The day turned fine but no movement of the Ice. We are therefore compelled to remain untill some favourable change opens a passage for us. In searching among the Rocks and low ravines among them Could not find any Specimens of Flora different from those already collected, and very few even of the Same. The country around appears to be very barren, and wood for fuel not abundant at this Spot.

**24 July 1838, Tuesday.** The Indians went out this morning. The day fine. Musquitoes pretty numerous. At 1 P.M. the Ice having considerably melted and moved a little, afforded a passage to the boats round the Point, light, by pushing through some narrow openings and raising them as much as the Crews could. The Baggage was Carried across the Point from whence about 2 miles of a very Steep Rocky coast was got over with great Risk of Crushing the Boats while the Ice was moving. The pressure was so great at one time that the Castor was injured. One of its planks Split, fortunately above Water Mark. A Small rivulet about 2½ miles from the Point, where the Ice was yet ashore outside of a point near us, ended our advance for the day. There being a Space free of Ice a nett was put down near the mouth of the Rivulet. The Indians only came in at midnight with a Small Deer Killed by LaRocque. They have seen many tracks of Esquimaux out from the coast when hunting, but not of a recent date. The Wild Sorel[51] is found about here, and the Dandelion[52] in full flower upon the Sandy Shore of the Small bend that the Rivulet falls into. Some hurtleberry bushes[53] are in Blossom, the earlier blossoms beginning to fall off.

**25 July 1838, Wednesday.** The refraction Shews a Chain of Islands out about 12 or 15 miles, Stretching towards Cape Barrow. But one Small Arctic Salmon was taken in the nett. At 8 A.M. we pushed off, but were Stopped at the Point about 3 hours when, assisted by the flow of the tide moving the broken Ice we advanced by pushing off some Pieces from Shore, with the Exertions of all hands, Cutting through others with hatchets where they rested on Small points of Rocks. In this manner we proceeded about 5 miles to the Second Main Point from the Cape we left yesterday.[54] Several Seals were seen. The Country appears very rocky and Barren. In low, wet places only does any vegetation appear in the Shape of mosses, grass & a few dwarf willows.

**26 July 1838, Thursday.** The Sun sett last night at 10.h 17'. At 6.15 this morning we pushed off and after much labour in some places, getting through the broken Ice, got a clear run where the Ice is removed about a mile from the Coast. Put ashore on an Island Eastward of Port Epworth[55] of Sr. J. Franklin to take breakfast, Where some drift wood Split by the Esquimaux was piled up for future use it seems. Con-

---

51 Arctic sorrel (*Oxyria digyna*), an important antiscorbutic.

52 Arctic dandelion (*Taraxacum lapponicum*).

53 Blueberry (*Vaccinium uliginosum*).

54 They camped near the mouth of the Tree River (Simpson, *Narrative*, 274).

55 The deep embayment at the mouth of the Tree River.

tinued on; had light Sail Wind part of the day. At a point of Rocks on the w side of Gray's Bay[56] saw another deposit of Esquimaux property put up as those already mentioned, but did not Stop to Examine it. The Ice in considerable quantity in the Bay, but in one part sufficiently broken to allow us a passage through. We landed at 8h.15' P.M. in the mouth of a Small River, nearly facing Hepburn's Islands,[57] about 4 miles from the Eastern point of the Bay, the land not so high or rocky as to the Westward. The Soil Sandy about the mouth of the River but Clayey bottom about a mile from it with high Banks, and Hills resembling those above the Bloody Fall in Copper Mine River. Some large Brown ducks[58] were Killed, and Seals very numerous in the Bay. Two were Shot in course of the day, but sank before they could be got hold of. Several Wolves prowling about. One of them with her Whelps was chased some time, but Could not get hold of any of them.[59] A nett was put down in the River which is about 120 yards wide, and pretty deep above a Sand Bar at its estuary on the Coast. This Stream we take to be Wentzel River of Sr. J. Franklin,[60] although not exactly corresponding to the description.[61] Some Small Stinted firs grow on hills about a mile from the Coast, Eastward of the River's mouth, about 150 ft high.

**27 July 1838, Friday.** Our nett was taken up without having taken a fish. We left our encampment at 10 A.M. and found the Ice closing in Shore as we proceeded. Between 3 and 4 miles from our Encampment we passed the Channel of another River, which by reference to the Chart and description is the one named Wentzel River. We had to follow so close in Shore that no real distance was made although we continued untill 8 P.M. toiling round the deep and narrow Bays and inlets that are frequent along this part of the Coast. Became again more Rocky as we approached Cape Barrow. The Rocks near our Encampment, which is Eastward of Inman Harbour,[62] consist of a Red Granite with many veins of Quartz intersecting in all directions. The beach principally formed of debris of those rocks broken up by the action of the Surf & Cold and laying to a Considerable depth in every little hollow. Two specimens of plants not yet Seen this Season were found.

---

56 Named by Franklin after Mr Gray, principal of Belfast Academy, on 24 July 1821 (Franklin, *Narrative of a Journey*, 366).

57 Named by Frankin after John Hepburn, one of the seamen on his first expedition (Franklin, *Narrative of a Journey*, 366). In fact there is only one island, although a low, narrow waist could make it appear like two.

58 Probably female eiders, either common eiders (*Somateria mollissima*) or king eiders (*Somateria spectabilis*).

59 But one male wolf was shot by McKay, according to Simpson (*Narrative*, 276).

60 Named by Franklin after Willard Wentzel, a clerk with the North West Company seconded to Franklin's first expedition (Franklin, *Narrative of a Journey*, 366) and father or grandfather of Ferdinand. But in fact this was not the Wentzel River; Dease and Simpson discovered their mistake next day. It was most likely a fairly major, unnamed river which reaches the sea some eight kilometres southwest of the mouth of the Wentzel.

61 The party's only thermometer was accidentally broken here (Simpson, *Narrative*, 278).

62 Named by Franklin on 25 July 1821 after Professor Inman of the Royal Naval College at Portsmouth (Franklin, *Narrative of a Journey*, 368).

**28 July 1838, Saturday.** It was 10 A.M. before we could proceed and laboured hard to get round Some points where the Ice was Closed on the Rocks. The Boats were much pressed in some places. The distance performed not great.
Put ashore in a Small Enclosed harbour. The Ice driving on the point Close to us rendered any attempt at rounding it dangerous. The Wind from NE. A Seal was killed and lost again.

• • •

## *2. Extract from letter from Murdoch McPherson to the governor, chief factors, and chief traders of the Northern Department, 28 July 1838*[63]

Portage la Loche, 28th July 1838

Gentlemen,
... 2. The Winter Express reached McKenzie's River this year unusually early (2nd April – the earliest it ever reached us before being the 15th) which enabled me to forward the Instruments, Books, etc. that came forward by that conveyance for the Expedition by land to Ft. Confidence where all was safely rendered 9th May and by the return of the bearers to Fort Simpson 18th June I was happy to learn that the operations of the Expedition were going on as successfully as could be desired ...

M.McPherson

• • •

## *3. Peter Dease's journal, continued*[64]

**29 July 1838, Sunday.** Left our Encampment at 10 A.M., and at 12 put ashore for Mr. Simpson to get an observation, which places the position of it 2 Miles South of Cape Barrow.[65] After leaving the Place some time a Deer was Seen & Killed on the Beach. Put up in a Bay that answers the Description of Detention Harbour.[66] A Swan[67] was killed of two that were seen in a Small Lake on an Island. A Shell of the Sea Urchin was found on a Rock about 40 ft from the Water. Many were afterwards found, also the Common Clamshells, muscles and quantities of Seawrack in every Bay where a low beach is to be found. More or less wood drifted upon the Coast, but very few trees of a large growth, appearing more like the Stinted wood of the Copper Mine, than the large trees of the McKenzie's. The twisted fibres of the wood in general tend to confirm me in that opinion. After having doubled Cape Barrow, the land does not appear so high as at Barrow nor the Coast so abrupt and

63 HBCA B.200/b/11, f. 1.
64 MCM M214.
65 Named by Franklin after Sir John Barrow, second secretary to the Admiralty, on 25 July 1821 (Franklin, *Narrative of a Journey*, 367).
66 Named by Franklin for the detention of his expedition here by ice from 26 to 29 July 1821, even although they portaged across a headland to try to make some progress (Franklin, *Narrative of a Journey*, 368–70)
67 Whistling swan (*Olor colombianus*).

Steep; the Islands also Show more vegetation as we advance in Bathurst's Inlet,[68] as the Ice is too compact to admit of any attempt to cross the Inlet among the Islands.
**30 July 1838, Monday.** Left our Encampment at 5 A.M.,[69] but a thick fog coming on obliged us to Keep along the main land[70] untill near Midday when it Cleared off; the weather Calm. The Oar was plied untill we got to Moore's and Marcet's Islands. On the latter[71] we landed and from its Summit the Ice appeared quite Solid and firm out to Seaward. A breeze Sprung up from ENE, which enabled us to hoist Sails to Cross Arctic Sound, some broken Ice floating in a long Stream but not to Stop us. Landed at Point Wollaston[72] for the night with a view to Cross over to Barry's Islands[73] Should a passage Clear in the morning, which will save a great distance, Should we have to go round the whole deep Inlet. The Indian LaRocque went out with his Gun and after two hours absence, came to inform us he had Killed a large buck Deer, but could not Cut it up having forgot to take his Knife. His Companion went with him to bring it & they returned with two instead of one; the other Shot one also (fine fat animals).
**31 July 1838, Tuesday.** In course of the night the Ice cleared and broke up Sufficiently to afford a passage to the nearest Island of which we took advantage at 6 A.M. and crossed to its northern point, where any further progress was compleatly obstructed by Ice. At Same time the wind from Eastward driving the broken Ice upon the point and filling up the Channel to westward of us, rendered it necessary to put ashore for the remainder of the day. Some drift wood was collected which from its appearance resembled in its twist & Stinted growth the wood we saw along the Copper Mine River. The upper Strata of this and most of the Surrounding Islands appear to be of the Trap formation,[74] with boulders of Granite along the water edge. Some of them appear to have a Granite foundation, covered with the Trap Rock. Shells as previously mentioned abundant. A large Sea Snail Shell was found to day. Nothing in the vegetation different from what occurs all along hitherto. From the Summit of the Rocks which rise to about 200 ft., the land on the Eastern Side the Inlet is plainly visible.
**1 August 1838, Wednesday.** The wind continued from the Same Quarter. The Ice driving westward, an opening appears to be forming on the East Side of the Inlet which affords us hopes of getting off tomorrow. Some Rein Deer were seen, but at a distance. Jas. McKay reports that this Island is not so long as we supposed it to be. A Channel that he saw to the South, He thinks if the Ice continues moving in the

68 Named by Franklin after Henry Bathurst, third Earl Bathurst, the secretary of state, on 10 August, 1821 (Franklin, *Narrative of a Journey*, 379).

69 Some new ice had formed during the night (Simpson, *Narrative*, 278).

70 Around Daniel Moore Bay (Simpson, *Narrative*, 279), named by Franklin after Mr Daniel Moore of Lincoln's Inn (Franklin, *Narrative of a Journey*, 371).

71 One of the Stockport Islands.

72 The tip of the long peninsula separating Arctic Sound from Bathurst Inlet *sensu stricto*.

73 Named by Franklin after Colonel Barry in August 1826 (Franklin, *Narrative of a Journey*, 394).

74 Dease's geological assessment is entirely correct; the islands in the northern part of Bathurst Inlet consist almost entirely of lava flows of the Akalulia Formation of Helikian age, consisting of dark grey to purplish basalts (Campbell, "Geology of the Helikian Rocks of the Bathurst Inlet Area, Coronation Gulf, Northwest Territories"). Some of them also display extensive diabase sills of Hadrynian age (Fraser, "Geological Notes on Northeastern District of Mackenzie, Northwest Territories").

same direction, will afford us a passage tomorrow. The Indians have gone out to look for Deer and did not come back. Geo. Sinclair also has gone out and not returned.

**2 August 1838, Thursday.** The Indians last night brought in a Deer they Killed, a large Buck. George Sinclair also Killed one and brought part of it. The Remainder was taken up after leaving our Encampment at Midday. Passing by the West Side the Island, after 4 hours' pull at the Oars and light Sailing we got to a narrow Inlet. At the Bottom of it over a narrow isthmus the Boats & Cargoes were carried. We reloaded & proceeded to another Island Eastwd. of us,[75] where we found the Ice packed on Shore. After pushing through about 300 yards distance of heavy Ice, we could advance no farther and landed. This Island had high Columnar Rocks of Trap on the N Side upon a Granite base. The upper Strata appears as pillars rising to the Summit and resembling the formation of the Copper Mountains below Kendal's River. As far as I can judge, Cliffs about 160 to 200 ft. high.

**3 August 1838, Friday.** This morning Mr Simpson, on looking about Some Debris or Masses of Rock that have been detached and fallen from the Cliffs found several pieces of virgin Copper,[76] and as the formation of most of the Islands appears the Same we have every reason to Suppose that great Quantities of Copper ore is contained in them. In Examining some of the Rocks of a dull Red Colour, some handsome veins of Quartz appear, in other parts Small, round & Bladder Shaped deposits, some of which about 6 inches diameter, were hollow inside a Shell of ⅔ of an Inch thick of the fine Clear Quartz. Sinclair Killed a fine Buck Deer.[77] McKay reports having seen a Channel South not very distant, by which he thinks we may get to the E Side the Inlet.

**4 August 1838, Saturday.** At 5 A.M. we pushed off to go round the South End of the Island; weather Calm and fine. Found a passage and although we found drift Ice in the Eastern Channel it did not Stop, although it retarded, our progress. Passing along an Island a Deer was Seen near the summit of a high Cliff on a Bank of Drift Snow. The Indian Macacoose fired at it, the distance and height such that the men laughed at him, when the animal made a few Springs to ascend the Summit. Our surprise was great when it suddenly came rolling down the Bank of Snow quite dead. It was taken in the Boat, being a Small one & the Ball found to have gone through it near the heart. The distance must have been near 250 yards. At 2 P.M. we landed on Point Everitt,[78] eastern Side the Inlet, and pleased in having at last crossed it with the hopes of finding a more open sea as we advance Eastward, we Sat down to a hearty breakfast, which was the more Welcome from the difficulty found in Collecting a Sufficiency of wood to Cook it. After 1½ hour we proceeded & found the Ice Still So close as to oblige us to follow the main land in most places. We Crossed over to an Island off Melville Sound and put ashore at 8.30 on an low point with a sandy beach. No part of this Island rises to any great elevation. Heard a distant Roll

75 Probably Eklalulia Island.

76 This is confirmed by Campbell ("Geology," 101).

77 On his way back to camp with the meat, Sinclair was followed by a female barren grounds grizzly with two cubs; they ran away when he stopped (Simpson, *Narrative*, 281).

78 At the west entrance to Baychimo Harbour, the site of the present settlement of Umingmaktok.

of thunder. The weather became Cloudy in the evening but no rain.

**5 August 1838, Sunday.** A Small Seal was Killed this morning by George Sinclair, which was got from the bottom after some search. Our advance was much impeded by Ice. At 4 P.M. after pushing through much broken Ice, we got the Cargoes Carried about 400 Yds. to get round a Point[79] with the boats lightened, after which untill 8 P.M. we had a pretty Clear run. We again put up on an Island about half way across the entrance of Melville Sound;[80] weather fine and Calm.

**6 August 1838, Monday.** The weather misty this morning. Crossed a Bay where new formed Ice was so thick as to cut through near ⅓ thickness of the Planks in our Boats' bows. Crossed to Cape Croker,[81] where the Ice was packed too close to allow us any chance. The boats were unloaded as the Ice was pressing on Shore. About 3 hours after, a narrow opening to a point near us formed. The Boats reloaded to proceed, when after getting into the Channel that has formed, the Ice closed upon us again, and with Some difficulty we got through to land again for the remainder of the day. The mist turned to rain & continued most of the day.

**7 August 1838, Tuesday.** This morning an opening formed by which we got round the Point. Passed the mouth of Walker Bay and in Coasting Round Riley Bay were again Stopped at a Small Point of Rocks which protects the Boats from the Ice. This Point forms a narrow Peninsula seperated from Walker's Bay only by narrow Sand bank near the bottom of the latter.

**8 August 1838, Wednesday.** Had some Rain to day; wind N by E. The Ice closed so round us that we can neither advance or retreat from our present Station. Could we get back to Walker's Bay which is narrow at the entrance, we might, by making a Portage of half a mile, get into the bottom of this one, where a Space appears Clear of Ice running across to the next point South of Porden Islands. 3 Deers were seen yesterday by Mr. Simpson while going towards the head of the Bay, but out of gunshot distance.

**9 August 1838, Thursday.** Wind freshened in course of the night with Showers of Rain. The Ice seperated in the Bay and at 7 A.M. we pushed off, hoisted sail with the Wind rather near, Sailed passed the S. End of Porden Islands, where the wind being too close, with the Oars got to Cape Flinders,[82] & after doubling it were again Stopped by Ice, at Pt. Turnagain.[83]

---

79 Cape Croker (Simpson, *Narrative*, 282); this is the southern entrance cape to Melville Sound.

80 Named by Franklin after Lord Melville, First Lord of the Admiralty (Franklin, *Narrative of a Journey*, 383). Here two sandhill cranes were stalking about; many female eiders flew over; and many old-squaws were moulting in the shallows (Simpson, *Narrative*, 283).

81 Dease appears to be in error here; this point must have been the west entrance cape to Melville Sound (unnamed) north of the Cockburn Islands.

82 The western tip of Kent Peninsula, named by Franklin on 16 August 1821 after his uncle, Captain Matthew Flinders, with whom he had sailed on his expedition in Australian waters in 1801–3 (Franklin, *Narrative of a Journey*, 386); the party was now leaving Bathurst Inlet.

83 Named by Franklin as being the point from which he was forced to turn back on 22 August 1821 (Franklin, *Narrative of a Journey*, 393). However, this point lay ten to twelve miles north of Franklin's last camp, and was reached by Richardson, Back, and Franklin on foot (Franklin, *Narrative of a Journey*, 387; Davis, *Sir John Franklin's Journals and Correspondence*, 166).

**[9 August 1838]** or near it, as by an Observation taken by Mr. Simpson he finds that our present position differs but about 2 miles from the last Encampment of Sr John Franklin in 1821. Two Swans were Killed to day. The bay near us appears to be a Sandy beach and very Shallow. The Ice driving on the Point has nearly enclosed us again. —

**10 August 1838, Friday.** The wind continues from the Same direction with increased force; the Ice packed in the Bay so that we cannot proceed. The Indians Killed two Laughing Geese to day. Saw no Deer but the tracks are numerous about not long Since. The land as far as we can See to Eastward is low: but few hills and those do not rise very high. Off the Coast some distance, but two Islands outside, about 5 miles distance to NE.[84] of us and 2 Small rocky Islets near our present Station.

**11 August 1838, Saturday.** The Wind Continued with Showers of rain occasionally through the day. Geo. Sinclair was out & came back at a late hour. He reports the Ice to be packed on Shore as far as he went, which must have been from 12 to 15 miles judging by the time of his absence, about 14 hours. Fired, he Says, 9 Shots at Deers without Killing any.

**12 August, 1838, Sunday.** Cold Wind and Showers of Rain now & then throughout the day. No change in the Ice. Wood getting Scarce for fuel. The Indians have been out but met with no success. Wind SW part of the day.

**13 August 1838, Monday.** Weather not so Cold or Cloudy. No change in the Ice. Rained a little in the night. The Indians each Killed & brought in a Deer. The rise and fall of the Tide is very little and irregular.

**14 August 1838, Tuesday.** Had Snow during part of the day. Wind continues from the same quarter. Weather too Cold to be pleasant. The Ice has been driven on Shore as far as we can See to Eastward. Alternate Showers of Rain & Snow Afternoon. Have to Collect fuel from a distance of three or 4 miles, therefore can Keep no regular fire. Only for cooking is fire made morning and Evening.

**15 August 1838, Wednesday.** Weather continues as Yesterday. LaRocque nevertheless went out to hunt and brought a Small Deer, the only one he Saw. The Wet does not allow taking Exercise and too Cold to remain in the Tents without additional Cloathes or Covering. A Seal Came very near the boats last night, the first we have seen about the Place. We are now entirely enclosed by Ice on all Sides. Wind Changed to the NW.

**16 August 1838, Thursday.** No change in the Weather, nor any appearance of the Ice clearing or moving.[85] A Swan was Killed by one of the Indians.

**17–18–19 August 1838.** We are Still surrounded by Ice. Wind Continues from the same quarter with very little change. Yesterday George Sinclair with William McDonald went out, and report having gone as far along the Coast as where a large Island appears Some distance out from land. Suppose Sir Humphrey Davies Island. The Ice too much packed in Shore to afford any prospect of getting from this for Some time. Under these Circumstances, as the Season is drawing on when we may Expect Cold

84 This must be a mistake for NW.

85 At 1 a.m. Simpson saw the first stars of the autumn (Simpson, *Narrative*, 288).

Nights, Mr Simpson and I think that our best plan will be to Employ the remainder of this month in gaining Some Knowledge of the Coast Eastward by a pedestrian voyage, which will prove at least of Some advantage towards ascertaining the trending beyond the Island mentioned, and to continue as far as may be practicable for the return at the end of the month.[86]

**19 August 1838, Sunday.** Should Wind & Weather produce a removal of the Ice so as to allow one of the Boats to follow or overtake them with the men remaining, the other would await their return at or near our present Station, the return not to be delayed beyond the last day of the month. On proposing the plan to our Steersmen both agreed to its being the best way of advancing the object of the Expedition that remains for us now, Seeing the uncertainty of how long we may yet be Kept at this place. Five men with the two Indians we therefore named to accompany Mr Simpson, and every thing prepared for their departure in the morning. One of the Canvas Canoes with its mounting or timbers to be taken, with Provisions for 10 days.

**20 August 1838, Monday.** This Morning Mr Simpson with George Sinclair, Geo. Flett, Wm. McDonald, John McKey, Laurent Cartier & the 2 Indians took their departure, assisted for the day by Pierre Morin, Frans. Felix,[87] who will come back tomorrow. The Day Mild but rather Wet fog in the morning. In the afternoon a Shower of Rain came on; the Evening became rather Cold. Wind as for some days past.

**21 August 1838, Tuesday.** At 4 P.M. the men came back, delivered a note from Mr Simpson Stating their days travel to have been rather more than 20 miles,[88] the Coast keeping a NE trending. A large Island appears out Some distance which they did not see the end of.[89] Several Deers were Seen by them but none Killed. The day has been fine; Wind NW as usual.

**22 August 1838, Wednesday.** Had a frost over night which formed Ice over a Small bite of the harbour we are in, but dissolved in course of the day. The Wind turned to SW in the morning but afternoon returned to NW. The Ice yet quite Steady: no move nor does it appear to dissolve at all. The Ebb and flow of the tide seems to occur but once in 24 hours. The former at lowest 8 A.M.; the latter 7 P.M. rises 10½ Ins. we Cannot in our present position form any Idea of the Sett of the Current.

**23 August 1838, Thursday.** The tide at lowest from 6 to 8h.30' A.M. during which time it was nearly Stationary. The flood at highest 7.30; rise much the Same as Yesterday. The night not So Cold as the preceding, but Ice was formed again. Wind light from NW to SW.

**24 August 1838, Friday.** The night was mild: White Clouds, light SW wind. The men go a great distance for Wood, and not much to be got now. The tide much the Same as Yesterday.

---

86 According to Simpson's version of events, it was he who proposed this overland foray eastwards (Simpson, *Narrative*, 289).

87 "And Hector Morrison" (Dease's original note).

88 Midway during this first day, Simpson and party had passed the extreme point reached by Franklin in 1821 (Simpson, *Narrative*, 292).

89 This could only have been part of Victoria Island.

**25 August 1838, Saturday.** Very little wind but from the Same quarter as yesterday. Clouds rising. Afternoon appearance of Rain.

**26 August 1838, Sunday.** Had a Shower of Rain in the night. The Weather Cloudy & mild all day. The Wind Sett round more to Westward afternoon but light. No change in the Ice about us.

**27 August 1838, Monday.** Cloudy & misty Weather all day with drizzling rain. The wind from SW to W in the afternoon. Saw some Geese and bustards flying Southwd.

**28 August 1838, Tuesday.** The Wind turned to the northward. A Clearance appears to be forming out near the Islands, but no move of the Ice along the land. Had a Shower of Sleet & Rain. The day Cloudy and Cold.

**29 August 1838, Wednesday.** Wind & Weather much the Same as Yesterday. Ice cleared in the offing with an opening approaching to Shore at this Point, but much heavy Ice remains about us yet. In the Evening we had the pleasure of greeting the Safe return of Mr Simpson and party, who proceeded to the 26th and reached to 104 ½ W Long., where a Deep Bay intervening, which would have taken Some time to walk round, made him determine upon returning from a Cape named Cape Alexander[90] which forms the Western Point of the Bay. A Continuation of the Outward Islands or Land was Seen out to Northward as far as they went,[91] but something like a point or termination to it Eastward, Mr Simpson thinks he saw from the Summit of Cape Alexander. The distance eastward from here he estimates upwards of 100 miles. The Land low with few Hills or Mountains near the Coast, and none but small rivulets which they Could Easily wade over were found in that distance. They brought in part of a fine fat Buck Deer Killed by Sinclair yesterday, affording us an agreeable Supper. John McKey and Geo. Flett have Suffered considerably from Swelled legs & feet from the frequent walking in Water which is at all times Cold along the Coast. Geese in flocks flying Southward.

**30 August 1838, Thursday.** A Strong NW Wind. The Ice driving upon the Shore prevented us from leaving this tedious encampment. Weather Cold & Cloudy. There is a Clear offing between us and the Islands but the Stream of Ice inshore yet too much to get through.

**31 August 1838, Friday.** At 10 A.M. there remained but a narrow Space of broken Ice upon the Point of Rocks forming our Harbour. This we pushed our Boats through after some labour, and after Clearing the harbour had to push through a considerable quantity of floating Ice with a heavy Swell, but this also was Cleared, and the Wind being favourable, we hoisted Sail on our return, taking a Straight Course Among the Islands[92] for Cape Barrow. Snow Showers at a distance appeared in course of the day. The Wind Shifting, we had to pass by the South of a large group of Islands[93] and finally, towards evening, to cross a Broad Channel between Islands, pulling at the Oar,

90 Named by Simpson after his brother.

91 This was named Victoria Land by Simpson (*Narrative*, 297), now Victoria Island. He named the most easterly visible point Cape Pelly.

92 Through the Wilmot Islands.

93 Chapman Islands.

and put up for the night on an Island a little north East of the one where a Swan was Killed on the 29th Ulto, in sight of Cape Barrow about 9 miles distant.

**1 September 1838, Saturday.** Wind being rather Strong and unfavourable, we left our Encampment at 9 A.M. and doubled Cape Barrow, where the Sails were hoisted and took us at Sunsett to Wentzel River, where a Buck Deer was Seen near the Beach and Killed. Put up for the night. Many Seals were Seen in course of the day. No Ice was seen to Westward of our last night's Encampment, but a considerable Quantity among the Islands in Crossing the Inlet.[94] The Wind fresh this Evening, NE.

**2 September 1838, Sunday.** Left our Encampment at 3h A.M.30' with a light Easterly wind which freshened to a Strong breeze. Put ashore on the Island where we took breakfast on the 26th July for the same purpose, and after continued with a Strong gale and Heavy Swells, and put up after dark on a Point nearly South of Lawford Islands. Had some light Snow Showers in course of the day but the Weather was not Cold. The Esquimaux have not yet returned apparently to their Deposit at Cape [name omitted] as it was Seen as we passed, and another on an Island Westward of our Encampment of 19/20th July.

**3 September 1838, Monday.** A fine Calm day. Had some light Showers of Snow in the morning but the Weather Cleared up again about 10 A.M. Had a chase after a Deer that was crossing a Channel between 2 Islands but Could not get near Enough to fire with any Effect. Continued with the Oar and at 6.30 landed at our Old Station in the Copper Mine River,[95] where the Esquimaux have come since our departure and taken away all their property but the Sledge, which remains, & have put up two marks pointing that they have gone out to Seaward. I am pleased to see that these poor people have not lost their property, as I feared they would not venture to return for it. Many Seals were Sporting near the mouth of the River and Deers were seen in Several places along the Coast in Course of the day. Whether the River mentioned as flowing from the Southward & falling in about 12 miles hence is the Tree river of Sr. John Franklin I cannot say,[96] but it is the same the Indians mentioned in July as having Wood growing along its banks, and appears at its Estuary to be a considerable Stream.

**4 September 1838, Tuesday.** This morning after having put up a present of a Kettle, hatchet, beads, Iron &c., &c. for the Esquimaux secured under the Sledge in a good deal Box, as a remuneration for any loss they may have Sustained, we embarked to proceed up the River as far as may be practicable in order to Secure our Boats from damage or Breakage by the natives.[97] The Water is considerably lower than when we descended it. On the first Island below the Fall[98] we landed and left our Masts,

94 Bathurst Inlet.

95 The party had thus covered the distance from Point Turnagain to the mouth of the Coppermine in four days, as against the twenty-five days which the outward trip had occupied.

96 It is not; it is the Napaaktoktok River.

97 Richardson had assessed the river to be unnavigable (Franklin, *Narrative of a Second Expedition*, 367) and had left his boats at the river mouth, where, as we have seen, they were broken up by the Inuit.

98 Bloody Fall, which now, with the lower water levels, was just a strong, steep rapid (Simpson, *Narrative*, 306).

yards and Spare oars. Proceeded to the Falls where in the Cleft of a Rock opposite the Rapid on the w Side, we deposited 10 Bags Pemican, the Canvas Canoes, netts, Corks & Canoe timbers, which was Covered over with Tarpaulin & Secured from Wild Animals being covered over with Turf and Stones. Some Crowberries and hurtleberries, and Cocks eyes were found about the Rocks in considerable quantities for this Barren Country. Many Bustards and White Geese[99] seen flying over, all going Southward. Put up for the night at the lower end of the Portage.

**5 September 1838, Wednesday.** The Boats were crossed over the E Side, launched over a point of Rocks and towed up the Remainder of the principal Fall whence, recrossing, they were towed up the Remainder of the Rapid with some danger in the upper Part. The Elbow Rapid, above which is very Shallow with Shoals between which the current runs very Swift, gave some labour, but the boats were got over without damage. Although the Stream in general is very Strong we got up about 9 miles from the Falls. Many deers were seen but none Killed. Put up for the night Some distance below Escape Rapid. Weather Cloudy & foggy but not Cold.[100]

**6 September 1838, Thursday.** Mr Simpson and I walked, the Indians also. At Escape Rapid the Castor in getting up a Strong rush between large Stones, Sheered, nearly upsett and drove from one Stone on another some distance, but was fortunately stopped, with but McKay & Felix in her at the time. Considerably injured & repaired, and proceeded untill Sunsett. The Indians Killed 2 Small Deers. Many were seen in course of the day, also large flocks of White Geese.

**7 September 1838, Friday.** Fine mild day. Got above the Rapid where we encamped 22nd June and put up for the night. This is the last of the Strong rapids, although many Strong ridges of rapidous Current occur in many Places. The tracking very difficult along Sandy Stone Cliffs.[101]

**8 September 1838, Saturday.** The Weather Continues fine and Warm. The tracking much better and the River not so Shallow. Fired Several Shots at Deer in Course of the day without Effect.

**9 September 1838, Sunday.** A Continuance of the Same Fine Weather. About Midday got to a bend above the Copper Mountains which affords the Shortest route direct to where Fletts' Station was on Kendal's River.[102] We therefore landed, unloaded and drew up the Boats for the Winter & put the property[103] in Security in a hollow, Covered over & loaded with Stones.

**10 September 1838, Monday.** After having secured every thing that remains, and mens' loads all settled, each having from 50 to 60 lb Wt. including Provisions to carry, we took our departure at 8 A.M. Our Course wsw for the Head of Kendal's River

---

99 Snow geese (*Chen caerulescens*).

100 This was contradicted by Simpson, who reported that it was snowing heavily and that ice an inch thick formed in the kettles, presumably overnight.

101 That night they camped among trees, admittedly quite stunted, for the first time (Simpson, *Narrative*, 308).

102 And about eight or nine kilometres below the confluence of the Kendall and the Coppermine (Simpson, *Narrative*, 309).

103 Including three bags of pemmican and two of flour (Simpson, *Narrative*, 309).

which we crossed about a mile below Flett's Station & Continued untill near Sunsett. Put up among the Hills to S & W of the Dismal Lakes. A Small fawn was Killed afternoon. Apparently in Search of its dam, came very near us while the men were resting and was Shot by Sinclair. The distance made about 20 miles. Saw great numbers of white geese feeding near Small Lakes.

**11 September 1838, Tuesday.** Continued the same general Course. Afternoon, along the Banks of the upper Parts of Kendall River[104] that falls in at the last narrows between the Dismal Lakes, found a Camp of 4 lodges (rather brush huts) of Hare Indian families with but one Elderly man who says two others are out hunting and the remainder of the Party gone to the House[105] with Provisions to get some Amunition. Have nothing to give us but a Small backfat of a Deer for which they were paid, and we Continued our route for about 3 miles farther and put up along a ravine running from the River towards the Valley through which Dease's River runs. Some Deers were seen near the Encampment. The Indians went after them but did not Kill any, although three Shots were fired at them.

**12 September 1838, Wednesday.** Travelling very bad most of the day, passing over hills covered with broken loose Stones; very fatiguing. Put up at Sunsett near a Small Lake where a barrier is made and Snares sett for deer. The fence as we could judge must Extend upwards of a mile in length, formed by Dry Wood with doors or passages where Snares are placed.[106] A Smoke appears at a distance on the Hills North Side of the River. Suppose it to be Indians out hunting. A large Signal fire was made.

**13 September 1838, Thursday.** At 1 P.M. fell upon the River.[107] Some of the party sett about making a raft to Cross over. McKay, Sinclair and others proceeded on along the banks in Search of a fording Place, which was found about a mile lower down, and of which we were apprized, after two of the Party had Crossed with the Raft made. We proceeded down to the ford & the parties rejoined after some time, firing Signals to Collect the Stragglers as we were Striking off from the River. 2 Shotts were fired not far from us, and Shortly after two Young men came running after us and informed us that they have been Sent off by Ritch on the 11th with Provisions to proceed to the Cop. Mine River should they not meet us Sooner. Two of their Companions they left at Some distance while they came on to overtake us, being Considerably out of our route. We kept them Company for about 3 miles, when seeing that it led us too much out of our Way, gave them directions to return as soon as possible to the House, and left them. A note delivered from Ritch informs us that all is well at the House. Put up for the night near a Small Lake on the N Side of which we were told an Old infirm Hare Indian (that we left in the Spring very unwell

---

104 Now Teshierpi River.

105 Fort Confidence.

106 The building of caribou fences such as these probably reached its peak of efficiency among the Gwich'in of the northern Yukon. For details see McClellan and Denniston, "Environment and Culture in the Cordillera," 377.

107 Dease River.

at the Camp) died. We therefore gave this Lake, that has 2 Islands in it, the name of Cholle's Lake[108] after him.

**14 September 1838, Friday.** Our usual rate of travelling brought us at 3 P.M. Safe to the house where all the party Excepting the Boy Frs. Boucher, are once more assembled. Thanks to the Almighty for all his mercies and the protection afforded us. Norquay is Station'd on the Island, Fishing, but with the Exception of what was dried during Summer no fish has been Collected or hung up. About 4000 lb. Dry Provs. has been stored, principally received from the different Parties of Hare Indians. Our Chipewyan Hunters have done very little in the hunting Way during the Summer. Nor has any news been received from any of the Posts of McKenzie's River since our departure in the Spring. The buildings have been covered with bark, and the Kitchen floored and the Chimney built in it. Three netts are down here that have been attended to principally by young Wentzel, who has not Conducted himself in a Satisfactory manner during our Absence.

• • •

## *4. Letter from Peter Dease and Thomas Simpson to the governor, deputy governor, and Committee of the Honourable Hudson's Bay Company, London, 15 September 1838*[109]

Fort Confidence, Great Bear Lake, Sept. 15, 1838.

Hon. Sirs,

It now becomes our duty to report the incomplete success of the expedition to the eastward, this summer, in consequence of the extraordinary duration of the ice. Much, however, has been done to prepare the way for another attempt next year; and our hopes, instead of being depressed, are elevated by the knowledge so painfully acquired this season.

On the 6th of June, our boats were conveyed on the ice to the mouth of Dease's River (then just open), the ascent of which was commenced on the following day. With some assistance from Indians we reached the portage leading to the Dismal Lakes (discovered by Mr. Simpson last winter) and carried the boats across it without accident. The ice on these lakes was still perfectly solid; and we were provided with iron-shod sledges for the passage. On these we fixed the boats; and the wind being fair, hoisted sail, which greatly aided the crews on the hauling ropes. In this manner, these frozen reservoirs, which are full thirty miles long, were passed in two days; and we reached our provision station at Kendall River on the 19th. There we had the satisfaction to find the two men (left there by Mr. Simpson in April) well; and their Hare Indian hunters successful in the chase. Two of these active fellows consented at once to accompany us along the coast, and proved not only good

---

108 Chollah Lake, according to Simpson (*Narrative*, 312).

109 Alexander Simpson, *Life and Travels*, 287–98; Dease and Simpson, "An Account of Arctic Discovery on the Northern Shore of America in the Summer of 1838."

voyageurs but, during our frequent detentions among the ice, killed so many reindeer as enabled us to save nearly half our summer's stock of provisions.

Next day (20th of June), we proceeded to Coppermine River, which was still fast. It, however, gave way on the 22nd, and we descended all its terrible rapids at full flood, while the ice was still driving.

Below the Bloody Fall the river did not clear till the 26th; and on the 1st of July we pitched our tents at the ocean. Two or three Esquimaux families were seen there; but they took the alarm and fled over the ice towards some distant islands. Here, and on various parts of the coast, a fine collection of plants was made by Mr. Dease.

We remained imprisoned in the mouth of the Coppermine River, awaiting the opening of the ice, till the 17th of July. Our subsequent progress along the coast was one incessant, we may say desperate, struggle with the same cold obdurate foe, in which the boats sustained serious damage, several planks being more than half cut through. At various points we saw *caches* of the Esquimaux placed upon the lofty rocks out of reach of beasts of prey; but we did not fall in with any of the owners, who seemed to have all gone inland to kill reindeer after their winter seal-hunt among the islands. Fragments of Dr. Richardson's mahogany boats were found widely scattered; and many articles left by his party at the Bloody Fall were carefully preserved in the native keepings.

On the 29th of July, we at length succeeded in doubling Cape Barrow. The northern part of Bathurt's Inlet was still covered with a solid sheet of ice; and, instead of being able to cross over direct to point Turnagain, we were compelled to make a circuit of one hundred and forty miles by Arctic Sound and Barry's Islands.

On the easternmost of that group, Mr. Simpson discovered, at the base of a crumbling cliff, several pieces of pure copper ore; and the adjacent islands had also the appearance of being strongly impregnated with that metal. A series of specimens of all the principal rocks along the coast was preserved.

In order to attain Cape Flinders, we had to perform a portage across an island, and several over the ice. On the 9th of August, we doubled that cape; and in a little bay three miles to the southward of Franklin's furthest encampment in 1821 our boats were finally arrested by the ice which encompassed them for twenty-two days. So different was the season of 1838 from that of 1821, when Franklin found a perfectly open sea there on the 16th of August. In June, the early part of July, and towards the middle of August, we had frequent storms accompanied by snow and frost; but during the greater part of July and the beginning of August calms prevailed, which, together with the severity of the preceding winter, we consider the cause of the tardy disruption of the ice this season.

On the 20th of August we were obliged to relinquish all hopes of advancing further with the boats. That our efforts might not, however, prove wholly fruitless, Mr. Simpson offered to conduct an exploring party on foot for ten days. It was at the same time arranged between us that, in the event of any favourable movement taking place on the ice, Mr. Dease should immediately advance with one boat. Signals were agreed upon to prevent our missing each other on the way; and should we unfortunately do so, the last day of August was fixed for the rendezvous of both parties at Boat Harbour. That unlucky spot is situated in latitude 68° 16' 25"N., long. 109° 20' 45"W. Variation of the compass 46° East.

Mr. Simpson's narrative of his journey and discoveries to the eastward is here annexed.

On the 31st of August we cut our way out of our icy harbour – the grave of one year's hopes – and, having the benefit of fair winds, crossed Bathurst's Inlet by Wilmot's Islands, and safely re-entered the Coppermine River on the 3rd of September. The following day we proceeded to the Bloody Fall and there secreted our superfluous provisions.

The ascent of the Coppermine (hitherto deemed impracticable)[110] to near the junction of Kendall River was accomplished on the fifth day. We deposited the boats in a woody bluff, where they can be conveniently repaired next spring. Then, taking our bundles on our backs, we traversed the "barren grounds" and returned to winter quarters yesterday.

We had the satisfaction to find everything in good order: the buildings rendered more comfortable, and some provisions collected. Our return so much earlier than we ourselves expected, on leaving point Turnagain, has enabled us to commence the fall fisheries in good time: and, though our stock of ammunition and other necessaries for the Indians is reduced very low, want no longer stares us in the face, as it did for several months after our arrival here last year. We are most happy to add that the natives have experienced neither famine nor sickness this season, the only death within our knowledge being that of a blind old man.

P.W. Dease.
Thomas Simpson.

### Mr. Simpson's narrative of his journey on foot to the eastward

On the 20th of August, the date appointed for the return of former expeditions from these desolate shores, I left our boats still hopelessly beset with ice, to perform a ten days' journey of discovery on foot to the eastward. My companions were five servants and two Indians. We carried a wooden-framed canvas canoe; and nearly the same other baggage as on the journey to Cape Barrow last year, with the addition of a tent for the nightly shelter of the whole party on a coast almost destitute of fuel. Each man's load at starting weighed about half a hundredweight; and our daily progress averaged twenty geographical miles.

About the middle of the first day's journey we passed the furthest point to which Sir John Franklin and his officers walked in 1821. Beyond that the coast preserved its N.N.E. trending to our encampment of the same night, situated on the pitch of a low cape, which I have named Cape Franklin.

From west to north-east a new land or crowded chain of islands of great extent, in many places high and covered with snow, stretched along at the distance, appar-

---

110 By Richardson.

ently, of thirty miles; and led to the apprehension that we were entering a deep sound or inlet. The mainland now turned off to E.N.E., which continued to be nearly its general bearing for the three following days. It is flat in its outline, our path leading alternately over soft sand, sharp stones, and swampy ground. At the distance of from one to two miles, the coast is skirted by a range of low stony hills (then partially clothed with dull verdure) which send down to the sea numberless brooks and small streams. None of the latter at that season reached above our waists, though the deep and rugged channels of many of them showed that in the spring they must be powerful torrents. Two leagues inland a hill (which I have called Mount George, after Governor Simpson) rises to the height of six hundred feet;[111] and forms a conspicuous object for a day's journey on either side. The ice all along lay immovably aground upon the shallow beach, and extended in every direction as far as the eye could reach.

The great northern land still stretched out before us, and kept alive doubts of our being engaged in exploring an immense bay, which even the increase in the tides, the quantity of sea-weed and shells, and the discovery of the remains of a large whale, and of a polar bear, could not altogether dispel. These doubts were converted into certainty as we drew near, on the fourth evening, an elevated cape, and saw land, apparently, all around. With feelings of bitter disappointment I ascended the height, from whose summit a splendid and most unlooked-for view burst suddenly upon me.

The ocean, as if transformed by enchantment, rolled its free waves beneath, and beyond the reach of vision to the eastward. Islands of various shape and size overspread its surface; and the northern land terminated in a bold and lofty cape, bearing E.N.E. at least forty miles distant, while the coast of the continent trended away to the south-east. I stood in fact on a remarkable headland at the eastern entrance of an ice-obstructed strait.

The extensive land to the northward I have called Victoria Land, in honour of our youthful sovereign, and its eastern termination Cape Pelly, after the Governor of the Honourable Company. To the promontory, whence I saw this welcome sight, I have attached the name of Cape Alexander, after an only brother who would give his right hand to be the sharer of my journeys.

The rise and fall of the tide there was about three feet, being the greatest observed by us in the Arctic seas. The coast here changes its character – the water becomes deep – the approach easy; and I have little doubt that the islands contain secure harbours for shipping.

Next morning, at the distance of eight or nine miles, we crossed another high cape composed of trap rocks.[112] The travelling had become more and more toilsome, our road, now, passing over some miles of round loose stones, and then through wet mossy tracts sown with large boulders and tangled with dwarf willows. At our usual camping hour we opened a large bay studded with islands,[113] which ran for five miles to the S.S.W., and then turned off in a bold sweep of rounded granite hills dipping

111 In fact its height is slightly over six hundred feet.

112 Trap Point.

113 Minto Islands.

to the sight in the E.S.E., at the distance of thirty miles. To walk round even the portion of the bay in view would have consumed three days – the time allotted for outgoing was already expired; and two or three of my men were severely lame from the fatigue of their bundles, the inequalities of the ground, and the constant immersion in icy cold water.

I, besides, cherished hopes that, by making the best of our way back, we might, agreeably to my arrangement with Mr. Dease, meet him bringing on one of the boats, in which case, with an open sea before us, we could still have considerably extended our discoveries before the commencement of winter.

I may here remark that we were singularly fortunate in the five days of our outward journey, the weather being so moderate and clear that I daily obtained astronomical observations; whereas, before our departure from the boats, and during our return to them, we had continual storms, with frost, snow, rain, and fog.

Close to our furthest encampment appeared the site of three Esquimaux tents of the preceding year, with a little stone fire-place apart; we passed the remains of a larger camp, and several human skeletons, near Cape Franklin; but nowhere, throughout the journey, did we find recent traces of that few and scattered people.

The morning of the 25th of August was devoted to the determination of our position and the erection of a pillar of stones on the most elevated part of the point; after which I took possession of the country, with the usual ceremony, in the name of the Honourable Company, and for the Queen of Great Britain. In the pillar I deposited a brief sketch of our proceedings for the information of whoever might find it. Its situation is in lat. 68° 43' 39"N., long. (reduced from excellent lunars at the boats) 106° 3' 00"W., the magnetic variation being 60° 38' 23"E. The compass grew sluggish and uncertain in its movements as we advanced eastward, and frequently had to be shaken before it would traverse at all.

Two miles to the southward of our encampment a rapid river of some magnitude[114] discharges itself into the bay, the shores of which seemed more broken and indented than those along which we had travelled.

Independently of Victoria Land and numerous islands, I have had the satisfaction of tracing fully one hundred miles of coast; and of seeing thirty miles further, making in all (after deducting Franklin's half day's journey already mentioned) about one hundred and twenty miles of continental discovery. This is, in itself, important, yet I value it chiefly for having disclosed an open sea to the eastward, and for suggesting a new route along the southern shores of 'Victoria Land,' by which that open sea may be attained, while the mainland (as was the case this season) is yet environed by an impenetrable barrier of ice. Whether the open sea to the eastward may lead to Ross' Pillar, or to the estuary of Back's Great Fish River, it is hard to conjecture, though the trending of the most distant land in view should rather seem to favour the latter conclusion.

The same evening, on our return, we met the ice at Trap Cape driving rapidly

114 Which he named Beaufort River.

to the eastward. As we proceeded, the shores continued inaccessible; but open water was, now, everywhere visible in the offing. At dusk, on the 29th of August (our tenth day), we regained the boats, and found them still enclosed in the ice, which the north-westerly gales seemed to have accumulated from far and near towards point Turnagain.

Thomas Simpson

• • •

# Chapter Eight
# The Second Wintering at Fort Confidence

## INTRODUCTION

As soon as the exploring party returned to Fort Confidence they set about preparations for a second wintering: setting fish nets, cutting firewood, and arranging with Indians to hunt and fish for the expedition. By comparison with the first, our information on this second wintering is much less detailed: Dease continued to make daily journal entries only until 20 September. Thereafter he confined himself to only some half-dozen entries per month and even these ceased completely as of 31 December 1838. There is no coverage at all for this time period in Dease's surviving field notebooks. Simpson's expedition narrative does cover this second wintering, but only in a fairly general fashion.

Illness and near-starvation were much more prevalent among the local Indians than in the first wintering, and the resources of Fort Confidence were stretched to the limit to relieve the situation as far as possible. To quote Simpson: "throughout the entire season, a large proportion of our purchases from one set of natives always went in charity to another."[1]

Initially this critical situation arose from the murder of two young Dogrib girls near the Coppermine River in late September, probably by Edahadelly, who on occasion had gone hunting with Simpson. Whoever the culprit was, the Indians became thoroughly terrified and initially congregated at Fort Confidence,[2] abandoning their hunting. To exacerbate the strain this placed on the station's resources, the fisheries yielded much less than in the previous winter, and the caribou had largely deserted the peninsula between the Dease and McTavish arms, having migrated south of Great Bear Lake and to the Coppermine valley.

---

1 Simpson, *Narrative*, 340.

2 Simpson, *Narrative*, 320.

To attempt to redress this situation, George Sinclair, along with Larocque and Macacoose, were sent to the head of McTavish Arm to establish a hunting outpost. On hearing this, a large proportion of the distressed Indians who had been distributed at fishing camps up and down the shores of Great Bear Lake secretly made their way there. The hunters, however, had had little success and both they and the group of Indians, who had arrived unbidden, almost starved. Dease and Simpson were forced to send a large amount of the meat intended for making pemmican for the next summer's travels as relief provisions for this group. Ultimately the only fatalities were an old woman and a newborn baby, abandoned by its mother.[3] Despite these problems, preparations for the summer's travelling went steadily ahead, and by mid-June Dease and Simpson were ready to head eastwards again.

## DOCUMENTS

### *1. Peter Dease's journal, continued*[4]

**15 September 1838, Saturday.** The young men we met came back but have Spent half a Bag of the Pemican they were Carrying for us. All hands allowed to rest after their voyage. Three Old Indians are tented about us.
**16 September 1838, Sunday.** The netts here took 24 fishes on the 14th, 28 yesterday and 38 to day. Some Young Indians have been Sent out Some days since to the Chipewyans, who are hunting towards Cape McDonnell.
**17 September 1838, Monday.** Employed 4 men cutting Wood to build a fish Store, 2 Collecting fuel, the others at different Employments. The Young men who were at the hunters Came with some other Indians, and brought 42 lb. Grease, 28 Tongues, 2 Backfats and 25 lb. Pounded meat.
**18 September 1838, Tuesday.**[5] The men Employed at different Jobs about the Place. Some of the men preparing to go and Establish a fishing Station in the Main Land at about 9 miles distance from Norquay's. Laurent Cartier, Hector Morrison and Morin to assist for some time to take 8 netts, and build a fishing House. Four other men to go with them in a Boat to assist in the building.

• • •

3 Simpson, *Narrative*, 323.
4 MCM M214.
5 A partial solar eclipse occurred. The local Indians were amazed by this phenomenon, but even more amazed that Simpson had been able to predict it and had set up his telescope in readiness (Simpson, *Narrative*, 215–16).

## 2. *Extract from letter from Thomas Simpson to George Simpson, 18 September 1838*[6]

September 18th, 1838.

It will be needless to repeat the history of our journey, which the official dispatch contains. All that has been done is the fruit of my own personal exertions, achieved under circumstances of peculiar difficulty. I speak not of the difficulties of the way – these I have never much regarded; but of those of opinions with which I have had to struggle.

My worthy senior, like Franklin and Back, was alarmed by the storms, the snow, and frost in August, circumstances which always attended the disruption of any great body of ice; and insisted that the 20th of that month, the date fixed for the return of the Government expeditions,[7] should also be that of ours. I contended that the clearest sea, and some moderate weather, will always be found in September. To the first of that month our final departure was, in consequence, protracted, and the intervening period enabled me to discover one hundred and twenty miles of the continental coast, a new northern land, many islands, and an open sea to the eastward. Had I been alone in the command, the latter circumstance, and the rapid disruption of the ice as we returned towards the boats, would have determined me to make a bold attempt to reach Ross' Pillar, or the Great Fish *this* season; but being somewhat in the situation of Sinbad in the tale, I was not altogether free to follow my own designs. That they would have proved, by this date, succesful, I have little doubt from the remarkable beauty of the weather which we still enjoy. *But my excellent senior is so much engrossed with family affairs that he is disposed to risk nothing; and is, therefore, the last man in the world for a discoverer. I write not in anger but in sorrow; I esteem Mr. Dease for his upright private character, while I cannot help regarding him and his followers as a dead weight upon the expedition.*

From Point Turnagain, by crossing Bathurst's Inlet direct to Cape Barrow, we reentered the Coppermine in four days, as I had maintained. Dease would have the time at least trebled – talked much of the heavy gales in September, and of the risk of being taken in the new ice – two things quite incompatible with each other.

At the Bloody Fall I had another contest against five weighty authorities. Franklin, Richardson, Dease, Mackay, and Sinclair concurred in pronouncing the further ascent of the river impracticable. I, however, carried my point against them all; and instead of abandoning our boats (and with them next year's voyage) to the Esquimaux, in five days we got them safely up to near the junction of Kendall River – the stream through which we entered the Coppermine in June. The opinions I expressed

6 Alexander Simpson, *Life and Travels*, 299–302.

7 Thomas Simpson is here referring to a clause in the Admiralty's official instructions to Franklin concerning his second expedition: "You are not to consider yourself authorized to risk yourself and party to the chance of being obliged to winter on the coast, but commence your return about the 15th or 20th of August to the established winter-quarters on Bear Lake" (Franklin, *Narrative of a Second Expedition*, xxiii).

regarding the Coppermine, its depth of water, the rapids, and precipices, and the probable formation of a narrow beach at their base by the subsiding of the spring flood, in every particular proved correct, *showing that ten years' experience well applied may be more valuable than that of a life-time.*

Since our return to this place, on the 14th inst., I have had the greatest trouble to keep the expedition together. Mr. Dease dwellt upon the probability of our recall – the insignificance of our remaining means – and the shattered condition of the boats; I argued that our honour was pledged in the cause – that we had provisions, if not goods, on hand – with nets to procure more; undertook, if necessary, to get our inland boat conveyed by one of the routes explored by myself last winter, to the Coppermine; reminded my senior that the sea boats were purposely left in a convenient situation for Ritch to repair them next spring; that we could get proper iron works for the purpose made at Fort Simpson; and recommended the instant dismissal of our expensive Chipewyan hunters.

Sinclair and several others applied for their discharge, and Dease again expressed his desire of relinquishing the further prosecution of the entreprise. Upon which, to bring the matter to a point, I proposed to remain and follow it up with *one* boat's crew of volunteers.

Had I acted like Parry, and others similarly circumstanced, I might have converted the erroneous notions of my senior to my own future aggrandisement; instead of which, sacrificing interest to honesty, I have been urging him on to his own advantage, reserving only the labour to myself. Would to God that I had again met with Esquimaux on my pedestrian journey, and could have obtained a *baidar* from them as last year,[8] I should then either have reached the Great Fish River, or Ross' Pillar – or have never returned to tell the tale.

• • •

### *3. Extract from letter from Thomas Simpson to Alexander Simpson, 18 September 1838*[9]

It was during ten days of that tedious interval that I performed the pedestrian journey to the eastward, of which I have already made mention. Its results are the tracing of one hundred miles of the continental coast, the seeing thirty miles farther, the discovery of an extensive "*terra incognita et nivea*" to the northward (distant about thirty miles from the main), which I had the honour of naming "Victoria Land," of an archipelago of islands, and, what I value more than all, of an open sea to the eastward. That open sea first greeted my astonished and delighted eyes on the fourth evening from the summit of a bold headland, which I have named Cape Alexander after an only brother, whom I love with an affection that the world, as it goes, knows little of.

8 The *umiak* which he borrowed to cover the final lap to Point Barrow.

9 Simpson, *Life and Travels*, 302–5.

Our journey was altogether a laborious one, owing to the roughness of the ground, the icy cold streams and swamps in which we were constantly immersed, and the severity of the weather on our return. Two or three of my men became completely lame, and one is still laid up from the fatigue then endured.

On the 29th at dusk, we rejoined Mr. Dease and our comrades at the boats. We had, on our return, found a great change in the state of the ice, which now only obstructed the shores, leaving everywhere a clear offing. Had I not been, like Sinbad the sailor, hampered with an old man on my back, I should have immediately turned eastward with both boats; but the apprehensions of my useless senior and of the crews overpowered my single voice.

We have made collections of minerals and plants; and on Barry's Islands in Bathurst's Inlet I had the good fortune to discover several pieces of virgin copper. I have three pet white wolves, which I found in a burrow at the height of land in June; then little brindled pups, but now fine stout fellows. They are remarkably docile, and as fawning as spaniels. I mean to train them to the sledge this winter; and (barring accidents) they shall have the honour of carrying me out at the close of the expedition ... Our remnant of trading articles for the Indians is very insignificant; but we have a snug little stock of provisions, and have now two or three good fall fisheries under way. Our little buildings, too, are much improved; and I hope to pass a quiet and comfortable winter, after the long and fatiguing journeys of the two last. *Luxuries* we have none, our only beverage being the Arctic tea-plant[10] infused without sugar.

For myself I am still – and I glory in it – but a clerk in their Honours' service, though I have won a distinguished place among Northern discoverers. I hope it may be as you say, that a wider field will be opened to me; though I confess I apprehend some slippery trick on the part of the concern on which my discoveries throw lustre. They cannot, however, bar the foot of the throne against me. Back, it appears, got "back" after doing nothing. My anticipations in reference to next summer are more sanguine than ever, for the knowledge so painfully acquired this season.

• • •

### *4. Letter from Thomas Simpson to a correspondent in Scotland and published in the Inverness Courier, 19 September 1838*[11]

Fort Confidence, 19th September, 1838

Our old enemy the ice has stepped cruelly in between us and the fulfilment of our hopes. I had a party of five men and two Indians on foot, and set out from this place on the 7th June. We successfully crossed the height of land with our boats, a portion of our undertaking long deemed particularly difficult and doubtful. But my

10 Labrador tea (*Ledum groenlandicum*), which produces a potable, if resinous, beverage.

11 *Nautical Magazine and Naval Chronicle* 8 (1839), 564–5.

repeated and fatiguing journeys of last winter had so thoroughly explored every route, that we were quite prepared for each obstacle that occurred. On the 30th we reached the Copper Mine, and found it still fast! It gave way a few days afterwards, and we descended all its terrific rapids, then swollen to their utmost height, along with the driving ice; grand but perilous running I assure you. Often had we to pull for life or death, to avoid the suction of the precipitous cliffs, along whose base the waters raged with overwhelming fury. The descent of Escape Rapid was the finest thing of the kind I ever witnessed. Below, the breakers made a clean breach over our little vessels; while above we were involved in a cloud of spray that dashed from an overhanging rock a hundred and fifty feet in height, and formed a magnificent rainbow around us in the bright sunshine. It was a gorgeous shower bath, however, that few would relish. On the 1st of July we encamped at the Northern Sea, which still glistened with ice as firm and impenetrable as adamant. It kept us imprisoned in the mouth of the Copper-mine till the 17th, and our subsequent progress to Franklin's point Turnagain (which was attained on the 9th of August) was one of incessant struggle with the same relentless foe, in which our poor boats got several planks more than half cut through. At Point Turnagain they were finally arrested, and remained beset for twenty-two days, so different was the season of 1838 from that of 1821, when Franklin found a perfectly clear sea on the 16th of August. The mild weather we experienced in July though very agreeable at the time, was in fact our bane; the gales which there would have cleared the ocean for us, reserved themselves till the following month, and brought down the ice from far and near upon the very part of the coast where we were so unfortunately confined. It was during ten days of that tedious interval (from the 20th to the 29th August) that I performed the pedestrian journey to the eastward; its results are the tracing of a 100 miles of the coast, the surveying 30 miles further, the discovery of an extensive snow-covered land to the north (distant about 30 miles from the main) which I have had the honour of naming "Victoria Land", besides many islands, and of an open sea to the eastward. At my furthest point we erected a lofty stone pillar, with a letter for the information of "whoever might find it". The march altogether was a most fatiguing one and the weather, on our return to the boats, very inclement. Five days carried us to the "Bloody Fall" of Hearne, as it is called; five more up the Coppermine to our boat deposit; and in a third space of five days we crossed the mountains on foot to this place, where we found everything in good order on the 14th instant. Some dry venison was stored up during our absence. We have now two dozen of nets set in the lake, and want no longer stares us in the face, as it did for several months after our arrival here last year. Luxuries we have none, our only beverage being an infusion of the arctic tea plant (a rather bitter but wholsome herb), without sugar.

My hopes of final success, instead of being depressed, are elevated by the check this year received, and the knowledge thus painfully acquired. The existence of an open sea to the eastward is no more doubtful, and should the main shore be again encircled by ice in August the southern coast of the great northern land before alluded to will afford an earlier passage into that eastern sea. Besides, I now regard September as the best month of the year for arctic navigation, and from the mouth of the Coppermine to this place is merely a hop, step, and a jump; an eight days' journey on foot. I traversed that ground over and over last winter, and again this autumn. By the way, that is the country for princely hunting; not a day but I had several chases after the rein-deer. A full grown buck, with his towering antlers, is a noble animal; in fact I now despise all meaner game. I have three tamed white wolves, too, of my own taking.

• • •

### *5. Peter Dease's journal, continued* [12]

**19 September 1838, Wednesday.** The men left for the Fishery. Some Indians Came in with 50 lb. fresh meat, 112 lb. half Dry, 8 lb. Grease and 10 Tongues. Deer is reported to be scarce this Season hitherto. Egoothi came from the Hunters & Says they are on their way here; but few deers in that direction (of Cape McDonnell). The netts that are down here take but few fishes.

**20 September 1838, Thursday.** This afternoon we had the Satisfaction of seeing a Boat coming under Sail from the Westward, which proved to be with our Supplies from the Southward in Charge of J.Bte. Fausseneuve, manned by 4 Servants & three Indians. The property according to list and Bill of lading correct and the Dispatches also, but no letter from the Governor & Council. The light Canoe was not yet arrived when the Portage Boats left Norway House. The Boat Came back from the Fishery with 820 fishes of which 560 from Norquay and 260 from Laurent's new fishery. The Hunters sent 50lb. fresh meat, 3lb. Grease, 4 tongues. They are to be here tomorrow or day after, and as our Supplies are not great it is my intention of sending them away with the Boat and take natives to hunt as they will be less expensive, although not so Efficient. But having Some dry Provisions I hope we will get through the Season without want.

• • •

### *6. Letter from Peter Dease and Thomas Simpson to Murdoch McPherson, 25 September 1838* [13]

Fort Confidence, 25th Septr. 1838

Dear Sir,

Our early return hither has enabled us to meet your boat, and to make satisfactory arrangements for another winter and voyage of discovery.

The supplies have all been rendered safe and we can only offer once more our sincere acknowledgements for your ready and kind attention to our various wants.

In consequence of a dangerous complaint one of our men, Pierre Morin, quits the service of the Expedition. In his stead we have been under the necessity of

12 MCM M214.

13 HBCA B.200/b/11, pp. 7–9.

engaging one of your people, James Hope, for Outfit 1839. Morin will perform the duty of the latter at Ft. Simpson till the departure of your March express when he will have to proceed to Athabasca to go out in the Boats of that District. We at the same time address the Athabasca gentlemen to send you back another man in the Spring. Please to let Morin have any essential articles he may require on his own account. We have the same request to make for young Wentzel on account of the Expedition from which he is dismissed.

Our Chipewyan Hunters have also received their discharge and go out in your boat. Have the goodness to afford them any facilities in your power to reach Great Slave Lake. They have a few furs which either at Ft. Norman or Ft. Simpson, should provide them with absolute necessaries and they have received from the Expedition a present among the four of 200 skins, payable in the Athabasca District.

Four of our own men, James McKay, James Hope, Charles Begg & John McKay accompany the boat and are to be employed as follows: McKay & Hope are to bring hither on the first snow with six dogs from Ft. Simpson our packet by the Athabasca Boats, including any instruments, parcels etc., some iron works now sent for [from] your blacksmith, Hope's clothes and some articles out of our men's orders which we are informed have all been rendered at Ft. Simpson. The other two men, Begg & McKay with 5 dogs are to proceed forthwith to Great Slave Lake with our dispatches to the Committee & Council. If you can still forward them by water to Ft. Resolution or even to Big Island it would be a most important advance. At Ft. Resolution they are to await the arrival of the ship letters which they will bring hither and thus save you an express.

Our next letters from hence will be forwarded to go out in your March packet and the bearers of them will bring back our letters etc. by the regular winter Express. It does not therefore appear at present that we shall have to trouble you to send hither at all this winter unless anything unexpected casts up, in which case we feel assured of your friendly offices. Indian guides, however, will be required to accompany our various envoys, whom you will also be pleased to supply with sledges and dog harness.

As for the remainder of our goods, after McKay has taken out the few things required by the people here, we do not wish them to be brought hither at all, being intended as Supplies for the Expedition at Great Slave Lake where we shall most probably have to spend the next winter. You will see from the general dispatch that we shall again require the assistance of a boat from you to take us all out to Ft. Norman next fall, but we do not wish anything to be sent in her beyond the provisions of the crew.

McKay is directed to consult with Mr. McBeath at Ft. Norman whether, on the event of his meeting a boat with the Expedition's goods coming down from Ft. Simpson, the latter (Mr. McB.) would be able to support him and his companions with six dogs till Bear Lake sets fast. If so, McKay is instructed to embark in that boat; if otherwise (and that it contains nothing of consequence to Ft. Norman) to turn it back to Ft. Simpson. Besides provisions to take the whole party thither (say 4 bags of pemmican, 600 lbs dry meat, and 300 lbs dried fish for the dogs) McKay has in charge 4 Bags of pemican to assist you in providing for our winter transport. And that Mr. McBeath may be fully acquainted with these arrangements this letter is inclosed open under cover to him to be transmitted forthwith after perusal.

With sincere regards, We remain, Dear Sir,
Yr. most obedt. hble. servts.
P.W. Dease, C.F.
Thomas Simpson

• • •

## *7. Extracts from letter from Thomas Simpson to James Hargrave, York Factory, 25 September 1838* [14]

*Private*

Fort Confidence, 25th Sept. 1838

My Dear Sir,

Your most friendly and welcome letter of 15th June came to hand here on the 20th Inst., a few days after our safe return from the coast. I am sincerely happy that you once more occupy your station at the Depot;[15] and feel truly obliged by your kind cooperation with Mr. Ross[16] in administering to our wants, which will enable us, God willing, to spend a much more comfortable winter than the last at this place of exile. Your opinion in reference to my exertions of last summer is too favorable. Those of the present season have been much greater, though attended with less complete success; for what are we poor worms when the force of the elements is arrayed against us! You will see the whole in our official dispatch; it would therefore be only wanting the few remaining hours to dilate on the subject here. We have practised such strict oeconomy that we are in a condition to perform another voyage of discovery next summer, and to pass the succeeding winter – probably at Great Slave Lake – without *any more goods*. We may safely challenge all the men in England to equal *that* ...

Back's expedition turned out much as I anticipated. My judgment was however founded more on moral than on physical considerations. Back I believe to be not only a vain, but a *bad* man; and his failure is retributive justice. I think with you that *his* adventures are closed; leaving a wide field, which cannot be fully explored by our *present* expedition. What can be done, however, we will do. No ship need ever try again except through Barrow's Straits. I cherish very strong hopes from the *open sea* discovered by me this season, to the south and east of my "Victoria Land". I shall do myself the pleasure of naming some point of my discoveries after yourself, in memory of the many happy days we spent together in the woods & marshes of York

14 Glazebrook, *The Hargrave Correspondence*, 265–7.
15 York Factory.
16 Donald Ross at Norway House.

factory; while I still looked upon the world with the heightened fancy of boyhood and held *theories*, some of which I dare say were crude enough ...

Assuring you of the unchangeable regard of
Your sincere friend
Thomas Simpson

• • •

*8. Peter Dease's journal, continued*[17]

**26 September 1838, Wednesday.** The Hunters Cast up on the 24th. Some natives also made their appearance. Received from them 346 lb. Fresh meat, 24 lb. Fat and 15 Tongues. The Chipewyan Hunters were settled with and a Present of 60 M. Brs.[18] given to each Eengheen & Nalhal – 40 each to Touzeyaze & Egoothi by a Bill on the Cos. Stores at Fort Chipewyan.[19] Pierre Morin has applied for his Discharge on account of Ill health, which was granted him.[20] James Hope one of the MKs[21] River men applied to, who agrees with C.F. McPherson's permission to be obtained to come back from Ft Simpson, with James McKay who accompanied them in order to come back with the Letters Expected by the Athabasca Boats. John McKey and Charles Begg are also to go up with this Boat and take our dispatches to Slave Lake, there to await the arrival of the Ship Letters[22] & bring them. The Boat left this in the afternoon. 9 Dogs taken by the men who are to come back.

• • •

*9. Letter from Thomas Simpson to Donald Ross, Norway House, 28 September 1838*[23]

*Private*

Fort Confidence 28th Sept. 1838

My Dear Sir,

With the utmost pleasure I now acknowledge your acceptable favors of 10th & 15th June, both received four days ago. You must, however, accept a stinted reply at present; as in that interval we have had to make all our arrangements for another northern campaign, and to write a history of that just closed, which is so full as I believe to leave little or nothing unsaid. If not entirely successful your friend has at least earned *some* fresh laurels this summer, and poor Back's disasters leave a clear field for further operations.

17 MCM M214.
18 Made Beavers: the standard unit of exchange in the fur trade.
19 The hunters were due to travel south with the boat which had just arrived from Fort Simpson.
20 "Young Wentzel also discharged" (Dease's original note).
21 Mackenzie's.
22 Mail arriving from Britain by the annual ship via York Factory.
23 BCA ADD.MS 625.

How can we sufficiently thank you for the trouble you have taken in administering to both our mental & bodily wants in this remotest corner of the earth. Such conduct proves "the friend indeed." I find plenty that are liberal enough of fine words "sed preterea nihil". You are of a different stamp.

Alas for Red River! Hurrah for the British Lion in Canada! I thought he would have to shake his mane soon. What a pity that the hyenas that led the poor beasts to the slaughter should have themselves escaped. If there be a *deepest* in the pit of perdition, it will surely be theirs.[24] The papers you have so generously deprived yourself of to send us will prove a rich fund of entertainment during the coming dark days & nights. Their Honors and his Excellency have totally overlooked us in that respect. We are *at length* certain that we live under a youthful Queen; and I have had the honor of naming a new land, discovered by me this season, after Her gracious Majesty. We have not heard a syllable of *English* news (in our letters, I mean, for the papers are yet unopened); from which we infer that all was quiet in the land of our fathers. This country budget seems much as usual – except his Xy's absence, and the smallpox among Rowand's plain troops. Here, thank God we are likely to pass a more comfortable winter than last; and after my very long and inclement snow shoe journies of the two preceding years, I may hope to taste some quiet and rest for a few months to come.

Your friend Adam McBeath and his nice little wife are our next door neighbours at Fort Norman; though at rather too great a distance to shake hands, even on New Year's morning. Where is daddy Charles going, now that he appears to be superseded by our worthy friend Hargraves, and what has the latter been about that he still stands a guarded and solitary oak, ungraced by some fair twining ivy?[25] Would I had his opportunities; not even midnight or shut toll bars should stop me. I must rub him up on the subject.

We now send for part of our men's orders to Fort Simpson or Slave Lake; and

---

24 Here Simpson is alluding to the near-synchronous but only minimally related rebellions of Lower Canada (now Quebec) and Upper Canada (now Ontario) in 1837–38. In the former case, widespread economic distress among the francophone majority and intensification of British immigration (which brought with it an outbreak of cholera) led the Patriotes, led by Louis-Joseph Papineau, to take up arms against the British authorities in 1837. Once this rebellion had been put down, twelve men were hanged and fifty-eight transported to Australia. In Upper Canada, discontent with the British authorities was particularly strong among American-born settlers who had moved to Canada prior to the War of 1812. Under the leadership of William Lyon Mackenzie, the radicals took advantage of the fact that Upper Canada had been stripped of troops to deal with the Papineau rebellion, and attempted to seize control of the government. Armed conflicts occurred in Toronto on 5 and 8 December 1837 and a few days later near Brantford. Raids across the border from the United States continued through 1838. As Simpson intimated, both leaders escaped. One outcome of these rebellions was Lord Durham's recommendation that the Canadas be merged into one colony; from this recommendation came the formation of the United Province of Canada in 1841.

25 But not for much longer. During a leave from York Factory in 1837, James Hargrave went to Britain with the specific aim of finding a wife. The outcome was that on 8 January 1840, during a second leave, he married Letitia Mactavish, daughter of Dugald Mactavish, sheriff of Argyll, and took her back to York Factory (Macleod, ed., *The Letters of Letitia Hargrave*, xxv–xxvi).

by McKay's return hope to receive further intelligence from yourself and other distant friends.

Meanwhile I always am
My Dear Sir,
Yours most faithfully
Thomas Simpson

P.S. Offer my kind regards to Mrs. Ross. Tell me how Jane gets on in the old country and whether Jessey & Robert are still with the revd. McCallum. Jones seemed on the homeward move as he writes me from the Stone Hermitage at Red River. Adieu!

• • •

## *10. Peter Dease's journal, continued*[26]

**1 October 1838, Monday.** The fisheries yielding very few fishes. A few Hare Indians brought in a little fresh meat on the 29th Ulto., also 316 lb. Dry meat, 58 tongues & 12 lb. Grease. [from] Nozanne and party. The men Employed cutting fuel & finishing the Kitchen, [and] putting up a Small building for the Fish, and a Round house. Larocque and Brother with Macacoose are employed as Ft. Hunters. The Deer not so numerous as last year about this place.

**4 October 1838, Thursday.** A Party of Indians came in with 320 lb. Fresh meat, 230 Dry'd, 84 tongues, 14 lb. Grease, 7 Fat. They were Settled with. A Party that are to the Northd. are Expected soon. No news from the fishery these some days.

**6 October 1838, Saturday.** Received 220 lb. Dry meat, 35 Tons.,[27] 24 lb. Grease. 12 fishes taken from 2 Netts put down at the Point to Eastward, attended by an Indian.

**13 October 1838, Saturday.** Nothing worthy notice has transpired the last week. Some Indians came in to day with 70 lb. Fresh meat, 84 lb. Dryd., 16 tongues & 3 lb. Grease. Report that Deer is Scarce. Many of the Indians Collecting towards the northern Lakes to Collect what fish they can.

**16 October 1838, Tuesday.** Sent a Boat to the Fisheries yesterday, from whence they brought us 1450 Sent by Norquay and 1080 by Laurent, who has Hector Morrison as Assistant, Young Boucher being with Norquay. 120 lb. Fresh meat was sent us by the Hunters. A Party of Indians in the Bay fishing. The weather has been fine and open this Season compared with last year.

**24 October 1838, Wednesday.** Nothing worthy notice has occurred during the Past Week. The Indians seem to be doing little or nothing. They are some of them about the Hunters' Tent towards the head of the South Branch of the River.[28]

**31 October 1838, Wednesday.** Two young lads were Called from the hunters' Tent to accompany Geo. Sinclair to the C M[29] River to see if the Boats & property are Safe. Provisions given them for 8 days. They are to leave this in the morning.

---

26 MCM M214.

27 Tongues.

28 Probably the East River, a tributary of Dease River.

29 Coppermine.

**8 November 1838, Thursday.** A few Indians casting up now & then but more for the purpose of begging from us than to bring any thing. But 6 Deer Ribs & 3 tongues have been brought this Week past. 104 fishes Received on the 3rd from Norquay's fishery. This evening a Party from Big Head's Camp brought 200 lb. fresh meat and 14 tongues. 118 fishes recd. from Norquay.
**9 November 1838.** George Sinclair and Companions came back from the Copper Mine River where they found all Safe. No Indians have been in that direction. LaRocque and his young Brother, who were with him, are preparing to go and hunt to the northward.
**16 November 1838.** The 2 hunters off on the 12th. This morning Sent off Sinclair with Nozanne hunting. The fisheries have given us 473 from Norquay & 181 from Laurent. A party of Starving Indians & Sick encamped on the Island, coming every day to beg something to eat & fish to make baits of, which we cannot well refuse. Deer, they say, is scarce and they have no netts to fish with. Lines are their only resource.
**19 November 1838, Monday.** Yesterday LaRocque came back unsuccessfull. This afternoon a party from Simpson's Bay, consisting of Little Chief with some of his followers, Cast up with some Provisions. They say Deer is not scarce in that direction but they are not successfull. The Parties about here are not successfull with their lines and very troublesome for food.
**21 November 1838.** The Provisions brought amount to 400 lb. Rib meat, 107 tongues, 6 lb. Grease. LaRocque with his Brother, accompanied by Gros Blanc, was sent off to hunt. Little Chief and party also left us to return.
**22 November 1838, Thursday.** George Sinclair brought 180 lb. Fresh meat Killed by Nozanne, who is removing.[30] The Indians have been Collecting in consequence of a murder Committed sometime in October near McTavish's Bay. 2 Young Girls were Killed, but the perpetrator of it not Known, therefore suspicion ascribes it to the Copper Indians, though in my opinion without truth, As in such a Case they would not have Stopped at the murder of 2 Children, but fallen upon the Camp.[31] Since the report has been Spread among them, their alarm is so great that nothing has been done, and they are Starving.[32] Our Fisheries failing. But 70 fishes received since the 13th.
**30 November 1838, Friday.** Nothing of any consequence has transpired during these 8 days. But 38 fishes received, but the Indians about us begging as usual and cannot be prevailed upon to remove for the present.

---

30 Moving camp.

31 The final verdict was that the murderer was one of the Hare Indians, Edahadelly (Simpson, *Narrative*, 318–19).

32 The results of these murders were far-reaching and severe. Most of the Indians abandoned their hunting and sought refuge at Fort Confidence. The task of feeding them placed an enormous strain on resources, especially since the caribou, which had been numerous on the peninsula between Dease and McTavish arms the previous winter, had deserted it for the Coppermine and the area north of Great Slave Lake (Simpson, *Narrative*, 320–1).

**9 December 1838, Sunday.** This Afternoon Laurent came from his fishery as they cannot, he says feed themselves. LaRocque came back and was Sent off again on the 6th. On the 3rd & 4th 760 lb. fresh meat, 156 lb. Ribbs & 20 Tongues were brought in, the latter by some Hare Indians; others from the Hunters.

• • •

## *11. Letter from Peter Dease to John Stuart,* [33] *Forres, Scotland, 18 December 1838* [34]

John Stuart Esqr. Great Bear Lake, 18th Decr. 1838

My Dear Sir,

I hope that the receipt of my addresses to you, would have convinced you ere this, that I should be sorry, willingly to incur the reproach of neglect or forgetfulness from my friends, especially those among whom it has been my fortune to pass those days when free from the cares and anxieties which time and the Vicissitudes of life are accompanied by, it was my endeavour to share with others the various pleasures that only youth and buoyancy of spirits can derive from surrounding objects, and the mutual interchange of kind offices. These last, when continued as in our case, leave the impression never to be eradicated and form the most pleasing retrospections of life, for although seperated almost from Pole to Pole as we may be, past scenes can always be recalled to memory. Such are almost the only enjoyments to me now in this country, where but few of the companions of those days remain, and these widely separated. I am sincerely pleased to hear that you enjoy that state of health without which no society or amusement can be agreeable; may a long continuance be granted you. For my own part, I have no reason, thank God, to complain of want of Bodily health, but whether during my Caledonia travels,[35] or since, I perceive a vast difference in my sight, which is much impaired, without however causing any painful sensations, symptoms which afford me but faint hopes of a restoration to former state of vigour.

With much pleasure I hear of the continued health & happiness of my respected Friend, Mr. Js. Leith, as also Messrs. A.N. McLeod and one of the oldest acquaintances in the Country, Mr. J. Haldane, and should feel greatly gratified by an opportunity of personally paying my respects to them, if spared to return in safety from this Expedition. I do not despair of having that satisfaction, it being very probable I may again cross the Atlantic. We were informed by our letters & Some public prints of the unfortunate result of Captain Back's expedition in the Terror, and the imminent danger they were in during most of the season.

Our last summer's campaign, I am sorry to say, was not attended with the success we hoped for, but still not entirely fruitless. The season, although fine in some aspects, was not favourable for the disruption of the Ice after a severe winter for cold.

---

33 See biography, p. 308.

34 Mr Warren Baker's private collection.

35 His sojourn in charge of the New Caledonia District.

The General winds we had tended to keep it on the coast; few, very few South or S.E. breezes, and calms prevalent. With the utmost toil along shore, pushing off & cutting through points of Ice, carrying etc. that we reached Pt. Turnagain of Sir John Franklin on 9th Augt., where again we were finally blocked up by Ice, seeing which a Coasting Pedestrian Excursion was performed by Mr. Simpson with a party. He proceeded upwards of 100 miles to W. Lon. 104 ½ from whence a Deep Bay which would have taken much time to perform the circuit of made him resolve upon returning, tho' the Sea was there open & had been for some space before reaching that. He rejoined us on the 29th and only on the 31st could we extricate ourselves from our hopeless encampment. The Ice still a considerable distance to the Eastward, precluded all hope of advancing for some time. The lateness of the season, without any prospect of seeing Natives (as we saw but one family at C.Mine River, who fled from us), made us think a return for the season most adviseable, with intentions of another Trial next summer, God Willing, when we trust that with the Experience gained of our route & the Coast, should the season be favourable, at least not more adverse, I hope we may succeed, the remaining distance but about 130 miles to Ross' Pillar I trust it may be granted us to accomplish. On our return we ascended the Cop. Mine with our Boats to near Kendals River, by which from the Dismal Lakes we had descended to it with our Boats & Cargoes – a Rapidous little Stream. Four strong rapids also occur in the Cop. Mine, above the Bloody fall, where we still found a Memento of the horrid massacre described by Hearne & to which he was an unwilling spectator. Our party are still in Spirits for a renewal of our Voyage. The Season has been extremely mild for this Climate, from which I entertain hopes, as we had a clear coast on our return Westward.

Mr. Robert Campbell last Summer has succeeded in reaching the heads of the Stikine, and a few miles below the Bridge where Mr. Jno. McLeod was, found a large camp of natives with a Party of Russians trading their furs, but his arrival was not looked upon with a favourable eye, although one of the Officers treated him to a Glass of Grog. There is a talk of Exploring Peel's River. With best wishes,

Believe me, yours truly,

Peter W. Dease.

1839, Jany. 20th. Since writing the above we have the satisfaction of the arrival of the men from Ft. Simpson bringing our letters by Atha. Boats, at same time Ship letters. The ship reached YF 28th Augt. and our news this place 27th Ulto. and you may suppose we were much gratified by the satisfaction expressed by their Honours at the result of our Westward voyage, and the generous manner in which they have testified their approbation of it.

I sincerely pray it may be granted us next season to be equally fortunate as we were to Westward. I see our report has been published & highly applauded. The merit is due to Mr. T.S. not me, as you know scientific pursuits are out of my reach, therefore do not aspire to undeserved distinction. Yours etc.

P.W. Dease

• • •

## *12. Peter Dease's journal, continued*[36]

**20 December 1838, Thursday.** Had 2 Netts put down on the 11th which have given 23 fishes Since. Larocque came home on the 12 unsuccessfull and was sent off again on the 17. George Sinclair sent with them.
**31 December 1838, Monday.** On the 27th we had the Satisfaction of the safe return of our men, viz. Js. McKay, Js. Hope, John McKey & Chs. Begg with 2 Indians, with the letters from the Country and those brought to YFactory by the Ship.[37]

• • •

Dease's journal breaks off at this point; no field notebook has survived for the remainder of the second wintering, nor is there any coverage in the fair copy of the journal. One suspects that the second wintering was so similar to the first that Dease simply tired of making daily journal entries. Both the field notes and the fair copy of the journal resume on 15 June 1839, at the start of the second trip eastwards.

We can, fortunately, gain an adequate picture of events during the second wintering from Simpson's account.[38] In March, Simpson made a trip south across Great Bear Lake to retrieve a cache of two caribou, left by some of the fort's hunters. But the cache had been opened and the meat eaten by wolverines. Simpson and his two companions returned to Fort Confidence on very short rations.

On his return to base Simpson found that prospects had improved noticeably. A group of eleven Hare Indians had arrived from a distant camp to the west with a supply of half-dried caribou ribs and with reports of the caribou being especially numerous in that area. This group made a later visit to Fort Confidence, again with a welcome contribution of meat. In the interim, Le Babillard and a small party arrived from the east; they had encountered a herd of muskoxen and brought some of the meat with them.

Dease and Simpson became increasingly anxious about Sinclair and his party, but that concern was relieved when the party returned to Fort Confidence on 15 April; they brought little or no meat with them, since their hunting efforts had been devoted almost completely to sustaining the large group of Indians who had opted to join them in their camp on the Coppermine.[39]

In late April, early preparations were put in hand for the summer's trip to the east. John Ritch set off for the depot on the Coppermine River with provisions for the summer and with planks for repairing the boats. He found that mice had got into the cache and eaten or spoiled some of the flour, and had chewed holes in the boat sails, but that otherwise the cache was intact.

The expedition received a very valuable addition during the course of the win-

---

36 MCM M214.

37 *Prince Rupert*, Captain John Grave, which, after being delayed by heavy ice in Hudson Strait, had reached York Factory on 28 August (Cooke and Holland, *The Exploration*, 165).

38 Simpson, *Narrative*.

39 Simpson, *Narrative*, 329.

ter in the person of Ouligbuck, the Inuk who had acted as interpreter on Franklin's second expedition in 1825–27. Thereafter he was employed at Fort Churchill and was dispatched to Ungava with Nicol Finlayson and Erland Erlandson, who established the post at Fort Chimo in 1830, and had been serving as interpreter there ever since. In 1836 a message had been sent to Fort Chimo to inform him that he had been assigned to join the expedition on Great Bear Lake. He had travelled south and west to Red River, and then west and north to Great Bear Lake; the trip from Red River alone had taken eleven weeks.[40]

Even as the winter eased towards spring, the food situation showed no sign of improving. In early May about thirty Indians from various parts of Great Bear Lake congregated, starving, at Fort Confidence. Dease and Simpson were unable to meet their requests for ammunition, since their own supplies were running low. Indeed on 1 May a party was dispatched to Fort Norman with an urgent order for powder and ball.[41] In the meantime the situation was relieved when the Indians were persuaded, on the 23rd, to move north to Kasbah Lake, which was known to have good spring fishing.

The express party returned from Fort Norman with the crucial powder and ball on 28 May. They had had a fast trip across the ice of Great Bear Lake on their outward journey, reaching Fort Franklin on 6 May. But there they found the Great Bear River open, the melt well under way and all the tributaries in flood, seriously slowing their progress. It was not until the 13th that they crossed the Mackenzie to reach Fort Norman, just before the river ice went out; this post was located on the west side of the river and some 30 miles above its present location from 1823 until 1844. On the return trip they had to cope with snow-free ground and the same flooded streams as far as Great Bear Lake; thereafter, for the first part of the trip across the lake, they were seriously hampered by melt pools on the lake ice, and it was only for the latter part of the trip that the dogs had a good ice surface over which to haul the toboggans.

The month of May was unusually cold: the temperature on one occasion dropped to -15°F (-26°C), and the mean for the entire month was +24.5°F (-4.2°C).[42] Inevitably, there was no sign of breakup on the rivers or the lakes. The beginning of June brought a striking and welcome improvement in the weather, with temperatures reaching 70°F (21°C) in the shade. The snow rapidly disappeared and Dease River broke up. Spirits at Fort Confidence generally rose. To quote Simpson: "Our men and the natives played at ball, and other out-of-door games. In the evening Mr. Dease's violin was oftener heard than during the long dreary winter, and to its enlivening strains the Indian youths danced and capered in the hall." [43]

• • •

40 Davies, ed., *Northern Quebec and Labrador Journals and Correspondence 1819–1835*; Simpson, *Narrative*, 334.

41 Simpson, *Narrative*, 335.

42 Simpson, *Narrative*, 338.

43 Simpson, *Narrative*, 341.

## *13. Letter from Thomas Simpson to Donald Ross, Norway House, 30 January 1839*[44]

*Private*

Donald Ross Esquire | Fort Confidence 30th Jany 1839

My Dear Sir,

I have now before me both your valued letters of 18th July & 7th Septr., which came to hand on the 27th Ulto. I should be ungrateful indeed were I not to acknowledge, and that in the warmest manner, the numerous obligations we lie under to your kind and unremitting good offices. The file of Canada papers arrived safe; but I beg that you will give yourself no further trouble of that sort on my account, as I expect to be more on snow shoes than at any thing else, next winter. The present winter has been the first and only "lull" I have enjoyed; nothing therefore could have been more acceptable than the mental instruction and entertainment with which you so liberally provided us. It prevents us from falling behind the rest of the world in the progress of ideas, and is doubly gratifying as a proof that, though far, we are not forgotten. I believe I mentioned to you in the fall, my opinions respecting the Canadian rebellions, (which is confirmed by a full perusal of the details) – that the feeble and ill-planned outbreak will tend to the *future* welfare and security of those fine countries. A Yankee war is of course, like all other wars, much to be deprecated, but I do not see that Britain has anything to apprehend from the result. She certainly stands in a more powerful and commanding attitude *now*, than she did during the last American war, which was so disgracefully forced upon her when she had her hands full in Europe.

I see that our Expedition *is* regarded as a *national* one; and we have every reason to be highly gratified with the manner in which it has been noticed, both by the Honble Company and by the British public. My *private* sentiments are of course never intended to go beyond the persons to whom they are addressed; and you yourself, my dear friend, are, I believe the only one to whom I confided my misgivings upon this subject, which are now happily removed. As to the tone of our *public* correspondence, neither Mr. Dease nor myself can find one syllable which we could wish altered, or by which we are not ready to abide. We therefore repose in peace.

The nautical almanacs, magnets and one of the two thermometers arrived safe – the other was broken. The "Dipping Needle" came, just as you sent it from Norway House; and it would have been well had it been so carefully packed in London. However, the united skills of Mr Dease, Ritch and myself, have been at work upon it, and I think that we have at last got it into something like a serviceable state. We are not far from the *magnetic* pole here. I am sorry to hear such gloomy tidings from Red river; but they are nothing more than I have long expected & prognosticated. I regret it the more on account of the valued friends whom we have there; but frankly confess that I have not the least sympathy with the depraved and worthless

44 BCA ADD.MS 625.

half breed population. Unless the Company maintains a tight little garrison at the Forts (say a score of determined fellows – Europeans, for what is the use of stone walls without people to defend them) – there will, I much fear, be *another* bloody reckoning ere long. Let not therefore an imaginary oeconomy lead to a false and dangerous security; for the "cure" depend upon it, will cost tenfold the "prevention." Our worthy friend Hargrave tells me that he is going to attend the Council at Fort Garry, where, if the multifarious business of Norway House do not prevent you, I have no doubt you will also be present. I would give something to meet our excellent Governor there this spring; and enjoy, were it but for an hour or two, the society of the circle that will be assembled around him. I was glad to learn from Mr. Christie that your daughter, Miss Jane, my old acquaintance, was well and making good progress at the seminary where she and Missey Christie were placed near London. He, worthy man, has suffered a trying loss in the death of his son James. I am much obliged to Mrs. Ross for her good wishes and beg to be very kindly [remembered] to her and the children.

We entertain confident expectations of more extended success this year than last; for the winter hitherto has been less severe; the knowledge of the coast already acquired is no small gain; and as we can have no occasion to take our worn out boats up from the coast, *next* fall, we shall thus have more time at our disposal. All these are matters of fact and calculation, which I do not think are likely to deceive us. We have certainly spun out matters wonderfully with our small means. As for the current winter, half of it, thank God, is now over our heads, and though nearly eaten out of house & home by starving Indians, we hope to get through the remainder somehow or other. I leave my esteemed friend Mr. Dease to give you his own news;

And remain, with unalterable regard

My Dear Sir,
Most faithfully yours,
Thomas Simpson

• • •

## *14. Letter from Peter Dease and Thomas Simpson to Murdoch McPherson, 31 January 1839*[45]

Ft. Confidence, 31st January 1839

Dear Sir,

Our packet from York Factory came to hand safely on the 27th ulto.; and the present one is intended to go out by your regular Express of 1st March as you will see by our letter to the Gentlemen along the communication.

The bearers Felix & Laurent Cartier are to remain at Ft. Simpson till the arrival of the Spring Express from the Southward when you will be pleased to furnish them with an Indian guide, and it will probably be the preferable plan both on account

45 HBCA B.200/b/11, pp. 19–20.

of provisions and what they will have to carry, to send them back by the same route via Marten Lake. Our men require by Felix some further articles out of their goods lying at Ft. Simpson, and to enable you to know exactly what belongs to each person the correct lists are now sent.

Subjoined are two orders on the execution of which we have to request your usual good Offices. The first is for orders to be forwarded hither by Felix, the second for some supplies that are absolutely necessary in case of our having to pass another winter here. These last you will be good enough to provide with the aid of the Athabasca gentlemen, and send them to this place with the fall boat regarding which we shall write you more particularly to Ft. Norman in May. Besides this we have to recommend your procuring from Athabasca ten bags of Pemican, five of which to be deposited at Ft. Simpson & five at Ft. Norman for the use of the Expedition, outgoing next winter.

With much esteem and regard, We remain, Dear Sir,
Your obedt. humble Servts,
Peter W. Dease, C.F.
Thomas Simpson

P.S. The two Ft. Simpson Indians not having made their appearance at the appointed time, James Hope has undertaken to act as guide via Marten Lake. You need not therefore send any Indian back with the party hither. A Net is sent to Beaulieu by this conveyance.

| LIST 1ST PR. FELIX | LIST 2ND PR. FALL BOAT |
|---|---|
| ½ Keg Gunpowder | 1 Keg Gunpowder |
| 2 galls. Salt | 1½ Bag Ball |
| ½ doz Files | ½ " Shot |
| ½ " Scalping Knives | 2 pc. Strouds,[46] 1 wh., 1 blue, incl. the one not sent this year |
| 1 Cent. large Gunflints | 2 doz. Scalpg Knives |
| 1 rudder iron lost by our men on their way from Ft. Simpson | ½ " Files |
| 1 pump spear broken do. do. | 1 Roll Twist Tobacco |
| 1 " " for sm. sea boat | 2 cents Gunflints |
| 1 roll Irish Twist Tobo. or an equivalent in Twist for Js. McKay pr. expenditure by him in purchasing fish at Marten Lake | 6 half axes |
| | 6 skeins Twine or net twine |
| | 2 Bales moose skins (say 30) |
| | Good wood for 5 dog sleys |
| 7 lbs. Souchong Tea (out of our remaing Stores) | 15 prs. snowshoe frames |
| | 5 axe helves |
| 1 sqr. Mustard | |

• • •

46 A densely woven woollen cloth, extensively traded by the Hudson's Bay Company.

## *15. Letter from Murdoch McPherson to A.C. McLeod and Colin Campbell, 27 February 1839*[47]

Ft. Simpson, 27th Feby. 1839

Gentlemen,

The Discovery Expedition has again made a demand upon us for supplies which are to be forwarded to Ft. Confidence next fall, in case of its having to pass another winter at that place.

The present demand is made separately on this district, but with an understanding that I should apply to you to furnish a share. I therefore hand you subjoined a list in the execution of which I have to request your usual good offices, and if possible that the whole may be ready to be embarked at Ft. Chipewyan or at Slave Lake on my return from Portage La Loche next August. I have the honor to be, Gentlemen,

Yr. Obedt. hble. Servt.
M. McPherson

LIST

| | |
|---|---|
| ½ Roll of Tobacco | 6 skeins Twine |
| 1 Keg Gunpowder | 1 doz. Scalpg. Knives |
| ½ Bag Ball | 30 large dressd. Moose skins |
| 1 pc. blue Strouds | 10 Bags of Pemican |
| 6 half axes | |

• • •

## *16. Letter from Thomas Simpson to Donald Ross, Norway House, 30 April 1839*[48]

*Private*

Fort Confidence, 30th April 1839

My Dear Sir,

Both your very welcome letters of 10th Septr. & Decr. came hither in the regular winter packet on the 13th Inst. – accompanied at long last by an Esquimaux Interpreter.[49] Better late than never, according to the proverb, and we are deeply indebted to our worthy friends Messrs. Charles & Christie in this most important matter. A thousand thanks for the magazines and newspapers, the former I now return agreeably to your desire. We have every reason to be highly gratified by the manner in which our lowly services have been appreciated by a generous public; although some of the prints seem to view them through the medium of party feeling, which, to say the least, is most absurd. Never mind, our discoveries will endure when the very names of the political sects of the day will be forgotten; and such conduct only

47 HBCA B.200/b/11, pp. 21–2.

48 BCA ADD.MS 625.

49 Ouligbuck.

inspires us with a sovereign contempt for the meanness and venality of these partisan scribblers.

You are now at Fort Garry – the seat of government of a vast but thinly peopled territory. Some strangers of distinction, it is said will be there on a visit to the buffalo plains; but we have been so abused by false reports of late, that we now disbelieve the gravest truths. I am however truly glad to hear that the people of Red River have at length stumbled, by chance, upon the only *natural* outlet to that isolated colony – the road to the Mississippi; and wish them a favourable market for their flour and cattle across the lines. It is to be hoped that no false principles of government may interfere to forcibly turn that trade down the *impracticable* channel of Hill river.[50] There has been too much of that quackery already. His Excellency may well have enough of travelling by this time, but the "Great Western"[51] makes play of running across the Ocean of Columbus. I pity the poor devils of scribes at your Council meeting, and from what I experienced of former ones had rather a thousand times be where I am – cheerless and frozen as all nature is. I should, however, like amazingly just to look in upon the turmoil of the scene. Last R.R. Council nobody had time to speak to anybody; and I don't know how many persons kissed the earth by coming in contact between the office and Government House; and when they got up again they hadn't time to ask pardon nor even to vent a single sacré!

In Canada, as we anticipated, the troops have kept all quiet, and if they but allow Sir John Colborne to retain Lord Durham's relinquished place, all will yet go well. The conduct of ministers, in abandoning Durham, when his measures were questioned, was weak and pusillanimous in the extreme, and deserves to draw down their dismissal, which cannot be very remote. Our "Bear Lake prophecies" on minor matters were not substantially erroneous. Logie *did* intend to splice the knot when he was hurriedly recalled to be the bearer of the express; and his *long* friend was prevented from taking the same course by bad health (which I sincerely regret) and philosophical *irresolution*. After so many years of "deliberation" we naturally supposed that the measures of these gentlemen would have been duly concocted? and promptly taken; we gave them credit, in fact, for more vigour than they possessed.

Ritch's affair I beg leave to waive, as what has transpired on the subject nowise alters the general impression here, respecting the injustice and impolicy of the fine. By the handicraft laws the boat moulds were the man's own property – the fruit of a long apprenticeship – and further, he has declared to us that the greater part of the *materials* of those moulds were actually used by him in boatbuilding, before his departure on the expedition, and that what remained could have been of no manner of use to any mortal. The man's services have been invaluable to us; he should not have been condemned absent & unheard, and the news almost drove him distracted. It is not the amount, it is the *public* disgrace that affected him; and we hope that the council will repair their hurried decision, as they did in the case of John

---

50 The route north to York Factory and out through Hudson Bay.

51 Brunell's new steamer.

Tod Esquire. I had a particular right to feel sore in the matter, having myself been formerly mulcted in the same sum, for an equally venial offence. But my returns were ever wrong in the eyes of the first personage in the country,[52] which was no doubt one of the *advantages* I derived from the relationship. I congratulate you, my dear friend, on your philoprogenitiveness, and beg to tender my kindest regards to Mrs. Ross, with a kiss to Donald [hole in paper] Your brother-in-law, our Fort Norman neighbour I hear has begun? betimes? in the same line. For myself, I live like an anchorite in these parts, which have never been gladdened by female charms.[53] The present winter has been altogether a miserable one, as we had not only our own party to provide for, but to rescue the natives from a repetition of the miserable fate of the Copper Indians at Fort Reliance.[54] We *have* saved them; and let those who would cast a stain on the humanity of the fur trader stand forward and say as much! There are no winter fisheries in Bear Lake; a murder and the withdrawal of the reindeer southward, caused the starvation which our exertions and sacrifices relieved. Nevertheless we stand ready and willing for a third campaign. My worthy friend Mr. Dease has had a twinge of gravel, and complains of his failing sight. He commenced his course of hardships too young, and now feels the approach of premature age. Otherwise we all, thank God, enjoy perfect health.

May all sublunary blessings attend you,
While I remain, My Dear Sir,
Yours ever faithfully,
Thomas Simpson

Donald Ross Esquire

• • •

## *17. Extract from letter from Thomas Simpson to James Hargrave, York Factory, 30 April 1839*[55]

*Private*

Fort Confidence, 30th April 1829

My Dear Sir,

Your valuable letter of 10th December is now before me, having reached this end of the world on the 13th Inst., together with a most unhoped for acquisition – Oulibuck the Esquimaux interpreter. As we have really nothing to say *officially* at present, having *no demands* to make, and our accounts being forwarded in February to your address, I may as well notice here your other (public) letter of same date. The foresight of the Committee has fortunately provided us with Nautical Almanacs for 1839 and 1840; but what Mr [Donald] Ross wrote you respecting the mountain barometer and Dipping needle is unluckily too true. The latter we have contrivd to

52 George Simpson.

53 Here Simpson is dismissing all the Indian women who frequented Fort Confidence.

54 During Back's wintering there in 1833–35.

55 Glazebrook, *The Hargrave Correspondence*, 294–5.

patch up so that it *may* answer, though doubtfully, as one of the spirit levels was smashed. The fault was bad package in London, but had Messrs. Banks & Wallace placed the little boxes in a Cassette,[56] the damage would not have happened. The disappointment is considerable, but it does not affect the principal business of the survey. The case would have been very different had not Mr. Smith lent me his super-excellent watch, seeing that London could not produce a chronometer suitable for the Hon'ble Coy's Expedition!

Many thanks for the Literary Gazette. Dr. Richardson has accorded us *easy* praise, for which we must feel grateful. The paragh. about Franklin's forbearance at Pillage Point is absurd, and your view of the case perfectly just. That affair about as much affected my reception at Point Barrow, (where perhaps you are not aware that in 1826 Messrs. Elson & Smyth with the Blossom's armed barge judged it *imprudent* to land) as Lord Durham's resignation of the Canadian Government is likely to influence the politics of the moon! We had our squabbles too on the coast, though we have hardly said anything about them, as the display of our arms was sufficient for Esquimaux stomachs. But friend Ross writes me that, at home, they have made quite a party affair of our expedition. Well, may God forgive them if they have, for I cannot, for the life of me, see what a purely scientific undertaking has to do with whig or tory, radical or conservative. I have strong hopes of making another decisive stroke this season, though we cannot expect to complete what Back *was* to have done in the Terror Bomb. Were there but a clear passage, we should perhaps astonish *you* with a visit some of these days. Alas! we must cut and carve our dreary way; we cannot say with the corsair,

O'er the glad waters of the dark blue sea,
Our thoughts as boundless, and our souls as free,
Far as the breeze can bear the billows foam,
Survey our empire and behold our home.[57]

We have had a miserable winter so far, on account of the Indians, numbers of whom we rescued from starvation; while the humane Capt. Back left his, under similar circumstances, to perish at the very doors of Fort Reliance. So much for the slandered fur trader ...

I ever am,
My Dear Hargrave,
Yours most faithfully
Thomas Simpson

James Hargrave Esquire

• • •

56 A trunk.

57 The opening lines of Lord Byron's *The Corsair*, this time quoted accurately.

## *18. Letter from Peter Dease and Thomas Simpson to the governor, chief factors, and chief traders of the Northern Department, 1 May 1839*[58]

Gentlemen,

A winter of intense anxiety is nearly over without mortality. But for our aid many of the natives must have perished from hunger; for a murder among themselves, last fall, paralysed their energies, and, to add to this calamity, the reindeer forsook their lands.

We have saved them through great sacrifices, and greater exertions. The environs of *our* winter quarters have been strewn with Indian corpses; yet, by the practice of rigid frugality, we are still in a condition to prosecute our discoveries on the northern sea.

Oolibuck, the Esquimaux interpreter, reached this place with the regular winter packet on the 13th ultimo. This valuable and unlooked for acquisition we owe to Chief Factors Christie and Charles, who engaged, to Chief Trader Todd, and the other gentlemen along the route, who so promptly forwarded him to his destination.

This winter has happily been less rigorous than the terrible one of 1837–8. In sheltered places indications of thaw already begin to appear at mid-day; and in another month we may expect to see the surface of the earth again.

Our boats are repaired; our provisions are on the banks of the Coppermine; and as soon as that impetuous river bursts its icy fetters in June, the rest of the party will be there, with hopes nowise damped by the hardships and languor of a second Polar winter.

We have the honour, etc.

P.W. Dease
Thomas Simpson

• • •

## *19. Letter from Peter Dease and Thomas Simpson to Murdoch McPherson, 1 May 1839*[59]

Fort Confidence, 1st May 1839

Dear Sir,

The bearers of this to Ft. Norman are Hope, McKay, Frammond and Boucher. The two first are directed to return immediately with some ammunition etc. in exchange for furs forwarded by them, of which a particular account is sent to Mr. McBeath. Boucher, who continues in pay of the expedition, is intended as last year to accompany you to Portage La Loche, and to form one of the crew of the fall boat to this place. Frammond is discharged by the expedition from 1st. prox.; but though not suited for us he is quite equal to the ordinary duties of the service, and would probably re-engage for McKenzie's river.

58 Alexander Simpson, *Life and Travels*, 307–8.

59 HBCA B.200/b/11, pp. 31–2.

With respect to the boat that is to bring the few supplies we asked for hither, the earlier she reaches this place the better, as the people will be requird to assist Ritch, Felix and Morrison in the fall fishery, and the drying of fish, so that we may have where with to pass, if necessary, part of another winter here. Three servants and three Indians will, we think, form a sufficient crew for a boat almost light. The Indians while here to be employed in hunting. With young Boucher, and a substitute for Frammond, there will thus be only one servant required from MacKenzie's river; and we beg that the steersman may be a good steady person.

If we do not get back here by 10th October Ritch will be instructed to forward your boat & crew with everything not absolutely required here, to Fort Simpson, and we shall make our way out the best way we can.

If we return from the coast before that date, we fully hope to reach you all together by open water.

Meanwhile with sincere esteem, We remain Dear Sir,

Yr. obedt. hble. Servts.
P.W. Dease
Thomas Simpson

P.S. Of the ten Bags Pemican from Athaba. four had better be deposited at Ft. Norman & six remain at Ft. Simpson.

• • •

## *20. Letter from J.H. Pelly to the Marquis of Normanby,*[60] *3 May 1839*[61]

Hudson's Bay House, 3rd May 1839

My Lord,

I have the honor to inclose for Your Lordship's information copy of a communication under date Fort Confidence 15/20th Sptr. 1838, received from Messrs. Peter W. Dease and Thomas Simpson, Officers of the Hudson's Bay Company's service, in charge of the Expedition for exploring the Shores of the Northern Continent of America, detailing their proceedings during the past season ...

My Lord,
Your Lordship's most obedt. servt.
J.H.P. Govr.

• • •

60 Constantine Henry Phipps, Earl of Mulgrave, was created first Marquis of Normanby in June 1838. Previously a successful lord lieutenant of Ireland, he gave up that position in February 1839 to become secretary of war and the colonies, succeeding Lord Glenelg. He remained as colonial secretary in Peel's administration until August 1839, when he was succeeded by Lord John Russell (*Dictionary of National Biography*, 15:1116–17).

61 HBCA A.8/2, f. 217.

## *21. Thomas Simpson's will, 11 June, 1839*[62]

In the event of my death on the present Arctic expedition, my worldly means, to the best of my knowledge, will stand as follows: Five hundred pounds sterling in the hands of the Hudson's Bay Company; my revisionary share as a Chief Trader in that concern, worth, at the utmost, Fifteen hundred pounds Sterling; whatever monies the British Government may award me for the acknowledged discovery, in the year 1837 of the long-sought North-west passage, or may arise from the publication of my maps and journal; and my half-share of a house and garden in the town of Dingwall, in the highlands of Scotland. Of these effects I will and bequeath the sum of one thousand pounds sterling, free of legacy duty and all other charges, to my mother, Mrs. Mary Simpson of Dingwall, aforesaid. To my uncle, Duncan Simpson, Esq., of Bellevue, near Beauly, in the same county of Ross, and his wife Mary jointly, or to the survivor of them, I bequeath the sum of one hundred pounds sterling; and to Mrs. Margaret Bain of Dingwall aforesaid, Twenty pounds sterling, in like manner free from legacy duty and all other charges. To my only brother, Alexander Simpson, now, or lately, in the service of the Hudson's Bay Company, and to his lawful heirs I will and bequeath the residue of my property as above stated; and all that may accrue to me from any other source whatsoever, saving and excepting my clothing, books, arms, or other moveables, not specified, at Great Bear Lake and Red River, to the value of one hundred pounds sterling or thereby, which I appropriate to satisfy any claims there may be against me at the said Colony of Red River. Should my mother, Uncle and Aunt, or Mrs. Margaret Bain, aforesaid, die before me, then, and in that case, it is my will that the sums bequeathed to them revert to my brother Alexander Simpson, and to his lawful heirs. I further constitute and appoint my said brother, Alexander Simpson, and my friend Alexander Christie, Esquire, at present Governor of Red River Colony, the executors of this my last Will and Testament.

In witness whereof I have hereunto set my hand and seal in duplicate, at Fort Confidence, Great Bear Lake, this eleventh day of June one thousand eight hundred and thirty nine.

Thomas Simpson

Signed, sealed and declared.
in the presence of us, place
and date *last* above written.
(Signed) Peter W. Dease, Chief Factor
John Ritch )
George Flett) Servants of the Hudson's Bay Company[63]

• • •

62 HBCA A.36/12, ff. 163–163d; Alexander Simpson, *Life and Travels*, 313–14.

63 Conspicuously absent from this will is any bequest to his "little fellows" at Berens River (see p. 286), certainly illegitimate and probably of mixed race, or to their mother.

## *22. Letter from James Hargrave to Peter Dease and Thomas Simpson, 11 June 1839*[64]

Fort Garry, 14th June 1839

Gentlemen,

I am directed by Governor Simpson to enclose to your address a Thermometer sent this season to the Indian Country by the Royal Geographical Society of London, likewise copies of observations by Dr. Richardson, on the objects for which it has been sent; and I am also to say that Govr. Simpson feels confident you will use your best endeavours to meet the wishes of the association.

I remain, Gentlemen,
Your mo. obedt. Humble servt.
James Hargrave

• • •

## *23. Letter from Peter Dease and Thomas Simpson to John Ritch, 14 June 1839*[65]

Fort Confidence, June 14th, 1839

Instructions to John Ritch

1. Your principal care during the summer, next to the safety of the establishment, will be to keep two fisheries a-going under the charge of Felix and Morrison, who, with the aid of Indians, are to dry as many fish as possible.

2. You will also trade their dried provisions from the Indians; but all who wish to withdraw to Forts Norman and Good Hope should be encouraged to do so, a very few of the best hunters being sufficient for our purpose.

3. About the middle of September a boat manned by servants and natives will arrive from Fort Simpson with some ammunition, leather, etc. (as per list 1), of which you will be very saving, in case of our having to pass part of another winter at Fort Confidence. The boat's crew to be placed as assistants to Felix and Morrison, till required to return to Mackenzie River.

4. We expect, God willing, to get back from the coast on or before the 10th of October. Should you not see or hear from us by that day, you will, on the following, despatch the Mackenzie River boat with her crew, the families at the establishment, private property, some of the dogs, and everything not absolutely necessary for subsistence at the place. You will put the whole in charge of Felix, in whose stead you can retain young Boucher; and to strengthen the crew of the boat, you may add a couple of active Indians not encumbered with families. This boat you will provision with dried meat (besides dried fish for the dogs) to Fort Simpson, deducting from the requisite quantity two bags of pemican, now at Fort Norman, which you will give Felix an order to receive. Should the fall be early, and the strait[66] threaten to freeze up before the 10th of October,

64 HBCA D.4/25, p. 55.

65 Alexander Simpson, *Life and Travels*, 309–12.

66 In front of Fort Confidence.

the boat must start earlier, as it is essential that it reach Fort Simpson by open water.

5. The pemican, etc., now here (as per list, No. 2) besides what you are to make, ammunition, nets, axes, sledges, some leather, snow shoes, and other articles pointed out to you must be left here in case of our having to perform a winter march to Fort Simpson, also the seven best trains of dogs (as per list, No. 3).

6. Should we unfortunately not get back to winter quarters this season, you must maintain the establishment till the 10th of October, 1840, and on the first ice, 1839, send an express to Fort Simpson (with a copy of these instructions), craving Chief Trader Macpherson's assistance and advice.

Peter W. Dease
Thomas Simpson

List No. 1. Articles expected from Mackenzie's River

1 Keg Gunpowder
1½ Bag Ball
½ Bag Salt
2 Bales containing 30 large Mooseskins
2 pieces strouds, 1 white, 1 blue
2 dozen yew handle knives
½ dozen files
1 Roll Twist Tobacco
2 Cents. Gunflints
6 Small axes
6 Skeins Twine, or net thread
Wood for 5 dozen sledges
15 pairs snow-shoes, and 5 axe helves

List No. 2. Dried provisions remaining at Fort Confidence

3 Bags Pemican
3 Bags Grease
3 Bags Flour
Also more than sufficient pounded meat to make, with two of the bags of grease, five more of pemican – in all, twelve pieces, upon which we depend on our return from the coast.

List No. 3. dogs to be kept

1st sledge: Brandy, Shanass, Toby
2nd do.: Sansbec, Captain, Castor
3rd do.: Cervolant, Lion, Boatswain
4th do.: Negre, Sauvage, Bob
5th do.: Carlo, Whiskey, Wabistan
6th do.: John MacKay's own train
8th do.: Laurent's own two, and Teyma
Felix's Tarriake and the Wolf to go out by the first boat.

• • •

# Chapter Nine
# Eastwards Again, 1839

## INTRODUCTION

The exploring party set off from Fort Confidence by boat as far as the mouth of the Dease River, on the afternoon of 15 June 1839, leaving John Ritch and three others in charge of the station. Then, packing their belongings on their backs, they set off overland. They reached the boat depot (where James McKay, George Sinclair, and John Norquay had been watching the boats) on the afternoon of the 19th. The latter group had launched the boats in the river on the 17th to allow the wood to swell to tighten the seams. On the 22nd they started down the Coppermine, running all the rapids as far as Bloody Fall, where they found the cache of pemmican and other supplies undisturbed, although two bags of pemmican were mouldy.

From the camp at the mouth of the Coppermine, Simpson and five men, including Ouligbuck, headed west to explore the lower Richardson River. Next evening (25th) Ouligbuck returned with two Inuit, members of a group whom Simpson and party had found camped on the Richardson. They, however, were unable to provide much information beyond the immediate area. Before they left next day the older man promised to have a pair of boots made for Dease and to deliver them when he returned to the mouth of the Coppermine in late September.

On the 28th the boats were launched and the party ran down the Coppermine to the sea. The sea was still icebound to the east, however, and hence it was the evening of 3 July before they were able to start eastwards. Even this was a false start; beyond the first point the ice was still solid and they had to land and camp again to wait for the ice to clear. One positive aspect of these delays was that by fishing and hunting they were able to replenish their provisions. Halted again by ice on an island halfway between Port Epworth and Gray's Bay over the period 11–15 July, the men amused themselves by building a cairn 15 feet high.

The two boats reached Cape Barrow on 18 July; beyond it there was some ice obstructing the channels between the islands, but they were able to hold a much more northerly course across the mouth of Bathurst Inlet than in the previous year and they reached Point Turnagain on the evening of the 20th. After a meal they pushed off and camped for the night at Cape Franklin. They were windbound at this point for several days (until the morning of the 25th) but then made excellent progress along the coast of Kent Peninsula, reaching Cape Alexander on the evening of the 28th. This was almost a month earlier than the date on which Simpson had reached this spot on foot the previous year.

Progress eastwards from here was handicapped by fog and ice; the latter, however, was broken, and they were able to take advantage of leads and polynias. One positive feature of the repeated delays was that the hunters were able to supply the party with a steady supply of caribou and muskox meat. They passed the Ellice River on 1 August. Next day, on a prominent headland, they found traces of an Inuit camp, including the bones and skin of a polar bear. This was the only polar bear, alive or dead, that the expedition encountered, and to mark this occasion they named the headland Whitebear Point. Pushing on eastwards, they successfully negotiated the bafflingly intricate scatter of islands at the east end of Queen Maud Gulf, and made their way north along the west coast of Adelaide Peninsula. On 10 August they found themselves at the west entrance to Simpson Strait, separating Adelaide Peninsula from King William Island, and headed east through the strait. Beyond it, Rasmussen Basin was totally free of ice. On the 13th Sinclair and McKay, who had been here with Back in 1834, thought they recognized Point Ogle,[1] which had been their farthest point on that occasion, but were not entirely sure, owing to fog. Two days later, while camped near Point Pechell,[2] they definitely recognized that section of coast and Montreal Island[3] to seaward. Next day, 16 August, this was definitely confirmed when they led Dease and Simpson to the cache left by Back and King on Montreal Island.

With this Dease and Simpson had fulfilled their instructions and might well have been justified in starting for home. But since it was still quite early in the season, the sea was ice-free, and they had ample food in hand, they canvassed the opinion of the men whether they should push on eastwards to try to solve the mystery of whether Boothia were an island or a peninsula. The men approved this suggestion and during the night of 16/17 August they ran across to the east shore of Chantrey Inlet, landing on a high cape which they named Cape Britannia. On the afternoon of the 17th they started northwards, but a strong headwind and high waves forced them to put ashore again after only a few miles. They were still windbound next day, and to pass the time the men built another impressive cairn 14 feet high. A message was placed in a bottle inserted about five feet above its base.

---

1 Named by George Back on 8 August 1834 after Vice-Admiral Sir Charles Ogle (Back, *Narrative of the Arctic Land Expedition*, 409).

2 Named by Back on 7 August 1834 after Sir J.B. Pechell, Bart. (Back, *Narrative of the Arctic Land Expedition*, 407).

3 Named by Back after the city of the same name, for the help he had received there (Back, *Narrative of the Arctic Land Expedition*, 309).

A headwind still impeded progress next day and the following day (20 August). After they had covered only about four miles on the morning of the 20th, the gale forced them to put ashore at the mouth of a small river, which they named Castor and Pollux after their boats. From what they could see of the land to the north and east, they were convinced that a strait extended eastwards from this point into the southern part of the Gulf of Boothia, i.e. that Boothia was an island rather than a peninsula. Despite the fact that the sea was ice-free to the north and east, they recognized that exploring the remaining coastline to Fury and Hecla Strait would require another expedition and decided that prudence dictated that they turn back. Hence, after building another cairn, this one only 10 feet high, they started back west at 12.20 p.m. on 20 August. With a fair wind and a heavy swell they easily ran right across Chantrey Inlet to Point Pechell next day. After exploring the bay between Point Ogle and Point Richardson[4] on the 24th, on the 25th they ran north across the eastern entrance of Simpson Strait to King William Island (which they labelled Ross's Land). Heading west through Simpson Strait, they surveyed the south coast of King William Island. When the coast began to swing northwards (on the 25th), they stopped and built another large cairn (14 feet high) on a headland they named Cape John Herschel, before heading south across the west end of Simpson Strait.

Progress back west was seriously slowed by adverse winds, which on occasion kept them windbound all day, but they reached Whitebear Point on 5 September. Next day they explored the shores of Melbourne Island, then started north for the south coast of Victoria Island, reaching it at Cape Colborne around midnight on 6/7 September. They coasted west past Cambridge and Wellington bays, surveying the coast as they went, then, at 8 a.m. on the 10th, with a stiff wind from east-by-north they steered south-by-east for Cape Barrow; the crossing took until 7 p.m. Even then they continued west until 10.30 p.m., landing for the night just east of the Wentzel River. Here they found plenty of driftwood and enjoyed the luxury of a blazing fire for the first time in weeks.

Further progress westwards was made unpleasant by snow showers and hence the party was relieved to reach the mouth of the Coppermine on the afternoon of 16 September. Dease was delighted to find that the old Inuk who had promised to produce a pair of boots had kept his word, leaving them hanging from a conspicuous pole. One boat was left at the river mouth; the other was tracked upriver to the location where the boats had been cached for the winter. Here the second boat and all superfluous food and equipment were turned over to the two Indians, Larocque and Macacoose, and the party began hiking overland to the Dease River. The ground was already snow-covered and many of the smaller lakes were frozen and bearing. They reached Fort Confidence on the evening of 24 September, undoubtedly very satisfied with their summer's achievements.

---

4 Named by Back after Sir John Richardson (Back, *Narrative of the Artic Land Expedition*, 420).

## DOCUMENTS

### *1. Peter Dease's journal, continued*[5]

**15 June 1839, Saturday.** The Ice in the Bay has melted along Shore leaving apparently sufficient open water to pass with a Boat to the River.[6] Every thing being prepared we took our departure, with Francois Felix, Hector Morrison and John (an Indian) who are to take back the Boat, they being the men appointed for the Summer under the Orders of John Ritch who is left in Charge of the Establishment, with written Instructions how to act in the Event of our not returning by a Specified time in the Autumn. Our Party, consisting of George Flett, Wm. McDonald, Chs. Begg, James Hope, John McKey, Laurent Cartier, the Esquimaux Oolibuck, and the 2 Indians LaRocque & Macacoose, Mr Simpson & I, At 1 P.M. embarked with the hearty wishes of those left behind us for our welfare & Success & pulled round the Bay along Shore, with Considerable obstructions from the Ice, which only Causes a little Retardment, to get to the River which is quite free of Ice some time. We proceeded to the first Shoal rapid where we landed at 4h.30' and took leave of our Friends. Each person took his bundle on his Back & commenced our march to gain the summit of the hills on the South side, the travelling tolerably good although some swamps occur. The water has had time to draw off in part & leaves them passable, with water about Knee deep in Some Places. We continued untill 9 P.M., taking rest at the end of each hours' walk. Put up for the night opposite the Old Man of Hoy (the Rock below the Rapid)[7] where we were visited Shortly by a young Indian from the Camp of LaRocque's Elder Brother, who is at a Short distance. This lad says they have Killed 4 Deers & Some of the party are gone to the House with meat. I should have noticed that our Guides, James McKay and George Sinclair, are with John Norquay in Charge of the Boats & property, at the Boat Station on the Banks of the C.M.[8] River.
**16 June 1839, Sunday.** At 6h.30' we resumed our march. The young Indian left us to go and apprize the hunters of our being on the Way. At 9 the Hunter L R's[9] Brother with another young man Joined us, each bringing a piece of fresh Deer's meat, which sufficed for breakfast. They continue with us for the day. Some Deers were seen; the Indians & Hope went after some. Shortly after Oolibuck went after others – also LaRocque & Macacoose – and 3 Deers were Killed, One by each party & brot. to the Place we put up for the night, being along a Creek running into the S branch of Dease's River,[10] near the hammock[11] of wood where boards were Sawed for the Boats. Another Indian of the Same Camp Joined in the Evening.

---

5 MCM M214.
6 Dease River.
7 See p. 170, n. 5.
8 Coppermine.
9 Larocque's.
10 East River.
11 Copse or clump of trees.

**17 June 1839, Monday.** This Morning Continued our route. One of the Indians accompanies us as far as the Boat Station; the others left us. Crossed the South Branch of Dease's River & Struck across the Hills which are here barren for the upper Part, for a Small branch forming one of the feeders of Kendal's River which we got to at 9 P.M. & put up on a low Island where both Channels are easily forded, being the only Spot for some distance where wood for fuel is to be found.

• • •

## *2. Letter from George Simpson to Peter Dease and Thomas Simpson, 17 June 1839*[12]

R.R. Settl. 17th June 1839

Dear Sirs,

I have the pleasure to acknowledge receipt of your welcome and much valued letter addressed to the Governor, Chief Factors & Chief Traders, of 15th September, with copy narrative of Mr. Simpson's jurney on foot to the eastward, likewise copy of your letter of 30th January to the Governor & Committee; all of which have been laid before the Governor and Council; while I congratulate you on your safe return from the late perilous voyage, I have much satisfaction in saying that one and all who have perused these papers feel that although the success of the expedition to the eastward has been incomplete, much has nevertheless been done towards accomplishing the survey, and that the exertions of yourselves and gallant little band in the causes of discovery and science merit our warmest commendations, in short, are beyond all praise.

We approved very much of your determination to make one more effort this summer to trace the coast round from Point Turnagain to Back's Great Fish River, and as the ice broke up early in this quarter, we think there is a probability that with you the present is a more favorable season for your operations than last, affording fair promise that before the Arctic regions resume their winter garb you will have ascended Great Fish River and got quartered in the Honble. Company's establishment at Great Slave Lake.[13]

We observe that whether successful or otherwise in accomplishing the survey to Great Fish River, you are not prepared to continue the operations of the expedition next year, which is to be regretted, as we were in hopes that after that section of coast had been surveyed, you would have been in a position to push your discoveries to the straits of the Fury & Hecla. That, however, we find cannot be done under any circumstances, you may therefore repair to the Depot, and take a winter's leave of absence, if agreeable to you, by way of recruiting after your severe & hazardous labours, during which, I have no doubt, plans will be matured for

12 HBCA D.4/25, pp. 75–6.

13 Fort Resolution.

completing this very difficult & interesting service, which cannot be allowed to fall to the ground, while a shadow of hope remains that there is a possibility of accomplishing it.

With esteem, I remain, Dear Sirs,
Your mo. obedt. hum. servt.
Geo. Simpson

• • •

## *3. Peter Dease's journal, continued*[14]

**18 June 1839, Tuesday.** The Weather rainy in the morning. After crossing the River we directed our Course for the Hills below the lower Dismal Lakes. The party seperated: Mr Simpson, G. Flett & I reached Kendal River about 2 miles below its run from the Dismal Lakes & found it too deep to Ford. Therefore lighted a Signal fire to Call the party who Cast up near 2 hours after us & put up for the night, as we cannot Cross without Constructing a raft tomorrow.[15]

**19 June 1839, Wednesday.** A Raft was Constructed which ferried us over in 3 trips. After 3 hours Walking we Joined our party at the CM.[16] in good health & Spirits at 5 PM. The Indians have Supplied them abundantly with fresh Deer. 17 have been Killed since Ritch left this. The weather became Cloudy in the Evening. The CM River began to break up on the 3rd. The Ice continued drifting down untill the 13th. On the 17 the Boats were launched to tighten them,[17] and are not yet Sufficiently Soaked, they say, after being so long dry.

**20 June 1839, Thursday.** The weather rainy & Windy all day which rendered it useless to empty the Boats of Water as they Could not Dry, if any Caulking is required, which the Guides think they may want.[18]

**21 June 1839, Friday.** The Weather Cloudy but dry; the wind NW and Nbw. Sent the Indians out to Kill Deer. They were Crossed over in a Boat. They came back after an absence of 3 hours with the meat of 3 Deers, two of which were Killed by André, one of the lads who was hunter here, and the other by his Companion Macacoose. Wrote a note for Ritch which will be taken by the 3 Indians who are to leave this tomorrow, As we intend to take our Departure also for the Coast. The water in the River rather lower that it was when we passed down here last year.

**22 June 1839, Saturday.** At 8 A.M. we embarked the Crews as follows: James McKay, Guide & Steersman, James Hope bowsman, Chs. Begg, John McKey, J. Norquay & Macacoose Midn., & P.W. Dease Officer, in the [blank]; Geo. Sinclair Guide & Steersman, Laurent Cartier bowsman, Geo. Flett, Wm. McDonald, Oolibuck & LaRocque

---

14 MCM M214.
15 Only three caribou were seen during the day (FNB C, f. 4).
16 Coppermine River.
17 To allow the wood to swell, thus tightening the seams.
18 The boats were baled out in the evening (FNB C, f. 5).

Midn. & Thos. Simpson Officer, and got down all the Rapids Safe T G[19] to the Bloody Fall. In running the Upper and Escape Rapids, the Boats took in some sprays in consequence of the heavy swells of these Rapids but no damage sustained. Being rather late the Tents were erected at the lower end. The property and Boats will be taken across in the Morning. Our Deposit in the Cleft of the Rocks was found quite untouched. Two of the Bags of Pemican were mouldy, one of them quite unserviceable, the other partially spoiled, but that I presume was from having been wet previously, as they did not appear to have got wet since they were left.[20] A Bag was left this morning about 3 miles below the Boat Station to serve as Provisions for our return in the Fall.

**23 June 1839, Sunday.** The property was carried over[21] this morning & the Boats left in a Small pool of water to prevent them from drying while we remain here, as an Excursion is to be made in order to Examine Richardson's River & trace it to the Coast from where it was seen last year by the men. The Eddy at the Foot of the Rapid seems to be alive with fish, rising to the Surface in great numbers, apparently trout & Salmon.

**24 June 1839, Monday.** This morning Mr. Simpson with J. McKey & Hope, Norquay, Oolibuck and LaRocque left us to go and Examine the River. We can see no vestiges[22] of any nature of a recent date nor does it appear they have frequented this place for some years back. Trying with a hook & line 2 fishes were caught, a Small trout and a Salmon. The day fine.

**25 June 1839, Tuesday.** In course of the day 4 fishes of the Salmon & Trout Species were caught with the line. Towards Evening Oolibuck cast up, accompanied by 2 Esquimaux,[23] An Elderly and a young man. They seemed to be under some alarm when approaching the Tent, but this was removed by a friendly shake of the hand, and a cordial reception. These, by a note received from Mr Simpson, are of a Small party that he found on Richardson's River last night a little Distance from the Coast, encamped on an Island and been seen by the party, before they were aware of the approach of Strangers. Oolibuck was sent alone to reconcile them to the appearance of Whites, which he partly Effected. An Old man[24] who was infirm & could not fly was, however, the only person who did not make a move when Mr. Simpson and party appeared. He with some of the men crossed to the Hut on the Island. Some women on their arrival covered their heads and began chaunting what seemed to be a dirge or death Song and imploring mercy. Some children in the hut were only Known to be there from hearing their cries coming from under a heap of Sundry baggage piled up in a corner to conceal them. They were however soon

19 Thank God.

20 Allowance had been made for half these provisions being spoiled, so this situation was a pleasant surprise.

21 I.e. portaged down past the fall.

22 Signs of Inuit.

23 Whom he had met on the Richardson River (FNB C, f. 6).

24 Named Awalook (Simpson, *Narrative*, 346).

assured that their visitors intended them no injury & the night was passed in tranquility. This morning, After sending Oolibuck with these two men, Mr. Simpson having been to the Coast previously, mentions the Estuary of the River to be in Back's Inlet,[25] and is to follow the Stream upwards part of this day and probably will be back tomorrow or the day following. Very little intelligence can be obtained from these people. They seem quite ignorant of the Coast any distance either to East or Westward of the Islands in Sight,[26] on which they pass the Winter & cross to the main land to fish & hunt during the Summer Season. This, by Oolibuck's imperfect interpretation, seems to be their regular routine of life. Our two visitors appear to be highly pleased with all they see and evinced some curiosity in Examining the different articles shewn them, without any disposition to steal or pry further than was permitted. The young man was highly pleased at the Cordiality shewn him by Macacoose, our hunter, who was no less gratified at having at length seen Esquimaux & talked to them.[27] They were left to pass the night together with Oolibuck.

**26 June 1839, Wednesday.** This morning several presents were made to the Esquimaux who in return offered what they had. Each presented a Copper Dagger with long handles of Deer horn, some arrow points & Seal Skin line. No Ivory was seen with them, from which I infer that this part of the Coast is not frequented by the Walrus.[28] Before leaving us to return to their Tents, I requested of our Esqx. friend to get a pair of Seal Skin Boots made for me & to come and meet us here in the Fall, sometime from 15th to 20th Septr., which he promised to do and took the measure of my foot. They left us about Midday. George Sinclair went out to hunt. In the Evening about 9 P.M. Mr Simpson & party came back have Examined the River[29] about 20 miles from the Coast to where it divides into two branches. The Inlet into which the River falls appears to be deep; the most distant point seen from the Estuary of the River about 15 miles to the northward. Sinclair not returned yet.

**27 June 1839, Thursday.** This morning Sinclair came back. Killed a Deer and left his load about 3 miles from here which was sent for. The Boats were launched to the River. Mr Simpson and I with a crew crossed over to try angling on the opposite Side, but after 2 hours trial returned without success. It would appear that the more violent agitation of the Eddy on this Side is the reason the Salmon takes the fish Bait.

**28 June 1839, Friday.** After Breakfast the Boats were loaded & we proceeded down. Took our masts, yards & oars at the Island, and passing at our last year's encampment landed to see if the articles left there had been taken, & found the Space Clear.

---

25 So named by Richardson in 1826 (Franklin, *Narrative of a Second Expedition*, 260); now Richardson Bay.

26 Couper Islands.

27 The young Inuk slept beside Macacoose that night (Simpson, *Narrative*, 349).

28 Dease is quite correct. The Atlantic walrus (*Odobenus rosmarus rosmarus*) rarely penetrates farther west than Boothia Peninsula, while the Pacific walrus (*O. r. divergens*) rarely reaches the Beaufort Sea from the west (Banfield, *The Mammals of Canada*, 364–5).

29 Richardson River.

This we were informed of by the Esqx. at the Fall. Encamped on the N Side the Island off the River. Put down a nett at 6 PM. out of which 10 Salmon and a fine white fish were taken in the Evening. The Bay[30] to westward is pretty free of Ice, but from the Outward Islands the Ice is firm to the Shore near our present Station. Some Bustards flying about the Island. 3 Eggs were found but had young ones in them. This morning the weather was fine & Warm, the Thermo. at +77, but in the Evening a wind from ENE refreshes the air.

**29 June 1839, Saturday.** The Weather Cold with a Strong Wind from NbE. Some pieces of Ice detaching from outside & seperating before reaching land. The nett took 21 fishes, most of them Salmon. Wind moderated in the Evening. Vegetation far more advanced than last year, I think at least a fortnight earlier, as Several flowers are in full bloom, that had not begun to open at this date last year.

**30 June 1839, Sunday.** This day fine & Warm. Wind Changed to W b S. in course of the day. Sinclair, Oolibuck & McKay went out to see if they may fall in with any Esquimaux about the River to the Eastward where they were seen last year.[31] The latter Came back in the Evening without having seen any. LaRocque and Macacoose were out. Fired several Shots at a Deer without Effect. The netts (another was put down) took 31 fishes, among which a Small flounder. Musquitoes begin to torment us; numerous this evening. A Bustard was Killed & Some eggs found but not Eatable.[32] Golden plover[33] has its nest with 4 eggs in it about 15 yards from the Tent. The Anxious bird comes running towards any one going in that direction, but has become reconciled to the Sight & has not abandoned the nest. It is truly pleasing to observe with what regularity the male comes to relieve its companion of the Charge while the other goes to seek its food, and the Chirping notice given of its approach, which is made very Cautiously, untill the female leaves its charge to him.

**1 July 1839, Monday.** Musquitoes swarming. The wind at 9 A.M. turned to NE and brought on Clouds with a few drops of Rain, but Cleared again. Sinclair came back at 10 A.M. Informs us that he & Oolibuck found a Camp of Esqx. yesterday, 8 Tents containing as far as they could see 16 men and about 34 Women & Children. They were Encamped on the borders of a lake, on the point of a neck of land Jutting out & before they discovered their visitors, these were on the narrow neck of land which prevented their Escape in any direction except over the Lake. On the first alarm some fled, but Shortly after returned. Among them were the man I saw last year, who found his property & what was left for him, which he Expressed himself well pleased with. 5 men of the party were yet out hunting. Some deers were Killed by them yesterday. They seemed to be in no want; had fish & meat fresh with which they entertained them hospitably during the night, & they left them this morning. Oolibuck has remained behind, Sinclair supposes being in want of rest, & made his appearance in the Evening. A Spot was Selected and a post put up to take the dip of the

30 Richardson's Bay (FNB C, f. 7).

31 Napaaktoktok River.

32 Because the embryos were already well formed (FNB C, f. 8).

33 *Pluvialis dominica.*

needle, which gives 86° 47'. Began to rain in the Evening.[34] Our netts gave about the usual Supply.

**2 July 1839, Tuesday.** The rain continued all night and ceased about 8 A.M. A thick mist followed which only Cleared late in the Evening. The Netts continue to Supply us abundantly, indeed more than we can consume, principally Salmon, one of which measured 3 feet in length & would weigh about 18 or 20 lbs. The Ice appears to be thinning in the Bay. A passage open for about 3 miles, running to Shore in the Bay to Eastward.

**3 July 1839, Wednesday.** At 5 A.M. this morning we crossed the Bay but found the Ice firmly adhering to Shore to Eastward of the point we had in view yesterday. Landed & had the Boats unloaded. Several of the men[35] went out to fire at Some Deers Seen on the hills. Only Macacoose Killed a Buck which was brought. McKey & Sinclair fired at wolves seen prowling about. Weather became Cloudy towards evening. Mr Simpson Killed a large duck. A Seal was fired at without effect.

**4 July 1839, Thursday.** A Cloudy & foggy day. Cold wind from the NE drove the Ice upon the Shore, pressing it up considerably. Different from last year at a later Season, the water is quite Salt whereas it was drinkable and fresh last Season in most places untill we reached Bathurst Inlet.

**5 July 1839, Friday.** Had rain in the night. A thick mist & Cloudy weather. Oolibuck out hunting. Morning wind ENE, changed to NE and NNE in the Evening. The Clouds Clearing to the Northward. The Ice pressing on the Shore.

**6 July 1839, Saturday.** This morning Oolibuck came back unsuccessfull; did not see either Deer or Natives. Weather Cleared about Midday. Wind turned to E, to ESE, and to SE. George Sinclair out hunting since morning.

**7 July 1839, Sunday.** This morning Sinclair came back; Killed nothing. The wind has Cleared a Space along the land of which we took advantage and proceeded about 4 miles, and had to land again. Rain came on and lasted part of the day. The Ice appears to be loosing from Shore this Evening. I hope it will Clear to allow us to proceed tomorrow.

**8 July 1839, Monday.** An opening formed which enabled us to Cross to the Islands of Berens group & pass by the Channels to opposite the E end, where our progress is again Stopped. Landed on the main land after pushing through the broken Ice some distance with great labour. Several of the men[36] & the Indians went out. Hope only Killed a Buck Deer which was brought between him & McKey. Mr. Simpson Killed 2 Bustards and a laughing Goose. The wind changed from NE to W which brought on rainy Clouds & Swarms of musquitoes that are very tormenting.

**9 July 1839, Tuesday.** Our Hunters, Oolibuck & Sinclair, Came back this morning. The 2 former Killed each a Deer which they have put up along the Coast. Oolibuck brought one. Sinclair Killed nothing. A narrow opening again formed which allowed an advance of about 3 miles and take up the 2 Deers Killed by the

---

34 Ouligbuck returned in the evening, having killed nothing (FNB C, f. 8).

35 Ouligbuck and the two Indians (FNB C, f. 9).

36 Sinclair, Hope, McKay, and Ouligbuck (FNB C, f. 10).

Indians. The Musquitoes as numerous & tormenting as I have seen in almost any part of the Country in the Interior.[37]

**10 July 1839, Wednesday.** A Southerly breeze coming on Cleared the Ice which is much decayed, from the Shore. At 1h.30' AM. we embarked, the breeze favouring. The Guide in Shipping the Rudder in a hurry, broke both pipes of it. We however proceeded under Sail for 1¼ hour when, the wind freshening we landed, repaired the Rudder & put off again. At 4h.30' the wind fell. We continued with the oars untill 9 P.M. and put up for the night at our Encampment of 24th last year. The Ice much decayed along Shore & not near so thick as last year. Several Seals were seen Sporting about the Boats. One was Killed by Sinclair but sunk before it Could be got hold of. Islands seen at a distance out[38] most of the day.

**11 July 1839, Thursday.** Embarked at 4 A.M. & took breakfast on an Island lying off Port Epworth of Sr. J Franklin. The Ice close in Shore. Had to push the pieces in several places, untill we got to an Island about half distance between our breakfast place & Gray's Bay.[39] The Ice too Compact to get on. We therefore Sheltered the Boats behind a projecting point of Rocks & landed for the day. Rainy Clouds Collected but the Shower, which appeared to be a very heavy one, passed to the Southward of us, but a light Shower passed over us. Weather Cleared afternoon. Wind SE.

**12 July 1839, Friday.** Wind Variable. Some dark heavy Clouds passed over. Had thunder, with some light Rain. The wind turned round to WNW at 6 PM. The Ice yet quite across a Small Bay near us, driving Eastward with the wind.

**13 July 1839, Saturday.** The wind continues from the same quarter. An Arctic Marmot[40] was caught & brought alive by one of the men. They are numerous about here. On walking some distance it was ascertained that instead of a point of the main-land with a Bay, as we thought, forming a peninsula, this is an Island. A narrow Channel passes to the Bay East of us. The tide rose 12 Inches; high water at 4 P.M.

**14 July 1839, Sunday.** Nothing particular occurred. The wind continues from the same quarter. The Ice filling up the Bay Eastward of us.

**15 July 1839, Monday.** The Ice continued driving to Eastward untill afternoon. In the evening a Clear Sky. At midnight the sun appeared but 9' above the horizon or ⅓ of its disk. Some Marmots, 1 old and 3 young. were brought by Hope & McKey, quite lively, but having no convenient place to Keep them all were liberated. In the Evening the Ice began to move out to the westward. The men, amusing themselves, Erected a pillar 15 ft. high here.

**16 July 1839, Tuesday.** At 10 A.M. a clearance to Westward enabled us to embark. We pulled round by the Channel to westward in hopes of finding a passage direct, leading into Gray's Bay, but found a bight, which we thought the entrance of a Second channel, to be but a small bay, therefore pulled back and pushed through the

---

37 Given Dease's years of experience in the northern bush, this is a very telling comment.

38 I.e. offshore.

39 Where they had had breakfast on the 26th the previous year (FNB C, f. 11); i.e. they were fifteen days ahead of their previous year's progress.

40 Arctic ground squirrel *(Spermophilus parryii)* or *sik-sik*.

channel, almost filled with broken Ice and got out along the coast eastward of our Encampment and proceeded about 5 miles, when we were again Stopped.[41] After 7 P.M. wind turned to W & S.

**17 July 1839, Wednesday.** At 1h 30 A.M. a Clearance appeared for Some distance along Shore, the wind SW. We pushed off, hoisted Sails and got on about 4 miles when a barrier of Ice stopped us for 2 hours. A move of the Ice again allowed us, by pushing through the floating pieces, to reach Gray's Bay where we had a Brisk Sail Wind for about two thirds of the distance. The Ice, driving fast with the Wind, gave some labour in pushing through. We were again Stopped about two miles W of Wentzel River at 9 A.M. Oolibuck, after landing, went out & Killed a fine Buck in good condition. At 7 P.M. a partial Clearance again Enabled us to push off; wind light. Passed Wentzel River about a mile & had to land again. The view to Seaward shews a lane of open water along the Chain of Islands outside. Our landing is on a Small Rocky Island near the main land, a Channel of about ¼ mile breadth between.

**18 July 1839, Thursday.** At 4 A.M. we again pushed off. Much impeded for Some time by drifting Ice. Wind ESE ahead. Got to the Extreme point of Cape Barrow, with much ice driving between it and the high Islands lying off at about 1½ mile distance.[42] Landed on a Small Island, and as the Ice was rather too thick, a Tent was Erected and an Observation for the Dip taken which gives 86° 47' as the mean result; Varn. 43 E. The Ice cleared a little in mean time along the point. Wind Changed to WSW. Pushed off at 8 PM. Crossed over[43] to the first Islands NE of the Cape,[44] to attain which we had to labour hard to get through pieces of Ice forming a floating barrier of upwards of a mile in Breadth.

**19 July 1839, Friday.** We landed only at 3 A.M.[45] The upper rim of the Sun only seen above the horizon at midnight. Our Crews were allowed some hours rest. The Bay in which we landed is formed by Several Islands near us. Left our resting place at 1 P.M. and with partial obstructions from Ice, Keep through Channels leading rather too much to SE. Came to an opening of the Islands entirely blocked up by Ice. On the Islands either side deer were seen, and some of the Hunters landing with Mr Simpson, who Joined LaRocque, two fine bucks were Killed. Sinclair & Oolibuck pursued the Chase and brought Another. Another was wounded by Hope but escaped. One of them was got with some difficulty from the Water that he had taken to cross an Inlet & was Killed crossing it. The remainder of the day was passed without any Clearance of Ice. The wind changed to SE.

**20 July 1839, Saturday.** No appearance of the Ice moving. We left our Camp & passing by the Channel we came through untill another Island. After Sailing about 8 miles opened a course leading more directly across the Entrance of Bathurst Inlet for Point Turnagain, outside most of the Islands that are off the Eastern Shore. Had

41 Here they fired at a caribou but missed (FNB C, f. 12).

42 Jameson Islands; Dease has understimated the distance to these islands, which is about 4 miles.

43 Under sail (FNB C, f. 13).

44 Chapman Islands.

45 Having covered only eighteen miles (FNB C, f. 13).

some Ice in the passage but at 7.30 P.M. reached that Point.[46] Landed to take Supper and proceeded with a SE wind.

**21 July 1839, Sunday.** At 2 A.M. we landed at Cape Franklin discovered by Mr. Simpson last year, a distance of about 25 miles from our Supper Place. After 7 hours rest embarked, but could not proceed more than 1½ miles when we had to land, the wind being too close & the Spray breaking over the boats. The Space of open Water along Shore from a mile to a mile and a half broad. The Ice Stretching along Victoria land as far as we can see with the SpyGlass. The land from Point Turnagain low & Sandy, Sloping gradually to the waters edge from the hills about a mile from the coast that are not more than 80 to 100 ft. high, to this Cape, which has a fine Hill commanding a fine view from its summit. The opposite or Victoria land about 20 miles distant to the NW receding to a greater distance, to the NE forming apparently a Bay which is lost sight of in the distance. Patches of Snow appear on the Hills which do not appear to be high. To the Eastward a handsome Hill about 15 Ms. distant, named Mount George in honour of our worthy Governor G. Simpson Esqre. under whose directions we have the Honour of acting. The Country is covered with Short grass & moss affording fine feeding for the Deer & Musk Ox.[47] No variety of Flora has been seen differing from the Specimens Collected last Season & Except the Small Birds of the Plover & Lark Kind, Common over the Country & the Small birds, nothing is to be found.

**22 July 1839, Monday.** The wind continues with the same violence as yesterday & Kept us all day.[48] The refraction Shews Victoria Land at a Considerable distance to the NE. On the other Side it seems to Stretch to one point South of West from this Spot. Three of the men were out hunting without success although fresh tracks of Deer are numerous, but with this wind, as they are not annoyed by musquitoes, it is likely they go from the Coast to find better feeding. Windy Clouds[49] to N & NW, and fleecy light Clouds to the E and SE. A clear Sky overhead.

**23 July 1839, Tuesday.** The Same wind Kept us all day. The Coast to Eastward appears to be Clear of Ice. But little drift wood found for fuel. One of the Indians yesterday while hunting found the Burial place of an Esquimaux about 3 miles forward, & as far as we can Judge by the appearance of some arrowpoints picked up by him near the Spot, I think it is not more than 2 years since. One of the arrowpoints is of Iron and not much rusted.

**24 July 1839, Wednesday.** A Continuation of the Same violent wind Kept us again all day. Some of our Hunters were out, and fired at Deer but without Success.

**25 July 1839, Thursday.** The wind blew very hard during the night but moderated about noon. We Embarked. The weather became quite Calm towards Evening. A Deer

46 The farthest point reached by the boats the previous year. The party had managed to take a much more direct route across Bathurst Inlet as compared to the previous year's route.

47 Simpson (*Narrative*, 356) was particularly impressed by the abundant Labrador tea (*Ledum groenlandicum),* which was in flower.

48 The wind filled the tents and covered their food with drifting sand (Simpson, *Narrative*, 356).

49 Most probably cirrus clouds.

was Seen along the Coast which Oolibuck & Larocque went after & Killed it, a fine fat Buck. After a pull of 15 Miles we landed at 7 P.M., took Supper & pushed off again. The night fine & Calm. Continued our route along the Coast similar to that described untill

**26 July 1839, Friday,** 3 A.M. when we landed to allow the crews some rest and food. After taking breakfast we proceeded & land upon a Small Rocky Island, a few miles frome Cape Alexander, named after Mr. Alexr. Simpson, brother to my friend and Companion. Ice becoming thicker as we approach that Cape. We landed at 7 P.M. on its Western point & ascended the high Rocky Hills to get a view from the Summit. From there the Ice appears to Eastward as far as we can see, but much broken and in heavier pieces than to westward, yet not in hummocks or Bergs, and is driven on this Point where we got the Boats round after great labour & risk in advancing about a mile, which took them an hour and a half to get over. Some Islands appear out between us & Victoria Land, the Continuation of which we can see to a Point running NE[50] with a Hill apparently terminating it, being the last we can see. The summit of this Point presents several ridges of Broken rocks, consisting of fragments of Lime, Sand Stone and Trap Rocks, with very little vegetation. A Deep Bay presents itself to Eastward & SE with many Islands, stretching out as far as to form a line E from this Point. Here for the first time along this Coast we have two regular tides, the rise & fall 15 Inches, from 6 to 12 morning & Evening.

**27 July 1839, Saturday.** This morning we embarked, pushing through the narrow spaces between the broken pieces of Ice, but could only proceed about four miles when we found the Ice packed upon the Point, and Kept us ashore the remainder of the day. Macacoose Killed a Small Deer.

**28 July 1839, Sunday.** This morning at 1 A.M. a space cleared, and we got under way with a light wind from WSW, but could only advance about 3 miles. The Ebb tide leaving the Ice packed on Shore Stopped us untill Midday, when the flow of the tide again gave us an opportunity of advancing. The wind fair, took us under Sail at 3 P.M. to the Pillar erected by Mr Simpson & party last year,[51] where after taking Observations again pushed off. The Bay takes a S & SW bend and several rocky Islands to SE of the Pillar Point, with projecting, rocky Points, which had the appearance of receiving a River, but upon Examination was found to be a narrow inlet. The Land become higher & rocky for some distance round the Bay, with many Islands, to one of which we crossed.[52] The Ice close on the N Side but allowing Space to pass in Shore. The main land trending to SE as far as we can see, but a Distant point appears to Eastward, whether mainland or an Island we could not ascertain from the Island on which we landed, rising about 200 feet high. We proceeded about 11 miles from Pillar Point along the Land & put up. Islands appear numerous outside as well as along the Bay.

---

50 Named Cape Pelly (FNB C, f. 17).

51 The top of the cairn had collapsed; they did not now take time to rebuild it, but did dig up the folding canoe which Simpson had buried here for safety the year before (Simpson, *Narrative,* 358).

52 The highest island of the Minto group (Simpson, *Narrative*, 359).

**29 July 1839, Monday.** Rain came on about midnight & continued by Showers with little intermission untill Midday, when we pushed off with a S wind. At 2 P.M. proceeded about 3 miles, where an Inlet runs in to Westward some distance[53] & seen from a high point of rocks to terminate in a narrow Bay. Crossed over among a group of low Rocky Islands with flat sunken rocks in many places between them. A thick fog coming on, Sailed to the bottom of a Deep Bay, the E Point of which runs out about three or 4 miles, which another inlet on the E of us forms into a narrow isthmus about 300 yards across.[54] We landed to walk across it & Sent the Boats round, where they Joined us after 2 hours pulling & Sailing. We put up for the night at this Place. Fog dispersed a little. The country as far as we can see again has the appearance of being low & flat with many Swamps & Small Lakes. A point seen ahead of us[55] seems to be mainland. Much Ice outside.

**30 July 1839, Tuesday.** The fog continued untill near Midday when it dispersed sufficiently to allow Observations to be taken for Lat. & Long.; Variation 54° by azimuth. Pushed off with a SE wind & Sailed to the Point mentioned yesterday, where the Ice is very thick along Shore but much broken. Proceeded by pushing through about a mile and a half when we could make no way. Put ashore for the remainder of the day. LaRocque Killed a Deer soon after landing. His Companion Macacoose fired 8 shots at others but only lamed one which he did not get up with. Reported that he saw the footmarks of two Esquimaux that had passed only two days since. If he is correct they are no doubt somewhere in a Bay that runs in to the Southward of us, and forms a Small peninsula with that near which we Slept last night, as both have a run, one to the E the other to Westward.

**31 July 1838, Wednesday.** We left our Camp at 2h.50. A.M. A dense fog came on Shortly after but not wishing to lose the advantage of a fair wind, continued on under Sail and found ourselves after some time in the end of a Bay. A group of Islands & Some rocky flats; a considerable quantity of Ice. We had to take the oars & poles to Extricate ourselves from this Situation, and get out from the Bay with a head wind. After getting round the Point, the Islands very numerous and the Ice very thick outside, and appears much heavier Ice than we had to the westward. No appearance of land out to Seaward. Ice near the points, but following the lanes of water among

---

53 Labyrinth Bay.

54 The combination of the intricacy of the coastline and the fog made route finding extremely difficult. In Simpson's words:

> It would be an endless task to attempt to enumerate the bays, islands, and long, narrow, projecting points that followed. The coast continued to stretch south and south-east, but lost its bold character, and became low and stony. This lowness of the land increased the intricacies and perplexity of the route, rendering it necessary to ascend every elevation that presented itself to ascertain where to make for next ...
>
> So confused were our people by the devious course we were obliged to pursue as to lose all idea of the true direction, few of them being able to indicate it within eight points of the compass. I even overheard one stoutly maintaining, in a cloudy day, that west was east! (Simpson, *Narrative*, 360–1)

55 To the east (FNB C, f. 19).

the masses of Ice, occasionally passing through channels among the Islands, Continued to advance untill 7 P.M. We landed on a point of Rocks at the end of a Sandy flat on each side, with some Rocky hills scattered in land. The Coast Shallow; Sand & Muddy Bottom. Many Deers seen about the Place, which our hunters went after. LaRocque killed 3 Bucks, Oolibuck one and James Hope one. The latter having pursued his Some distance brought but a Small part of it. The others all brought to the Boats.[56] The Animals, being in fine Condition afforded the men a fine feast of Fat meat. Hope says he saw a large herd of Musk Cattle grazing but did not go after them. This part of the Coast, although some considerable Hills are Seen, appears in general to be pretty level with fine feeding for the animals that frequent it. The morning was Cold & damp but towards evening, the weather became fine & Clear.[57]

**1 August 1838, Thursday.** Left our Camp at 9 AM. Wind N & W. A Small Bay which we had to pass after pushing off was nearly buried by Sand flats; the water much discoloured & fresh. Obliged us to keep out a considerable distance to avoid the Sand Shoals, which continue running out towards some Small Islands about 3 miles out. These flats were accounted for by James Hope telling us that while pursuing the Deer he Killed yesterday he fell upon the Banks of a River much broader than the Copper Mine[58] with a Deep Stream & Strong current of which he Judged from the Deer that he Saw Crossing it being carried some distance down the Stream before reaching the opposite shore. He was about 3 Miles then from our Camp Inland & that a small Channel from the Main Stream run to the Westward of our Camp. We were too far to ascertain the accuracy of his Statement at the time, but from the water being fresh 1½ mile from the land leads us to believe it must be a large Stream. The land now appearing less rocky off the Coast or any distance we can see, the Bays in several places presenting broad sand flats. Put up for the night at 7 P.M. on a projecting point of Rocks at the End of a Broad Sand flat on each side. The Ice before us to Eastward driven Close in Shore.

**2 August 1839, Friday.** Could only leave our Camp afternoon; the wind NE. Much impeded by the Ice, and Stopped after proceeding with much difficulty about 8 miles with oars & pushing through the Ice with poles. The Bones of a Whale were seen about 5 miles E of our Camp, about 100 Yards from the Water's edge.[59] Also, an Esquimaux camp of a single hut of this Season, with several pieces of the Skin & bones of a polar Bear.[60] Some twigs of the Black dwarf Birch, tied together in the

---

56 Macacoose wounded another but it escaped (FNB C, f. 20).

57 Two Inuit sleds were found and numerous old stone caches, suggesting that this was a location where Inuit regularly left their winter gear (for life and hunting on the sea ice) before heading inland for the summer (Simpson, *Narrative*, 361–2).

58 This was the Ellice River, named after the Right Honourable Edward Ellice (Simpson, *Narrative*, 362).

59 Very large areas along the arctic coast and throughout the arctic islands have emerged from the sea during postglacial time because of glacio-isostatic rebound. Whale bones, especially the larger bones such as the skull, jaw bones, and ribs, particularly of the bowhead whale (*Balaena mysticetus*), are commonly found in the raised marine deposits, often high above sea level and far inland.

60 The point was hence named Whitebear Point.

shape of a Mat, to put under them, which served us very well as fuel, particularly as it had been well oiled, burned finely.

**3 August 1839, Saturday.** The Ice packed ashore Kept us all day. The Wind from NW b W. Oolibuck was out and brought part of a Deer that he Killed. Observations were taken which gives our present position to be in Lat. 68° 7', Long [omitted], Dip 86° 27' & Varn. 54° 45'E. The Tide 18 inches.

**4 August 1839, Sunday.** The Ice continues close inshore. The wind changed at 1 P.M. from W to NE, which Keeps the Ice that was moving with the tide this morning ashore & continued from that airt[61] all day. Some Deers came near the camp & were fired at without Killing any. Bustards numerous in the Small Lakes about us. Wood Scarce, Although some was found yesterday & day before, both Fir & Poplar. The latter is not found along the Copper Mine as far as we Know, but the twisted fibres of the Fir very much resemble the wood found along that Stream & on the Coast near it. Lunars were taken by Mr Simpson to day & yesterday. The tide's rise is at 6 and 12.

**5 August 1839, Monday.** The Ice moved this morning. At 9 A.M. we got under way, working through the floating masses for some time. Islands very numerous in the Bay, with inlets running no distance in land. Many rocky flats & Shoals. Our Course S of E most part of the day. Many old sites of Esquimaux camps & caches seen along the Coast. Indeed We have seldom landed without seeing some old vestiges of them. 2 Swans were shot to day; being in the Moulting Season, Could not fly but gave us a long pursuit before they were Killed. Some Musquitoes seen in course of the day but not vigorous, consequently did not annoy us much. Landed for the night on an Island about 18 miles from our last camp.[62] The Ice heavy out to Seaward & bordering along the numerous Islands which serve as a Barrier between it and the main land. Fortunately for us it is so, otherwise we Could not Expect to advance much.

**6 August 1839, Tuesday.** At 7 A.M. we pushed off from the Island with a SE wind rather near. My boat had to pull up to windward to Keep our course to the nearest point. Islands numerous and Bays pretty deep. Our course S of E generally since morning. Landed to take breakfast about noon. Wind turned round to S & freshened towards evening. Landed at 9.30 P.M. took supper & pushed off again at 1. AM. A Deer was Killed in the water crossing from an Island to the main land.

**7 August 1839, Wednesday.** The wind favourable untill 10 A.M. Landed on a low Point of Rocks. The Coast appears Studded with low Islets of Granite Rock; low points running out from Narrow Indents. Observations were taken. The Compasses very Sluggish; cannot well be depended on. In course of the day a Buck Deer was Killed on an Island. The weather delightfully fine & warm. A light air of wind from the Southward served merely to cool the air of a hot Sunshine. Put up for the night on a Point of the mainland. A Chain of Low Rocky Islands continue to Skirt the coast. Bays & inlets as in the morning. The Ice outside along the Outer Islands, beyond which nothing but Ice seen to Seaward. After putting up, some Deers were seen.

61 Direction.

62 It was a clear night and the first stars were visible (Simpson, *Narrative,* 363).

Hope and Oolibuck went after them and each Killed one. Part only of each was brought to the Camp. We are now nearly due South of Captn. James Ross's point Franklin[63] in the 100th parallell of Long. The Coast line & Islands very deceptive to the eye, shewing from the want of height and conspicuous objects more distance, than is found to attain them. From our present Position a distant Point or Island appears to be running out to NNE at a considerable distance.

**8 August 1839, Thursday.** The morning fine. At 8 A.M. pushed off. A light Southerly wind helped our progress untill 11h.40' when we landed to take observations: Varn. 25° 45'E. Pushed off again & Crossed a Bay, the Extreme point of which runs out to Northward, which we got to at Sunsett. Some Islands larger & higher [than] those we have passed, outside of us. Landed in a Sandy bay. The Land higher & more vegetation. Lat 68'.8'.30'.

**9 August 1839, Friday.** This morning after leaving our Camp some time, we saw a Channel running to Eastward of our last night's Station.[64] Leaving that an Island running along the Coast to northward & NNW untill 9 P.M., we landed upon a point from [which] the mainline of the Coast appears to trend more to the East of north. Our present position about 68° 24'N Lat. and 98° 30'W. Lon. The Azimuth to day only gave 15° 50'E. Varn. Two Seals were Killed & Sunk to the Bottom before they could be got hold of, and although we could see them had too great a depth to get them up. Our principal Object in Killing them was to use the blubber as fuel, wood being very Scarcely found & merely in Small pieces or Chips about old Esquimaux Camps. Many Deers were seen in course of the day. Since Midday the Soil in general appears to be a Yellowish Sand & fragments of Limestone with Boulders of various Coloured Granite along the edge of the Islands as well as the Mainland. The Country tolerably level; many Small ponds & Lakes. Vegetation scanty. Few or no Rocks to be seen in the distance.

**10 August 1839, Saturday.** Much Ice among the Islands & driven on Shore near us. Left our encampment about noon. Our route among a number of Islands which are so situated as to have the appearance at a Short distance of being points of the main land. We Consequently went round several Bays in Expectations of finding Channels, where we were actually along the main land. We proceeded to 9 P.M. and put up on an Island. From its Summit, about 200 ft. high, a line of Sand to the northward appears to be running East & West, at Some distance from us. Several Islands near us in the same line,[65] with a Strait or Inlet, beyond which appears a broad Space with Some Ice broken & open sea in the distance. These appearances lead us to hope

63 Cape Franklin, on northwestern King William Island, seen by James Ross to the southwest from his farthest point, Victory Point, on 29 May 1830 (Ross, *Narrative of a Second Voyage*, 418).

64 "On an Island this morning found a Deposit of Deers' paws with the first Joints or Marrow Bones unbroken, put up in the Spring by the Natives, merely covered over with a few Stones" (Dease's own note in original).

65 What Dease is describing is the partially submerged drumlin field which forms the mainland and offshore islands in the western entrance to Simpson Strait (the strait between King William Island and the mainland); drumlins are rounded hills, teardrop-shaped in plan, whose long axes are parallel to the last direction of ice flow, and hence to each other.

that it may be a Strait leading to Captn. Backs Great Fish River.[66] A Deer was Killed on the Island by the Indians; being a young and lean one they brought but a Small Part of it. Along the Coast fragments of Limestone, Sand, and Granite boulders.

**11 August 1839, Sunday.** At 6 A.M. were afloat. Passed by several Islands with some Ice driving. The Space between us & the land to Northward, which untill we landed shews a Continued line, has narrowed to between 2 & three miles,[67] with Islands in the midst. Our position about 68° 30'N Lat & 98° 30'W Long. Several Deers were seen on the Hills and Islands. The Coast similar to yesterday. The wind turned to SE, heading us. A Strong current or tide running to Westward this morning. Weather fine and warm. The Ice found, in passing among the Islands, very thick & Several Salmon seen as we happened to Stop near some masses of Ice, sporting themselves in their Element. As we advanced the Northern Land receded allowing an open view to NE with the Sea open & free of Ice as far as we can See.[68] Put up at 9 P.M. in a Small Enclosed bay well sheltered. Several Lakes are seen from the Hills, one of them about 2 miles in length South of our Encampment. Many old vestiges of Esquimaux seen on the Islands & Mainland wherever we land. Some deers were seen when we landed and pursued by the Indians. They had, however, been roused & Could not be approached.

**12 August 1839, Monday.** In course of the night the weather became Cloudy with a Strong Easterly wind. At 6 A.M. came on a Heavy Shower of Rain with thunder and vivid flashes of lightning; the thunder peals from the SWest. The rain did not Cease untill towards evening. No fuel found about here. Our Economy of that article hitherto has been the most rigid. When any was found, merely using it to Cook our Victuals or Boil the Tea Kettle. But this rainy weather has deprived us of that Comfort during the day. The trending of the land seems to be to the Southward of East. The Dip of the Needle was taken by Mr Simpson, which Shews 89° 27', while that taken by Captn. Back on the Same date, 1834, at Point Sr. Chs. Ogle shewed 89° 26'. Lat 68° 13' 57"; Long. 94° 58' 1" by that Officer's table of the Varn. & Dip.

**13 August 1839, Tuesday.** This morning we got under way with a fine Sailing breeze. The Weather Dense and foggy, prevented our having any distinct or Distant view of the Land along which we Sailed & which it was necessary to Keep sight of. We, however, Crossed a Bay or Channel and Sailed along a point or Island whence another opening being passed, the remainder of the Day was along a low land. Rounded a Point, where McKay put a Question to Geo. Sinclair in the Cree language, asking if this did not resemble their Point.[69] An answer of uncertainty was given.[70] At 9 P.M. we landed on a low Shore with a Shallow beach.[71] Stones scat-

66 These hopes were well founded; they had reached the western entrance to Simpson Strait between King William Island (which Dease and Simpson named Ross's Land) and the mainland.

67 They were now in the narrowest part of Simpson Strait.

68 This was the open water of what has since been named Rasmussen Basin.

69 Both men had been members of Captain George Back's party which descended the Back River in 1834. Their farthest point had been Point Ogle, but they had also seen Point Richardson father west. It is not clear which point Dease is alluding to here.

70 Simpson also admitted to being uncertain, because of the fog (Simpson, *Narrative*, 369).

71 South of Point Pechell, according to Simpson (*Narrative*, 369).

tered out from Shore made us look out Some time for a good landing Spot, but none very convenient. Therefore the Boats were unloaded[72] to prevent any accidents.

**14 August 1839, Wednesday.** A NE gale came on in course of the night, with such violence as to Keep us wind bound all day. Some of the men were out hunting, and Came with the meat of 2 Deers. Weather Cold & Cloudy. Land Seen to the NE opens as 2 Distant hills or Islands about 20 to 25 miles off. No observations could be got to day. This is the first Storm of Wind we have had this Season with open sea. Some Willow roots[73] & moss was all that could be found for fuel, & with much difficulty a fire was Kindled to Cook our Supper.

**15 August 1839, Thursday.** The Clouds Cleared or dispersed, but the wind Kept up too Strong to allow us to push off. A heavy Surf breaking up the low beach would render it a dangerous trial to load the Boats. The Indians and some of the men were out. The latter Killed a Swan & brought a young one alive to the Camp. The Indians brought a young Arctic Fox. Their Freedom was given to both. Observations were taken which place us considerably to Southward of Pt. Richardson of Captn. Back. The land trending to the Southward. Mr Simpson was some distance along the Coast & Saw a point about 6 miles a head of us to the Southward. In the Evening our Conjectures as to the precise Spot we are now on, were Joyfully put an End to, by the return of George Sinclair who informs us that he has been at his Old Encampment near Montreal Island where he was in 1834 with Captn. Back, & that our present position is near Pt. Petchell of that Officer, to us a most gratifying piece of intelligence. This, as may be supposed, brought on observations from his companion on that voyage (McKay) that he was sure he recognized the Point named in honour of Sr. C. Ogle, as we passed it. After a thing is found out, it is certain to bring on recollections, as in this instance. However that may be, we were not the less pleased at being so near one of the main Objects of this Expedition,[74] and I believe few of the Party but felt, and Expressed a gratefull Thanks to that Mercifull Being who has afforded us His protection and enabled us so far to meet the views of our honourable Patrons & friends by visiting a Spot so conspicuous as cannot be mistaken and as a further proof we are told that a Deposit has been left on that Island,[75] which God Willing we will soon see.

**16 August 1839, Friday.** The wind fell, but a fog arose in the morning. Our Course, however, being Known we pushed off at 4h.30' A.M. and with Sails and oars in 7 hours landed on Montreal Island, precisely at the same place where our Gallant predecessors' Tent had been on this date in 1834. Some of the men went to Satisfy their curiosity about the Cache, which was found about 150 yds. from us, near the Cleft of a Rock, under a large Stone. They removed the Stones & found 2 Bags Pemican,

---

72 I.e. unloaded some little distance offshore.

73 Arctic willow (*Salix arctica*).

74 The mouth of the Back River.

75 The cache left by Captain Back and Dr King on Montreal Island. There is no mention of this cache in either Back's or King's account of the expedition.

alive with maggots, 2 Japanned[76] Canisters of Gunpowder, 1 lb. ea., 1 Empty one, 1 Small Japanned Vasculum[77] with 3 Cod hooks in it & 1 Box percussion Caps, 3 Bricks of Cocoa, most part of which was Spoiled. Finding that the principal articles were useless in any future Case, I trust it will not be considered improper that the other articles, being of little consequence & easily restored, were Kept as a memento of our voyage having succeeded in Joining what was unknown between Point Turnagain & the discoveries of Captn. Sr. Geo Back. The Season having been so far favourable and in hopes of a continuance of fine weather which might be employed to our advantage yet, We opened our views to the Crews of both Boats of Continuing to the Eastward for some days in order, if we could, to ascertain whether or not the Sea we are now on is connected by Straits or otherwise with the Gulph of Boothia or whether what is named by Sr. John Ross Boothia Land is a part of, or seperated from, the Continent of America. To this proposal they all replied much to our Satisfaction that they were ready to Continue with us wherever we thought proper to go. The Fog Cleared about noon, Observations taken & we pushed off, Keeping along the Island to its Extreme N Point, and Crossed over to some Small Islands off from it, with intention of Crossing to the Eastern Shore beyond Cape Hay[78] if we Could. The fog, however, coming on again, we remained ashore untill 9 PM. when it cleared up once more and with a delightfull Calm night we Struck out merrily with the Oars for the most distant point in view, our crews considering their labours being once more on new unexplored grounds as a second discovery in the same season. After six hours hard labour we got to land which presents a Rocky and uneven Coast, with many Small Bays.

**17 August 1839, Saturday.** We landed at a high Rocky Point to obtain a good view of the surrounding Country, and a most noble one presented itself a few miles a head of us, the coast seen trending to the Eastward with an open Sea to the North, the Estuary of the Great Fish or Backs River, as far as Could be distinguished with a glass, shewing the points named by that Enterprising & Gallant Officer, with Montreal and other Small Islands in the Distance, an indistinct view of Land on the W Side of the Estuary, & Several Islands out from this Point, particularly Ripon Island.[79] This led us to the Animating hope of finding an open Passage, should circumstances permit our advance, into the gulph of Boothia, as every appearance leads to the probability of its Existence.[80] We advanced to a high Cape from where the land trends to Eastward and landed to allow the Crews refreshment and rest at 4h.30' A.M. At noon an Observation was taken, and on fixing the Azimuth Compass to obtain the Variation, found after several trials that it retained the true meridian line.[81] Our best

76 Varnished.

77 A tin box for collecting plants, etc.

78 Named by Back after Mr Hay, under-secretary for the colonies (Back, *Narrative of the Arctic Land Expedition*, 408).

79 Named by Back after the Earl of Ripon (Back, *Narrative of the Arctic Land Expedition*, 424).

80 In this, Dease and Simpson were to be disappointed. The only passage running east to the Gulf of Boothia is Bellot Strait, some three hundred kilometres north of their position.

81 I.e. they were directly south of the North Magnetic Pole.

Compasses were set to ascertain its correctness, when precisely the same results were given. The name of Cape Britannia was given to this Cape in Honour of our Country. The wind arose from NE and blowing fresh. At 2 PM pushed off. The swells rising and being directly a head could not use the oars. Had to put up after proceeding but a few miles under Shelter of the Points. Our landing place in a Small Bay, secure from the wind, on a Sandy Beach near a high Point of Rocks. The Tent was Erected in the Bay to take the Dip. The Horizontal needle of the Instrument here Shews the variation to be W. Many Old vestiges of Esquimaux seen, but none of recent date. Sinclair Complains of a pain in his breast. Having no other medicine but a few Emetics and Jalap powders,[82] one of the latter was given him. Oolibuck, Hope and the 2 Indians went out hunting. The former and Larocque each Killed a Deer and brought a load. In the Evening the weather became Cool & Clouds rising.

**18 August 1839, Sunday.** The wind more to Eastward than yesterday, too Strong to make any head way against it. At noon the Azimuth Compass gave the Variation 18° W. In course of the day the men put up a Pillar or Pile of Stones on the highest point of the Rock near us, as it is uncertain how long we may be detained, or if another equally conspicuous place may be found for the purpose. A brief account of our Voyage and Discoveries was written & put in a Bottle, Sealed, and Deposited in the Pile about 5 feet from the Base, the Pillar being 14 feet high.[83] Possession taken of the Country in the name of our Gracious Sovereign, Queen Victoria the first. The British Flag hoisted and a Salute fired. Nothing found here in the botanical way. Specimens of the Rocks have been taken by Mr Simpson.

**19 August 1839, Monday.** The wind moderated, though Still blowing fresh. We left the Cape at 6 AM. and pulled against heavy swells for about 9 miles. A Deep bay after some time presented. The wind changing more to the Southward, allowed us to use the Sails. After doubling the Cape some points of Rocks appear in the Bay mentioned, eastward of which the land again becomes low & flatt, covered with Short grass & mosses with London pride,[84] which is to be found all along from the Cop. Mine. At 5 P.M. landed for Breakfast,[85] being our first meal since 7 PM. yesterday, after which we advanced about 4 miles & put up for the night. The Coast here again, as along the Strait to Westward, has fragments of limestone all along with Boulders of various Coloured Granite, & some large blocks of Sand Stone. The interior has the appearance of a Sandy Soil. The Coast trending ENE. Put up for the night.

**20 August 1839, Tuesday.** The wind directly a head, blowing fresh. We pushed off and pulled against heavy Swells for about 4 miles. The gale increasing, could make

---

82 A purgative made from the tuberous root of a Mexican plant, *Exogonium jalapa*.

83 In 1962 a party led by R. Cundy descended the Back River from Beechey Lake by kayak and reached this location, in hopes of finding this cairn, since Cundy believed it would have been a logical place for any survivors of the Franklin expedition to leave a message (Cundy, *Beacon Six*, 237). Although they found what they interpreted as the remains of the cairn, they were unable to find the bottle left by Dease and Simpson, or any message. However, they did find a message left by a party from the Geological Survey of Canada in 1960, to the effect that that party had been no more successful.

84 The saxifrage, *Saxifraga umbrosa*?

85 On a cape named Cape Selkirk after the Earl of Selkirk (Simpson, *Narrative*, 375).

no head way, we therefore landed in the entrance of a Small River[86] about 60 yards wide and ½ fm. water inside the Sandy Shoal at its Mouth. Walked with Mr Simpson to the highest hill in sight, about 3 Miles eastward of us, from which a Cape or Point of high Land appeared NNE about 30 to 40 miles distance apparently running out to Northward or NE, as we lost sight of it in the Distance. To the E of it appeared to be open Sea; to the Westward the land, after running some miles, also is lost sight of. About 5 miles from our position a low Point of Sand rises out, with 2 low Islands off the point, but affording a passage between apparently. Round the Point the land, as far as we Could Judge, takes a trending to the S of E with two blue hills having the appearance of Islands seen over the point, about 20 miles distant, which we both take to be Islands, and the land trending off from the Point. The Western end of the Gulph of Boothia being in the 93 or 4th parallell[87] of longitude & 69th of Lat., it is hardly probable that an Isthmus can intervene in the Short Space between us & Victoria Harbour or Gulph of Boothia. Even the open Space seen to the NE would lead to that conclusion, independent of the trending of the Land beyond this. However that may be, seeing the time arrived which we had fixed upon for the limit of our advance, our present detention, and no Chance even Should we persevere of reaching the Straits of the Fury and Hecla, with any reasonable prospect of having the means of return, which besides would involve the Whole party in the greatest risk, by the certainty of being set fast by Ice along the Coast with no provisions or means of passing the Winter along this dreary Coast, the little dependence that could be placed, in a case of necessity in obtaining any Assistance from the natives, whose habits are such as to lead them, when conscious of a Superiority in numbers or power to gratify their cupidity at any risque, These Considerations, with a further view of Exploring such parts of Ross's land[88] and Victoria Land on our return, as Circumstances might admit, made us determine, since we could not advance to the Evening, of making this the Point of our return. Therefore on our return to the Boats Another pillar was Erected, but being principally limestone of a yellowish Colour & Small size, it was raised about 10 feet high and 6 or 7 Square at the base. Possession taken with three hearty Cheers in the name of our Sovereign, Queen Victoria 1st, a Meridian Observation taken and at '20 P.M.[89] Embarked on our return. The Sails were Spread to a Strong E breeze. A very heavy Swell which rendered our Esquimaux Oolibuck quite Sea Sick. Our little Boats Stood the Gale Nobly and at 8h.45 P.M. we landed for the night at our Camp of the 18th at Cape Britannia from which is the Shortest distance to Cross the Estuary of the Great Fish River for Point Ogle.

**21 August 1839, Wednesday.** The Wind Continued fresh but more northerly. At 5 AM we again hoisted Sail, the wind being rather near[90] with a heavy Swell which drove us Considerably to leeward of our intended Course, wind heading us. We had

86 Named Castor and Pollux River, after their boats.

87 A slip of the pen; Dease clearly means meridian.

88 King William Island.

89 12.20 p.m. (FNB C, f. 35).

90 I.e. they were forced to sail close to the wind.

to land under Shelter of two Small Islands between the Points of Sr. J Petchell and Sr. C Ogle,[91] after a run of about 25 miles, to Beach and unload our Boats. Our Esquimaux again Sea Sick,[92] but after landing took his gun for Exercise. The Indians also went out to look for Deer but came back with 5 young bustards, which is all the game they saw. The Esqx. not returned. The country round a low Sandy Soil with very little vegetation. No hills of any magnitude; gentle rising grounds & hollows; some Lakes.

**22 August 1839, Thursday.** The Wind Still a head with too heavy a Surf to load the Boats in the morning. At 8 A.M. no appearance of Oolibuck rendered us anxious about him, particularly having been Sea Sick these 2 last days. Sent off 3 men with the Indians and gave them directions to divide into two parties & Endeavour to trace his footsteps on the Sandy hillocks, wherever they might fall in with them & return as soon as they might find him, at all Events to be back by Sunsett, should he not be found previously. After an absence of 2 hours we had the Satisfaction of Seeing them safe back with him & the meat of a Buck Deer he had Killed. On being mildly reproved for causing us such anxiety he said he had broken a leg of the animal at the first shot and did not wish to lose it. Continued the pursuit untill too late to come back yesterday, but at Same time his Attention to the Wind, that he thought was too Strong for us to gett off. The wind turned more to Westward, the Surf abated and allowing our Esquimaux a hasty breakfast, the Boats were loaded and we pushed off. Reached Point Sr. C. Ogle at 2 P.M. and a portage made over the Sandy bar Connecting the Island with the main land at low Water, after which, the wind being pretty free, hoisted Sail to Survey the Deep Bay between it & Point Richardson. We Continued Coasting the E side for about 20 miles. Passed a Strait formed by an Island lying between the mainland & McConochie Island[93] about half the distance down the Bay which is of Considerable Breadth.[94] At 10 P.M. got to the bottom which is a low flatt, as far as Can be seen. The Evening turned fine & Clear. Moss[95] was gathered to cook our Victuals. We enjoyed the comfort of a warm Supper of fine fresh venison.

**23 August 1839, Friday.** This morning after taking a Sketch of the bottom of the Bay which Could not be distinctly seen yesterday, We embarked and Coasted the W Side of it under a Close NW Wind part of the way to Point Richardson, which bears to N of West from Pt. Ogle, the wind being too Close a head to Cross over the Strait to Ross's land[96] about 15 miles distant running NE and SSW as far as we can Judge from here, and appears to be the two Points named by Sr. G. Back after Captn. James Ross and Sr. Felix Booth. We Erected the Tents for the night. Many

---

91 This landing spot was about ten to twelve miles south of Point Pechell (FNB C, f. 36).

92 Hope was also seasick (Simpson, *Narrative*, 378).

93 Named by Back after his friend Captain Maconochie, RN (Back, *Narrative of the Arctic Land Expedition*, 423).

94 They had hoped that this bay connected with Wilmot and Crampton Bay, on the west side of this major peninsula, which they named after Queen Adelaide (Simpson, *Narrative*, 378–9). It does not.

95 Most likely arctic heather (*Cassiope tetragona*) rather than moss.

96 King William Island.

Deers were seen in course of the day. Larocque & Hope went out and each Killed one. Observations taken.

**24 August 1839, Saturday.** The night was Clear with a Hoar frost appearing on the Stones this morning. At 5 A.M. with a fine Calm we Struck across the Strait in a NNE direction. Landed on an Island about 10 miles from the Point & hence about 5 miles to Ross's land. Put ashore at 10 A.M. for breakfast & to take observations. The Azimuth Compass gave 14° W Varn. Many Old vestiges of Esqx. Camps seen, about which the men Collected Chips of their wood work, Sufficient with the addition of a little moss to cook breakfast. At 1 P.M. continued our route to Westward. A light wind from SE arose, but did not give us any advance. The Coast along this, as on the mainland, is covered with Limestone in Blocks & fragments with Scattered boulders of various Coloured Granite.[97] The wind in the afternoon freshened to a moderate breeze, with which we got into the Strait[98] N of our Encampment of 11/12. at 6 P.M. Continued untill 9h.20' and put up for the night. A Strong Sett of tide or current running to Westward, whether on account of the Ebb cannot say. Full moon appeared at Sunsetting. No fuel found this evening.

**25 August 1839, Sunday.** At 6 A.M. this morning were on our way & Continued coasting the North Side of the Strait untill near Midday, when we landed to get Observations & the Dip, being at the Western Extremity of it. Lat. 68° 45'N; Long. [blank] West; the Dip [blank]. The Azimuth Shews between 45 & 50 E. Varn. From this Point the land trends to the Northward; an Island about 5 miles distant; beyond it an open Sea with a distant point of land to the NWestward, but lost in the distance & cannot be traced distinctly, which with the heavy Ice found between this and Cape Alexander, and some Still in Sight to Westward, leads us to suppose another Strait turning out to Northward from this point.[99] More particularly do I draw this inference from the Circumstance of Captain James Ross having (from his Pillar which is but 57 miles from this) an open Sea to Westward & no land in Sight. Although desirous of ascertaining that Point the lateness of the Season, with the distance we have yet to perform, renders it necessary to lose no time at present which might ultimately endanger the Safety of the whole Expedition. We therefore, while the necessary Observations for time, Dip and Varn. were taking, Told the Crews to Erect a Pillar, which was done. The Stone about the Point being in larger pieces than any we saw previously of limestone, a Monument or pile was raised 14 ft. high, possession as usual taken of the Country with our Flag unfurled and a Salute fired[100] before taking our Departure to cross the Strait which is about 8 miles to our encampment of the 10/11. The wind, which had turned to Westward, Shifted round to Northward & took us over under Sail in less than two hours. Passed on the E Side of the

97 Several caribou were seen grazing on shore (FNB C, f. 38).

98 Simpson Strait.

99 This assumption was quite correct. Queen Maud Gulf lay to the northwest, with Victoria Strait running north from it.

100 This point was named Cape John Herschel after the astronomer (Simpson, *Narrative*, 380).

Island[101] to ascertain if any Channel Exists Eastward of that one & followed what appeared to us as one, which only led us into the Bight of a Bay, where we put up at 9 P.M. for the night, and from the highest part of the Point, distinctly traced it to be a Peninsula formed by a Deep bay to the Southward with many Islands. The tide mark here about 3 ft. rise but the precise time not noted.

**26 August 1839, Monday.** A very thick low fog came on, although the sun appeared through it. Varn. by Azimuth 46°E. After Midday the Fog Cleared enough to allow us to see our Way. Some broken Ice appeared out to N of us. No land from here can be seen Westward of our Point yesterday. Continued our route, Keeping outside the Islands as much as our Course will allow. The fog Cleared entirely; a light wind from SE a head. Put up at 9h.30' a few miles to the SE of our Camp of the 9/10th

**27 August 1839, Tuesday.** At 6 A.M. this morning got under way, pulling against a fresh breeze a head. At 11.30 landed on an Island where a Camp of 3 Tents of Esquimaux have been this Season. Some of their mats of the dwarf black birch and a broken Snowshovel were found, which afforded us a Seasonable Supply of Fuel to cook breakfast. Several pieces of broken lamps[102] & Kettles of Pot Stone, with fragments of Deer Skins Strewed about. Leaving this Island, we struck out for the western end of the Islands off our Camp of the 8th but the wind freshening too much to get round them, [and] boats Shipping Water, we hoisted Sail to pass through one of the Channels, which was Effected after taking in a reef of the Foresail. The wind Shifting & Snowy Clouds Coming on, At 6 PM considered it most prudent to land & let the gathering Storm pass before taking the wide Bay before us, which affords no Shelter for about 15 miles. Put up for the night, and a most violent wind came on with threatening Clouds. The sea in a Short time appeared as if covered with foam. The Boats unloaded and drawn up.

**28 August 1839, Wednesday.** The wind moderated towards morning. At 5 AM pushed off, hoisted our Sails to a SE wind & crossed the Bay. Landed at 11 A.M. The men, looking for pieces of wood, fell upon a cache made by the Esqx. & brought some blubber found loose near it, which was used to Keep up our fire, which served at the Same time to Keep those who had not Strong Stomachs at a respectable distance. The wind turned westerly, and Came on violent. The boats making very little way, the men labouring hard. Put up near our Camp of the 7th. 2 Deers were Killed by Hope & Oolibuck; one of them lost.[103] Some Snow fell; Cold weather.

**29 August 1839, Thursday.** A very Strong head wind with frequent Snow Showers and Sleet rendered it a laborious duty to make any progress, which was continued from an Early hour to 3 P.M. The Islands occasionally affording a momentary Shelter, allowed us to advance, but the wind becoming too violent, landed at that hour for the day. The weather Cold with thick Snow Clouds.

**30 August 1839, Friday.** A Continuance of the Same weather Kept us ashore untill after breakfast, when we pushed off again during a lull of the wind, which Shifted

101 On which they had camped on 10/11 August (FNB C, f. 40).

102 The Inuit blubber lamp, or *kudlik*, was usually made from soapstone or steatite.

103 It swam across a river channel where they could not follow it (FNB C, f. 46).

round to Southward and the Sails were put up with a Close wind for about 3 hours. Passed the Point on which we took Supper the 6th and put up a few miles to the Westward of it for the night.

**31 August 1839, Saturday.** We left our Camp with a Strong wind a head. Frequent Showers of Sleet and Snow during the day. Weather Cold & Cloudy. Our progress not great in such weather, with heavy Swells. The labour was continued untill after Sunsett. Put up on an Island, where a Sufficiency of fuel was Collected to Cook our Victuals.

**1 September 1839, Sunday.** The night was Cold & Stormy. The Small pools of water in the hollows of the Rocks frozen this morning ½ an inch thick. The morning indicated no appearance of a favourable change. We however pushed off at 9 A.M. and with hard labour reached a point where we landed on the 5th Ulto. The wind coming in such violent gusts with Snow Showers, we had to land at 2 P.M. The Boats Sheltered by a point of Rocks forming a Cove. Vestiges of Esquimaux seen of the Spring.[104] A Deer was Killed by Geo. Sinclair after landing & brought.

**2 September 1839, Monday.** No alteration in the Weather.[105] Small pieces of Drift wood found, Sufficient to Cook our meals. Some of the men were out to day but had no Success.

**3 September 1839, Tuesday.** We were Confined all day to the Same spot. The Indians were out and brought each a load of meat. They tell us they Killed 5 Does and 2 Young Deers, and that they found the Whortleberries[106] and Cranberries[107] [in] plenty, some distance from the Coast. The fine Season we have had will account for these berries having ripened. Last Season none were to be seen any where. Sinclair caulked & pitched parts of his Boat.

**4 September 1839, Wednesday.** The Weather Cleared during the night, but Cold. Ice formed on the Small pools of water an inch thick. The wind moderated and at 4 P.M. the boats were loaded, and we pushed off. A fine Calm followed. We pulled about 12 miles and put up for the night on an Island about 3 miles out from the main Shore.

**5 September 1839, Thursday.** Embarked at 4 AM. Calm morning. Continued with the oars untill breakfast which was taken on one of the Outer Islands. While we were at that necessary occupation, a fine Easterly breeze came on, of which we took advantage advancing at a Steady rate. Landed to take observations at White Bear Point. Pushed off, and the wind freshening, we put in the River Mentioned 31st July[108] and advance 3 miles up its Stream, which flows with a Strong current and 4 fms. water in its Channel. The breadth greater than the Cop. Mine at that distance from the Coast. Put out again and continued under a Strong Wind and heavy Swells about

104 I.e. dating from the spring.
105 They woke to a wintry landscape, the country snow-covered and the ice on the pools thick enough to bear a man (Simpson, *Narrative*, 381).
106 Blueberries (*Vaccinium uliginosum*).
107 *Vaccinium vitis-ideae*.
108 The Ellice River.

15 miles Westward of the River, when, darkness coming on, put up on an Island. Weather turning Cloudy. A little Rain fell in the Evening.

**6 September 1839, Friday.** The wind not so Strong as last night but Still favourable. At sunrise we pushed off & hoisted Sails, Steering our Course for a large Island[109] that was Seen at a great distance from Cape Alexander. The wind being too near to get to the N End, we landed on it to take Observations. While breakfast was preparing, Mr. Simpson went up to the highest point near us to have a view, but the breadth of this Island did not afford a favourable one. A point of the opposite land[110] was seen at a considerable distance. We embarked and Coasted along its Eastern Shore to the Southward point & passed through a Channel Seperating it from other Islands running westward towards Pillar Point, and Steering up a few miles along its western Side, Struck across a little before sunsett for Victoria land about 25 Ms distance. The Sail served us to get across; the wind moderate, rather Close. Near Midnight we reached a Bluff Shore[111] lined with a wall of Ice which rendered landing unsafe. We therefore Coasted along these Icy Cliffs for about 3 miles, when a little Point being rounded presented a fine Sandy beach to land on, with water enough to run the Boats aground on. Our Crews were highly Amused with the luminous appearance of the agitated water round the Boats, some of them not having previously noticed it. The Weather Cloudy & threatening. The land covered with Snow. Sandy Soil, as was the Island we took breakfast on, with Small Limestone Shale.

**7 September 1839, Saturday.** At 1 P.M. left our Camp. The day turned out fine & Calm. After having proceeded about 5 miles landed at a high Point to take a view of the Surrounding Land. A Bay from the Westward[112] runs back of this Point to near the base of a Conspicuous Mountain[113] about 12 miles from us, the coast Continuing W b NW for some distance. An Esquimaux Cache with several of their implements, as Lamps, Shovels, Hoops and other articles was found about half way to the Summit of the Hill, to which I added some pieces of Iron hoop, needles and awls & Closed all securely as before. Proceeded on our route. Crossed the Bay mentioned which is near 20 miles Wide from E to W and Continued untill 11 P.M. when we put up for the night. The Coast for the most part presenting much the Same appearance of a level Country covered with moss & Short grass, with many Small Lakes. Deers seem to be pretty numerous. Several were Seen in the morning and 2 Snow Geese[114] Killed by Hope.

**8 September 1839, Sunday.** Started this morning with a fine Wind. Saw many marks of Esquimaux Caches on a point of Rocks but did not land. At 9.30 A.M. landed to take our breakfast on a Point of Sand Stone Rocks, that has many large blocks &

109 Melbourne Island.

110 Victoria Island.

111 Cape Colborne (Simpson, *Narrative*, 382).

112 Named Cambridge Bay, afer the Duke of Cambridge (Simpson, *Narrative*, 383).

113 Named Mount Pelly after John Henry Pelly, governor of the Hudson's Bay Company (Simpson, *Narrative*, 383).

114 *Chen caerulescens*.

masses of the same Scattered over the Point. A wide and Deep Bay[115] before us with Several High hills in the Distance along it, about 20 miles deep apparently. 2 Islands were passed E of this. Found several Deposits of Esqx. property on this Point, and at some distance, where Mr S. went to have a view of the Bay, he found the vestiges of one of their Camps which, from the Situation & length of time apparently occupied, seems to be one of their winter Stations. After Embarking the wind fell. Crossed the bay about 12 miles from E to W. Wind rose again, NE. Cloudy. Had a light Sail Wind. A Picturesque Mountain in view on the W Side of the Bay. Put up at 11 P.M. for the night.[116] Except for a few Points of Rocks mentioned, the country shows no difference. It is now covered with the Snow lately fallen and appears to have very few hills or mountains off the Coast. Limestone is the principal Rocks seen along the Coast; the Soil Sandy in most parts with a fine Bold Shore. No wood but what is found about Old Esquimaux huts, and then merely Small Chips. These people themselves, it appears, Economise it, as Several pieces of worked, but not in use, as well as some very rough planks of Knotty wood, has been found in their Caches.

**9 September 1839, Monday.** The Weather Cloudy in the morning, with a light wind from E & NE. Embarked at 8 A.M.,[117] hoisted Sails & proceeded to near Midday when we landed to get a Meridian Observation. Had Cape Franklin in Sight about SW of us. Proceeded untill 10 P.M. The land trending somewhat to the N of West, renders it necessary to leave this Coast in order to Strike across to Cape Barrow, as the distance by the apparently trending of this land will be increased, and the Season advancing may prevent our Departure from B Lake[118] by any present retardment, and cause a Season's delay in our Return from thence, as the winter transport from Ft. Confidence would be attended with great labour & Expense. We accordingly fixed upon this as our Point of departure from Victoria land.

**10 September 1839, Tuesday.** This morning a Stiff breeze from E b N. At 8 A.M. hoisted our Sails & Steering S by E course with a heavy Sea, Carried on for about 3 hours before any Object appeared, when we got a Sight of the outward Islands off the entrance to Bathurst Inlet and some time after appeared the Chain of Islands, running towards Cape Barrow, which last was Steered for. The wind became Stronger as we neared it, with very heavy Swells. We passed the Cape at 7 P.M. and continued on untill 10.30 P.M. when, it being too dark to distinguish the many Shoals along this part of the Coast, we landed for the night a few miles to the Eastward of Wentzel River where, although from darkness little Choice of a Spot could be made, a Sufficiency of wood was found to make a fine blazing fire, a luxury we had not enjoyed for a long time, and we had once more the prospect of the same Comfort, wherever we might land, a pleasing one in this Cold Climate.

115 Named Wellington Bay after the Duke of Wellington (Simpson, *Narrative*, 383).

116 Since it was a gently shelving beach, the men had to wade a long way ashore with the camping equipment (FNB C, f. 46).

117 Having first eaten breakfast (FNB C, f. 47).

118 Great Bear Lake.

**11 September 1839, Wednesday.** A fine Steady breeze this morning, rather Strong with heavy swells, took us to Port Epworth. The wind died away; weather became Cloudy with some rain. Doubled the Cape of Esqx. Cache with oars. A dead Calm for 2 hours followed, dark Clouds rising, and a Westerly Wind came on which, increasing in violence, obliged us to look out for a place that might Shelter our Boats from a heavy Surf. This we at last found at 4.30 P.M. in a Small Cove Sheltered by a point of Rocks. Showers of Snow & Rain came on alternately for the remainder of the day. We here, however, had fuel enough.

**12 September 1839, Thursday.** A Continuance of the Same bad weather Kept us ashore all day. Some of the men went out. But one Deer was seen & the tracks of musk Ox, but nothing Killed.

**13 September 1839, Friday.** The wind moderated. Afternoon we embarked, & pulled for some hours with a Calm, when the wind again Came from W a head and we put up at 11 P.M. in the mouth of a Small River we landed at on the 8th July.

**14 September 1839, Saturday.** Wind Continued untill afternoon. Embarked at 4 P.M. The weather Cloudy. Calm of a few hours was followed by a breeze Coming on from NE with rain. The Sails put up. Wind freshened to a Stiff gale and brought us into the Channel of the first group of Islands near the main land E of the C. Mine River at Midnight; the rain pouring all the time. Landed in a very wet mossy place,[119] Some snow falling with the rain. Passed a very wet & unpleasant night.

**15 September 1839, Sunday.** The wind Strong a head. Pushed off at 8 A.M. and proceeded about 3 miles under Shelter of the Islands but were forced to land at a point opening the Coast to the full force of the wind. On the Point of Rocks found a Cache made by the Esquimaux in Spring, also 2 of their Winter Sleds, Put up on large Stones. Snow and Rain by Showers part of the day. Some hurtleberries found about the Point and Cocks Eyes.[120]

**16 September 1839, Monday.** The wind moderated. Embarked at 8 A.M. Cold Cloudy Weather; the Hills covered with Snow. Saw 4 Swans, 2 Old with 2 young at a Spot where two old were seen passing in the Spring. They took flight to the Southward. Entered the Copper Mine at 2.30 P.M. and got to the Bloody Fall at 6. Here I found to my Surprise that our Esquimaux visitors in June have been back. The old man from whom I bespoke a pair of Boots, has punctually performed his promise. They were tied at the End of a Pole and put up in a Conspicuous place with some Seal Skin line. The Water in the River in a higher State than it was last year when we passed here. One of the Boats which is to be left, was drawn up; the Sails, rigging, 2 hatchets, Hoops and 6 Bags Pemican put up for our Esquimaux friends who may Just Chance to visit this Spot. I have no doubt that the Old man is not far distant; his tracks were not of an old date. James Hope with Larocque were appointed to proceed in the morning direct to See if any Indians have come to meet

---

119 Some Canada geese were seen, and Hope killed some snow geese (FNB C, f. 48).

120 I have been unable to identify this plant.

us, as promised in the Spring, at the Boat Station.[121] If not they are to go to the Dismal Lakes and make Signal fires, in case any of them Should be about there & come back to Join us on the 20th themselves at that Place.

**17 September 1839, Tuesday.** The Man and Indian left us early this morning. The Boat was launched and the property carried across the Portage by 9 A.M. Crossed the River. There being 2 reliefs of men for the towing line[122] Mr Simpson and I walked along the Hills. A little below Escape Rapid we got sight of a fine large Grizzly Bear in a hollow plain. After taking precautions to get a Shott at him, the Animal was alarmed by the noise made in the Boats a little below us. We however had 3 Shotts but rather too distant for good Effect and he escaped. The Esquimaux Oolibuck, being a heavy man, was Exempted from the Duty of the towline in Consequence of an alarm he had in wishing to follow the Indian Macacoose scrambling to get up a Steep Cliff along the River.[123] After making his way about half the ascent he got upon a narrow ledge of Rocks, beyond which he thought it dangerous to proceed, and Equally so in his now alarmed mind to retrace his way back. He therefore laid down and began to bawl out lustily for assistance which was rendered him after the Indian had run about ½ mile after the Boat. Some of the men went back and with a line Extricated him from his awkward position, to his great Joy, at finding himself again in Safety. The Boat, without other accident than the Breaking of the towing line once, got about 5 miles above Escape Rapid at 8 P.M.

**18 September 1839, Wednesday.** The most rapidous & Shoal part of the River being this day's labour, the Boat could not advance so well. Walking over the Hills I got rather farther a head of the Boat than was intended and passed the night in Alfresco opposite the upper rapid, having, however, the means of making a fire to Dry my Shoes and Socks. The only Consequence was going Supperless through the night. I cannot say to bed, as the Cold air without any Covering, did not allow the enjoyment of more than a few minutes Sleep occasionally, while the fire Kept off the Cold. Some deers Seen but none Killed.

**19 September 1839, Thursday.** This morning at an Early hour, I was joined by Mr. Simpson accompanied by 3 men whose Anxiety at my absence brought them on as soon as the light would allow, with the means of Shifting[124] my Cloathes & Shoes if required & Something to Eat. Their Kind attention was duly appreciated, and we proceeded to a point above the Rapid where we waited the coming of the Boat for breakfast. They had encamped about 9 miles below at Island Rapid without accident. Several deers were Seen. We proceeded after breakfast at a quicker rate; the banks being low affords better walking. A fine fat Buck Deer was Killed by Sinclair. Put up for the night at the foot of the Copper Mountain. The day tolerably fine. Ground

121 The depot on the Coppermine River where the boats had been cached the previous winter.

122 I.e. two towing crews of three men each, who took it in turns hauling the boats upstream.

123 The ground was snow-covered and the rocks coated with ice and hung with icicles (Simpson, *Narrative*, 390).

124 Changing.

covered with Snow, which there is very little appearance will dissolve before ensuing Spring, being pretty deep in hollow Places, from 12 to 18 inches. No wild fowl but a flock of Swans seen flying to the Southward.

**20 September 1839, Friday.** Many Deers seen to day Crossing the River in large herds. Mr. Simpson with Oolibuck and the Indian passed over the Hills. The 2 latter each Killed a Deer but too distant from the River to be taken in. One was Killed from the Boat. We got to the Bend of the River whence we intend taking our route overland, at 6 P.M. Hope and Larocque joined us in the Evening. Have been to the Lakes[125] without any vestiges of the Indians seen by them. They made Signal Fires, without any appearance of answering ones. They brought the meat of 2 Deers Killed near hand. Mr. Simpson, Crossing the Copper Mountain Saw a fine Wooded Valley to the Westward, but too distant to see if any Stream or River ran through it, but many Deers were Seen near it.

**21 September 1839, Saturday.** The Property to be carried across was divided among the men & Such Articles as can be dispensed with, together with the Boat, made over to our two Indians, who put all in Security, covered with a heap of Stones; nearly 2 Bags Pemmican also, which may be of use to them. We began our March near Midday. Crossed Kendal's River and put up for the night after Sunsett about 2 miles South of it at the last woods from its Borders. Shortly after 6 young men Cast up from Fort Confidence, with a note from J. Ritch dated the 14th, by which we have the Satisfaction of hearing that all was well at the House. The McKenzie River Boat, sent by our Friend M. McPherson Esqre. with the Supplies required, got to Ft. Confidence on the 8th and are waiting our return. One of these young men appointed with Macacoose to go back & bring on our Tent, which was left this morning.

**22 September 1839, Sunday.** The two young men as mentioned were Sent for the Tent.[126] The Party continued at a Steady pace, through much Snow in many places. Passed many Small Lakes on the Ice and put up at 7 P.M. for the night in a Hammock[127] of Pines in the upper part of Kendals River, near the Junction of its two branches.[128] Many hurtleberries found near this Spot.

**23 September 1839, Monday.** This morning a Strong wind a head with Snow. Our course directed for the Southern branch of Dease's River,[129] which we reached after Sunsett & put up where the Boat wood was Sawed in the Spring. Larocque's Brother & family encamped but a Short distance a head of us. Less Snow than yesterday.

**24 September 1839, Tuesday.** Started at 4.30 this morning, and after a hard walk of 30 miles got to the River where we found a Boat waiting for us, and got to the House after Sunsett, with the Satisfaction of finding all our people well. We have great reason to offer up prayers of Praise and Thanksgiving to the Almighty for all

---

125 Dismal Lakes.

126 And one Indian was sent on ahead to Fort Confidence (FNB C, f. 52).

127 Clump.

128 I.e. the junction of the Teshierpi with the Kendall.

129 East River.

the mercies bestowed upon us, and the protection afforded us through all our Trials. We received letters from many of our Friends but none yet from the Governor,[130] as no Express had come to Norway House previous to the MKs[131] River brigade leaving that. Hector Morrison with the Boy Boucher are yet on the Island[132] to take up their netts preparatory to our leaving this. Some Indians about the House. But a Small Collection of provisions has been made during the summer. Animals were Scarce & the best hunters removed to Some distance from the place, who have not yet made their appearance since the arrival of the Boat.

• • •

130 Governor George Simpson.

131 Mackenzie.

132 Ritch's Island.

# Chapter Ten
# Back South to Fort Simpson

## Introduction

Packing up for the trip back south took only a short time. On the morning of the 26th of September 1839, *Goliath* and a boat which had arrived from Fort Simpson were loaded, and in the afternoon both boats pushed off. They reached Cape McDonnell late on the 27th but were held windbound there for two days. Finally, on the 30th, the wind swung into the southeast, and the party risked attempting the long open-water crossing to Etacho Point. There they were windbound for a further two days but on 3 October they got under way again. By this time temperatures were well below freezing, and, as they crossed the mouth of Narre Ella Bay (Deerpass Bay) and continued along the west shore of Keith Arm, the boats, oars, dogs, and even men became encrusted with ice from the waves and spray. They passed the site of Fort Franklin on the 4th and on the 5th the passengers and dogs walked down the portage route while the boats ran the rapids on Great Bear River. Next morning, on the lower reaches of Great Bear River the ice was driving down the river in impressive quantities, but when they reached the Mackenzie they were pleasantly surprised to find it free of ice, and with the men tracking they made good progress upriver. They reached Fort Norman on the morning of the 7th. While the explorers enjoyed the great luxury of fresh vegetables from the post's garden, the boats were hauled out and cleared of ice by the expedient of lighting fires under them.

The party started upriver again next morning (8 October) with the men tracking, but using the sails whenever the wind was fair, and thus made excellent progress. As they approached Fort Simpson, ice began to drift down the river and to spread out from the shores. This would have made tracking extremely unpleasant, if not impossible, but fortunately the wind was fair and they were able to use the sails. At 11.00 a.m. on 14 October they reached Fort Simpson, to be warmly welcomed by

Murdoch McPherson. "Considering this as the termination of our third coasting voyage," Dease ended his journal at this point.

## DOCUMENTS

### *1. Peter Dease's journal, continued* [1]

**25 September 1839, Wednesday.** Settled with the Indians who have given any thing and gave a liberal reward to our Hunters Larocque & Macacoose who with his Companion Cast up to day. Ritch and the men Employed preparing Every thing for departure as the Cold Season is Coming on. The men from the Fishery came & have left the netts, ready to be taken up when we pass there.
**26 September 1839, Thursday.** After having taken all the Ironworks that may be usefull[2] & finished packing up, all our property was divided between the 2 Boats,[3] 10 men & Indians as Crew to each, and in the afternoon we parted with our 2 Indians & numbers of the tribe, amidst loud Shouts and a Deafening noise of the Cries of men Women, Children & Dogs, witnessing our Departure. Pulled round the E end of the Island & hoisted Sail with a fine light wind which took us to the fishing House where we put up for the night.
**27 September 1839, Friday.** Embarked this morning with a SE wind, and got to Cape McDonnell where, being too late to attempt Crossing put up for the night.
**28 September 1839, Saturday.** A Strong northerly wind Kept us ashore all day. Had some snow; the weather Cold; the Small Lakes frozen over.
**29 September 1839, Sunday.** The weather continued the same as yesterday. Some of the men were out with their Guns, but saw no Deer.
**30 September 1839, Monday.** Cloudy threatening weather. The wind Shifted round to SE and we took advantage of it & hoisted our Sails to Cross over to the Scented Grass mountain. As we proceeded the wind augmented to a Strong gale. The other Boat not bearing the Swells so well as ours, we Steered for land on the E Side of the Point and landed in a Bay that Sheltered us from the Wind at 4 P.M.
**1/2 October 1839, Tuesday/Wednesday.** A violent wind with Snowshowers Kept us these two days ashore. Some of the Indians employed as part of the crews of the Boat from Ft. Simpson, were out hunting & Killed 2 Deers. Most of the meat was brought. Weather Cold.
**3 October 1839, Thursday.** The Wind Moderated towards morning. We embarked at 7 A.M. Had a Strong Breeze crossing Narre Ella Bay[4] which worked our Boats considerably, thick Coats of Ice forming on the oars when used, which the men were

---

1 MCM M214.
2 I.e. iron fittings, such as door hinges.
3 The two boats were *Goliath* and the boat from Fort Simpson. The party totalled twenty-six and they took twenty dogs with them (Simpson, *Narrative*, 395).
4 Deerpass Bay.

obliged to Break off with the Hatchets occasionally, & when the Sail was up the Spray dashing over the Gunwales on the mens Cloathes encrusted them in Ice.[5] An Indian Canoe was seen on the beach a few miles Westward of that Bay, and though a Shott was fired and loud Calls to any one that might be ashore, no answer was made. We proceeded untill it was getting dark & landed for the night. Had to Clear away the Snow to get the Tents Erected. Ice forming along Shore.

**4 October 1839, Friday.** A northerly breeze in the morning filled our Sails, which turned afterwards very violent by Sudden Squalls So that both Boats Shipped a great deal of Water while Crossing the last Bay[6] westward, and all hands were highly pleased when we once more were Sheltered from its violence by Coasting near the land. Much Ice forming along Shore. In Crossing the Bay to the River a Shott was fired by an Indian,[7] who followed us about 6 miles down the River to our Encampment, and informed us that two Indians have passed on 7 days past with a Packet for us. I think it must have been their Canoe we saw yesterday, as they Know of no Indians in that direction. One of the Company's Servants is Stationed at the Old Place to carry on the Fishery, with whom this young man is Employed. This is a sad disappointment to us, as we can Spare no time at this Cold Season to await them here. We appointed James Hope to accompany this Indian and go after them, as it is Stated as most probable they will wait along the Lake for our arrival which is daily Expected. He will, in the event of finding them, follow us to Ft Simpson with the Packet as soon as possible. They were provided with provisions, Gun Amunition & Shoes, in Short all that they might require for a long voyage at this Season.

**5 October 1839, Saturday.** This morning the Thermo. outside the Tent was at -14°. We Sent off Hope & the Indian and Embarked at 5 A.M. As we proceeded we found the Ice augmenting in the River. All the passengers & Dogs of which we have 19 landed & walked from the head to below the Rapid, which was found very Shallow. The Crews landed below & made a good fire after their long Wading in the Water. Proceeded, after Joining them, untill Dusk & put up about 9 miles from its Confluence with the McKenzie. The Ice thick in the lower part of the River but not Strong yet.

**6 October 1839, Sunday.** A Cold night. The Ice driving very thick untill we got to the McKenzie in about 2 hours, which is as Clear of Ice as in August. The men once more took the towing line and advanced at a good Step on a fine Beach. Put up for the night about 9 miles below Fort Norman.

**7 October 1839, Monday.** At 9 A.M. this morning we got to Ft. Norman, where we met with a hearty Welcome from Mr. Adam McBeath, who has been Expecting us for some days past. The Boats being very heavy with Ice were drawn up to be Cleaned of it[8] & pitched where required, one of them being rather leaky. The day fine and dry, rather cold. We had here the luxury of vegetables, a great rarity to

---

5 The dogs also became heavily encrusted in ice from the freezing spray (Simpson, *Narrative*, 396).

6 Russel Bay, the last bay before Great Bear River.

7 At Old Fort Franklin (FNB C, f. 54).

8 Fires were used to melt the ice from the boats (Simpson, *Narrative*, 397).

us. This Post has had a good Crop of Potatoes, with some Barley, and other Culinary vegetables, a great improvement to the ordinary fare of the Country.

• • •

## *2. Fort Norman journal, 7–8 October 1839*[9]

**Monday 7th October 1839.** ... Early in the morning we was much gratified by the arrival of the gentlemen of the Discovery Expedition with 2 Boats on their way to Ft. Simpson, having completed the object of their pursuit much to the satisfaction of all.
**Tuesday 8th.** ... the gentlemen of the Expedition took their departure early this morning to Fort Simpson.

• • •

## *3. Peter Dease's journal, continued*[10]

**8 October 1839, Tuesday.** This morning at 8 A.M. we took leave of Mr. McBeath, and with a fine day for the men tracking, put up for the night on the Second Island below the Point of Old Ft Norman,[11] near Clarke's Rock.
**9 October 1839, Wednesday.** Nothing occurred during this day's travel but meeting 2 Indians in their Canoes on their way to the House. Put up some distance above Large Lake River.[12]
**10 October 1839, Thursday.** An Early Start this morning with the usual route, tracking untill towards evening a light NW wind came on which we took advantage of to hoist Sails. Put up about 12 miles below the Dipping Rock.[13]
**11 October 1839, Friday.** The wind increased this morning to a fine breeze. Saw a party of Indians at the Rock but being under Sail did not Stop. The wind Continued all day & took us about half way from the Rock to Willow Lake River. Some Indians with their families that were near the place came to us & gave us a few pounds of fresh moose meat, for which they were paid. Some of them passed the night chatting with their Countrymen that are with us.
**12 October 1839, Saturday.** A Sail Wind most of the day took us to an Island near the River Nahanny coming down through the Mountains. The evening Cold with Clouds & Some Snow.
**13 October 1839, Sunday.** The Same favourable wind continued all day. Found Ice driving in Considerable breadth along the Sn. shore of the River Since morning. In the Evening Crossed over to encamp near the 3 Islands about 15 ms. below Fort

---

9 HBCA B.152/a/18, 11–11d.
10 MCM M214.
11 Old Fort Norman, located on the east side of the Mackenzie near the confluence of the Redstone; the post was moved downstream in 1823.
12 Also known as Rivière du Grand Lac.
13 Or Roche qui Trempe à l'Eau.

Simpson & find that the Ice is forming on both Sides, but not so much on the North as on the South Side, the latter being apparently coming down from the South Branch or Liard River. Evening Cold.

**14 October 1839, Monday.** A great increase of Ice this morning in the River. The wind Still favouring enabled us to Keep outside of it, otherwise we could hardly have made any head way by tracking. At 11 A.M. we drove through a broad Space of driving ice to land at Fort Simpson, where we were received by our Friend M. McPherson Esqre. with that Cordial Welcome that none can better appreciate than friends who have from Circumstances been long Seperated, and fully enjoy the moment of a happy meeting. Considering this as the termination of our third Coasting voyage, I Shall from henceforward only make Such Occasional remarks regarding our further proceedings as circumstances may require.

• • •

## *4. Fort Simpson journal, 14 October 1839*[14]

**Monday 14th October 1839.** We were agreeably relieved today from much anxiety by the arrival of Messrs Dease and Simpson. Its most satisfactory to learn that they have succeeded to the extent of their most sanguine expectations in exploring the hitherto unknown extent of Country between Back's Great Fish River and their discoveries of last year. The People are all in excellent health & spirits.

• • •

## *5. Letter from Peter Dease and Thomas Simpson to the governor, deputy-governor, and Committee of the Hudson's Bay Company, 16 October 1838*[15]

Fort Simpson, October 16, 1839

Honorable Sirs,

We have the honour to report the completion of all the primary objects of the expedition – the entire fulfilment of Governor Simpson's original instructions, under which it has been our good fortune to act, and something more; though, as we plainly told your Honours last winter, it was quite out of the question to think of reaching the straits of Fury and Hecla from the Coppermine River.

On the 22nd of June, we descended that impetuous stream to the Bloody Fall, where we remained until the 28th. This interval was employed by Mr. Simpson in exploring Richardson's River, discovered in 1838, which discharges itself, as we then supposed, into the bottom of Back's Inlet, in lat. 67° 53' 57", long. 115° 56'w. A party of about thirty Esquimaux were encamped there, all of whom fled precipitately to

---

14 HBCA B.200/a/22, f. 8.

15 Alexander Simpson, *Life and Travels*, 315–29; *The Times*, Monday, 20 April 1840, p. 5; Dease and Simpson, "Narrative of the Progress of Arctic Discovery on the Northern Shore of America, in the Summer of 1839"; *Nautical Magazine and Naval Chronicle* 9 (1840), 439–45.

the hills, except one family, whose tent was placed on an island in the stream. With these last a communication was opened through our interpreter, Oolibuck; but the circle of their little lives being confined to the Behren's Isles and the borders of Richardson's River, they had no information to impart of any value.

On the 3rd of July, the first slight opening occurred in the sea ice, of which we took instant advantage; but our first week's journey did not exceed twenty miles, and it was the 18th, after sad work, before we could attain Cape Barrow. From its rocky heights we beheld, with equal surprise and delight, the wide extent of Coronation Gulf partially open, whereas, long after the same date in 1838, the whole party might have crossed it on foot. At midnight on the 20th, we landed at Cape Franklin, just one month earlier than Mr. Simpson's arrival there, on his pedestrian journey of the year before. A violent easterly gale arrested our progress for the next four days, and on the 27th and 28th, we encountered great peril in doubling Cape Alexander, amidst very heavy driving ice.

From Cape Alexander, situate in lat. 68° 33'N., long. 98° 10'W., the Arctic coast may be comprised in one spacious bay, stretching as far south as lat. 67° 40', before it turns off abruptly northward, to the last mentioned position. This vast sweep, of which but an inconsiderable portion was seen by Mr. Simpson last year, is indented by an endless succession of minor bays separated from one another by long narrow projecting points of land, enclosing an incalculable number of islands.

From this description, it will be evident that our route was an extremely intricate one, and the duties of the survey most harassing; but, whilst perplexed beyond measure in finding our way through these labyrinths, we derived great advantage from the protection afforded by the islands from the crushing force of the seaward ice, and the weather was generally clear. In fact, the most serious detention caused by ice on this part of the voyage was from the 1st to the 5th of August, on a point that jutted out beyond the insular chain. White Bear Point, as it was called, lies in lat. 68° 7' 35", long. 104° 36' 45"W., variation 54° 45'E. These bays and masses of islands present a distinct succession of Geological features, which can be best illustrated by our series of specimens of the rocks that compose this wild and barren coast. Vestiges of Esquimaux, mostly old, were met with wherever we landed. They appear to subsist in single families, or very small parties, and to travel inland for the deer hunt in the month of June, not returning to their sealing islands till the ice sets fast in October. A river twice the size of the Coppermine, which falls into the sea in lat. 68° 2'N., long. 104° 15'W., is much resorted to by reindeer and musk oxen in the summer season.

Finding the coast, as already remarked, trending northerly from the bottom of the great bay, we expected nothing less than to be carried round Cape Felix of Captain James Ross, contrary to the conjecture hazarded by Mr. Simpson, in his narrative of last year's journey. On the evening of the 10th of August, however (at the last point already given), we suddenly opened a strait running in to the southward of east, where the rapid rush of the tide scarcely left a doubt of the existence of an open sea leading to the mouth of Back's Great Fish River. This strait is ten miles wide at either extremity, but contracts to three in the centre. Even the narrow channel is much encroached upon by high shingle islands, but there is deep water in the middle throughout.

The 12th of August was signalised by the most terrific thunderstorm we have ever witnessed in these regions. Next day it blew roughly from the westward, with a very dense cold fog, but we ran rapidly south-east, passed Point Richardson and Point Ogle of Sir George Back, and continued on till the darkness of night and the increasing gale drove us ashore beyond Point Pechell. The storm shifted to the north-east and lasted till the 16th, when we directed our course, with flags flying, to the Montreal Island. On its northern side our people, guided by Mackay, soon found a deposit made among the rocks by some of Sir George Back's party, but, as Mackay seemed to think, without that officer's knowledge. It contained two bags of pemican, and a quantity of cocoa and chocolate, all perfectly rotten, besides an old tin vasculum, and two or three other trivial articles, of which we took possession as memorials of our having breakfasted on the identical spot where the tent of our gallant, though less successful, precursor stood on his return from Point Ogle to the Great Fish River, that very day five years before.

The arduous duty we had, in 1836, undertaken to perform, was thus fully accomplished; and the length and difficulty of the route back to the Coppermine would have amply justified our immediate return. We had all suffered more or less from the want of fuel, and the deprivation of warm food; and the prospects grew more cheerless as the cold weather stole on apace; but having already ascertained the separation of Boothia from the American continent,[16] on the western side of the Great Fish River, we determined not to desist till we had settled its relation thereto on the eastern side also. A fog which had come on dispersed towards evening, and unfolded a full view of the picturesque shores of the estuary. Far to the southward, Victoria Headland stood forth so clearly defined that we instantly recognised it by Sir George Back's exquisite drawing. Cape Beaufort we almost seemed to touch; and with the telescope we were able to discern a continuous line of high land as far round as north-east, about two points more northerly than Cape Hay, the extreme eastern point seen by Sir George Back.

The traverse to the furthest visible land occupied six hours' unremitting labour at the oar, and the sun was rising on the 17th when we scaled the bluff and singularly shaped rocky cape, to which our course had been directed. It stands in latitude 68° 3' 56"N., longitude 94° 35'W. The azimuth compass, by Jones, settled exactly in the true meridian, and agreed with two others, by the same maker, placed on the ground. From our proximity to the magnetic pole, the compass had latterly been of little or no use; but this was of the less consequence, as the astronomical observations were very frequent. The dip of the needle, which at Thunder Cove (12th of August) was 89° 29' 35", had here decreased to 89° 16' 40"N. This bold promontory, where we lay wind-bound till the 19th, was named Cape Britannia, in remembrance

16 Simpson and Dease were assuming that King William Island was attached to Boothia Peninsula to the east, and that the latter was an island.

of our glorious country. On the beetling rock that sheltered our encampment from the sea, and forms the most conspicuous object on all this part of the coast, we erected a conical pile of ponderous stones, fourteen feet high, that, if not pulled down by the natives, may defy the rage of a thousand storms. In it was placed a sealed bottle containing a sketch of our proceedings, and possession was taken of our extensive discoveries in the name of Victoria I, amidst the firing of guns and the enthusiastic cheers of the whole party.

On the 19th, the gale shifted from north-east to east-southeast, and after crossing a fine bay, due east, with no small toil and danger, the coast bent away north-east, which enabled us to effect a run of forty miles. Next day the wind resumed its former direction, and after pulling against it all the morning among the shoals and breakers and gaining only three miles, we were obliged to take refuge in the mouth of a small river.

From a limestone ridge, about a league inland, we obtained a view of some very remote blue land in the north-east quarter, in all probability one of the southern promontories of Boothia. Two considerable islands lay far in the offing, and others, high and distant, stretched from east to east-northeast.

Our view of the low main shore was confined to five miles in an easterly direction, after which it appeared to turn off greatly to the right. We could, therefore, scarcely doubt our having arrived at that large gulf uniformly described by the Esquimaux as containing many islands, and with numerous indentations stretching down to the southward, till it approaches within forty miles of Repulse and Wager Bays.[17] The exploration of such a gulf, which was the main object of the Terror's ill-starred voyage, would necessarily demand the whole time and energies of another expedition, having a starting or retreating point much nearer to he scene of operations than Great Bear Lake; and it was quite evident to us that any further fool-hardy perseverance could only lead to the loss of the great object already attained, together with that of the whole party.

We must here be allowed to express our admiration of Sir John Ross' extraordinary escape from this neighbourhood after the protracted endurance of hardships unparalleled in Arctic story.

The mouth of the stream which bounded the last career of our admirable little boats, and received their name, lies in lat. 68° 28' 27"N., long. 97° 3'W., variation of the compass 16° 20'W. The strong wind that had forbidden our advance gave wings to our retreat. The same night, the 20th of August, we landed once more at Cape Britannia, and next morning we crossed the inlet direct to Point Pechell, with a heavy sea. On the 22nd we explored a long narrow bay on the west side of Point Ogle, which extends to the sixty-eighth parallel of latitude. The north wind blew roughly, with sharp frost; and next day we got no further than Point Richardson. Thence we

17 The reference is to the Gulf of Boothia and its southern continuation, Committee Bay. Of course, in reality, the party was still separated from that inlet by the full width of Boothia Peninsula, extending all the way north to Bellot Strait.

crossed over on the 24th to what had, from the continent, appeared like two islands, but which we rightly conjectured to form part of the southern shore of Boothia, or, to speak with greater precision, of that land on which stands Cape Felix of Captain James Ross. This shore we had the satisfaction of tracing for about sixty miles till it turned up to the north in lat. 68° 41' 16"N., long. 98° 22'W., only fifty-seven miles from Ross' Pillar. The dip of the needle was 89° 28' 45"N., the magnetic pole bearing north-northeast, distant ninety miles. The variation, as shown by both the azimuth compass and the horizontal bar needle, was 45° east. The objects seen on this coast are easily enumerated – a low uninteresting limestone tract abounding nevertheless in reindeer, musk oxen, and old native encampments. To the westward a good deal of ice appeared, and vast numbers of snow-geese passed high overhead in long triangular flights, bound for milder skies.

Whilst engaged in taking observations, our men constructed another durable monument of our discoveries, which was saluted in the usual manner.[18] Then recrossing the strait on the 25th, we resumed for some time our outward route, only keeping more along the seaward verge of the islands so as to shape a straighter course.

The weather, from being threatening and unsettled, soon became unequivocally severe. On the 29th of August a snow storm began that lasted for seven days, during four days of which we were fixed to a single spot by the violence of the northwest gales, while the frost was so keen that the pools among the rocks on which we lay became solid enough to bear a man. A more moderate interval succeeded this fierce outbreak. Quitting the continent again at the large river already mentioned, we struck north-northwest for an extensive island twenty-two miles off, which we coasted northwest for twenty miles; and shortly before sunset on the 6th of September stood out thence due north for the nearest point of Victoria Land, which proved equally distant. We have never seen anything more brilliant than the phosphoric gleaming of the waves when darkness set in. The boats seemed to cleave a flood of molten silver, and the spray, dashed from their bows before a fresh breeze, fell back like showers of diamonds into the deep. It was a cold night, and when we at last made the land, cliffs, faced with eternal ice, obliged us to run on for a couple of leagues before we could take the shore with safety. The coast of Victoria Land, which we explored for upwards of one hundred and fifty miles, is incomparably the boldest we have met with in these seas. Often near the shore no bottom could be found with thirty-five fathoms of line, and the cerulean blue of the water everywhere indicated a profound depth. There are several noble bays, the largest of which, northwest of Cape Alexander, is twenty miles wide and equally deep, backed by snow-clad mountains. It attains to 69° 40' north, the highest latitude of this voyage.

At length we reached the extreme point seen by Mr. Simpson from Cape Franklin in 1838, where the coast of this large country begins again to trend northward of west, Cape Barrow lying by computation south-southwest, distant fifty miles. On

18 At Cape Herschel.

the 10th September, we crossed this magnificent strait with a strong east-southeast or side wind and a rough sea in which our gallant boats, old and worn out as they were, acquitted themselves beyond our most sanguine hopes. Our return from Cape Barrow was miserably retarded by furious northwest winds and severe stress of weather. Winter permanently set in on the 15th September, and next day to the undisguised joy of the whole party, we re-entered the Coppermine River, after by far the longest voyage ever performed in boats on the Polar Sea.

Leaving one of our little craft, together with the remains of the pemican (which through age and long exposure was become quite mouldy), and various other articles, as a prize to the first Esquimaux who may visit the Bloody Fall, we ascended the river with our double crew in four days, abandoned our tents and everything but absolute necessaries, crossed the barren grounds up to the knees in snow, having unluckily left our snowshoes on the coast, and safely reached Fort Confidence at dusk on the 24th. The fisheries had failed sooner than ever, and we had good reason to congratulate ourselves on not being doomed to pass a third winter within the Arctic Circle.

After settling with the Indians, liberally rewarding the most deserving, and supplying all with ammunition gratuitously, we took our departure on the evening of the 26th in two inland batteaux: one belonged to the expedition, the other came from Fort Simpson sixteen days before our arrival.

Our passage of Great Bear Lake was most boisterous and inclement; in crossing the body of the lake and other considerable traverses our boats, with everything in them, and even the very clothes on our backs, became converted into shapeless masses and concretions of ice. It was high time for us to escape from Great Bear Lake, for the temperature, which was at 4° below zero when we landed at the head of the river on the evening of the 4th of October, fell 10° lower in the course of the night, and next day we descended the rapid stream in the very midst of the driving ice. On entering the Mackenzie we experienced a mitigation of this excessive cold; but we should most assuredly have stuck fast above Fort Norman had not the northern gales again arose in their strength, and while they shattered and dispersed the rapidly forming ice, enabled us to stem the current under close reefed sails. At noon, on the 14th October, after forcing our way with no small risk through the torrent of ice forced out by the rivers of the mountains, we reached this place, and were cordially welcomed by our valuable friend Chief Trader Macpherson, who had for some time given up all hopes of our arrival.

Most of our people are still afflicted with acute pains and swellings in the limbs caused by cold and exposure, and we are assured by Mr. Macpherson that he has never known or heard of so early or rigorous a commencement of winter in Mackenzie's River; on the other hand, so fine a spring as that of 1839 seldom visits these frozen regions, and to this favouring circumstance, under Providence, ought our signal success to be partly ascribed.

*October 30th.* The state of the ice at length enables us to despatch couriers to Slave Lake. In the meantime Governor Simpson's highly valued letter of the 17th of June, which unfortunately missed us in our way hither, has cast up overland. We rejoice in having anticipated the Russian expedition[19] and secured to our country and the Company the indisputable honour of discovering the north-west passage,[20] which has been an object of search to all maritime nations for three centuries. When our expedition was planned at Norway House, in 1836, it was confidently expected that Sir George Back would have achieved the survey of the Gulf of Boothia with the Terror's boats, and that our meeting at the mouth of the Great Fish River would have left no blank in the geography of North America. That officer's failure, the exhaustion of our men and means, and the necessity of a new wintering ground, renders a fresh expedition indispensable for the examination of the Gulf of Boothia, the circuit of which to the Strait of the Fury and Hecla, according to the Esquimaux accounts, cannot be less than four or five hundred miles. It only remains for us to recommend to your approbation the plan proposed by Mr. Simpson to perfect this interesting service, which, as he has no wish to avail himself of the leave of absence granted, he is prepared to follow up whenever the limited means required are placed at his disposal.

We have the honour to be your most obedient, humble servants,

Peter W. Dease
Thomas Simpson

• • •

19 Here the allusion is to the expedition mounted by the Russian American Company in the summer of 1838. The Aleut-Russian Kashevarov sailed north through Bering Strait in the ship *Polifem* as far as Cape Lisburne. Kashevarov reached Point Barrow with five three-man *baidarki* (kayaks) on 23 July 1838. Continuing east, he was forced to turn back by the threatening attitude of the Inuit at Dease Inlet. For full details see VanStone, *A.F. Kashevarov's Coastal Explorations.*

20 For a discussion of this puzzling claim see Barr, "Richard Cyriax's Note concerning Thomas Simpson's Claim of Having Discovered the Northwest Passage."

# Chapter Eleven
# Aftermath

## Introduction

On completion of the expedition, Dease was posted to Fort Chipewyan for the winter of 1839–40, but, on being granted a leave of absence, he decided to spend it in Britain, which he had never visited. But first he had some very important business to attend to: at Red River, in August 1840, he formally married his wife of many years, Elizabeth Chouinard. They had four sons and four daughters and were already grandparents. The marriage was registered on 3 August 1840.[1]

At about this time, too, Dease heard about the public recognition of his and Thomas Simpson's achievements. On 3 July 1840 the colonial secretary, Lord John Russell,[2] had written to Sir John Pelly,[3] governor of the Hudson's Bay Company, to inform him "that the Queen has been pleased to signify Her gracious intention of granting to each of those Gentlemen a Pension, on the Civil List of Her Majesty, of One Hundred Pounds a year, for their exertions towards completing the discovery of the North West Passage."[4]

In the same letter Lord John reported that George Simpson was to be knighted and Sir John Pelly to be created a baronet for their services.

By the end of August, Dease was at York Factory waiting for the ship to England. In a letter dated 2 September, Letitia Hargrave, wife of James Hargrave, reported of Dease: "I like him much better than any of them, he looks like a gentleman & is

1 *Red River Register of Marriages,* 3 August 1840, HBCA E.4/1.

2 Lord John Russell, first Earl Russell, had become colonial secretary after the Marquis of Normanby in 1839. Later, in 1846, he became prime minister (*Dictionary of National Biography*, 17:454–64).

3 See biography, p. 292.

4 Letter from Lord John Russell to Sir John Henry Pelly, 3 July 1840 (HBCA A.13/2, p. 165); see p. 292.

pleasing in every respect."[5] But she could not resist the chance for a rather snide bit of gossip. She had heard of Dease's annuity of £100, but also of the persistent rumour that Thomas Simpson and he "expect the offer of knighthood wch. diverts the people here as they say Mrs. Dease is a very black squaw & will be a curious lady."[6]

Dease sailed from York Factory on board the annual supply ship, *Prince Rupert*, on 6 September, and on 21 October was presented to the Governor and Committee at one of their regular meetings at Hudson's Bay House.[7] A letter from Dease to Dr John Richardson dating from this period has survived and throws some interesting light on his field notebooks: "I send you my Short Diaries of the Different voyages; the 2 last you will find have not the winter transactions which I considered superfluous altho the season of 38/9 was a hard one for the natives and we had to send them provisions three days Journey to Save a party of them who would inevitably have died."[8]

Later Dease was awarded a medal by the French Société de Géographie.[9] Partly as a token of their appreciation, and partly in the light of his poor health, the Governor and Committee extended Dease's furlough until the date of his retirement on 1 June 1843.[10]

Early in 1841, Dease retired to a farm at Côte Ste Catherine near Montreal, where he was joined by his family from Red River.[11] Here he built a substantial house, which (with some modifications) is now the Hôtel de Ville of Outremont.[12] He appears to have fulfilled the functions of justice of the peace, since on 22 April 1850 he wrote to Sir George Simpson asking him to attend the Montreal Court House to give an opinion concerning a particularly irregular marriage.[13]

Throughout his career, George Simpson had acted as an investment councillor and broker for many of the officers of the Hudson's Bay Company, including Dease. This service did not end with Dease's retirement. On 10 January 1851 he wrote to Sir George asking him to sell some railway stocks on his behalf.[14]

Peter Dease died on 17 January 1863 at the age of seventy-five, and is buried in Mount Royal Cemetery.[15] His contribution to the exploration and history of the Canadian North is commemorated in a range of place names: Dease Lake and Dease River (a tributary of the Liard) in northern British Columbia, named by Robert Campbell; Dease Arm, the northeast arm of Great Bear Lake, and Dease River flow-

5 Macleod, *The Letters of Letitia Hargrave*, 76.
6 Macleod, *The Letters of Letitia Hargrave*, 76.
7 *Minutes of the Governor and Committee*, 21 October 1840, HBCA A.1/62.
8 Letter from Peter Dease to Dr John Richardson, 3 December 1840, MCM M2715.
9 Letter from Peter Dease to Sir George Simpson, 23 January 1843, HBCA D.5/8, ff. 46–7.
10 Sampson, "Dease, Peter Warren."
11 Letter from Peter Dease to Sir George Simpson, 23 January 1843, HBCA D.5/8, ff. 46–7.
12 Mr Warren Baker, personal communication, August 1997.
13 Letter from Peter Dease to Sir George Simpson, 22 April 1850, HBCA D.5/28, f. 127.
14 Letter from Peter Dease to Sir George Simpson, 10 January 1850, HBCA D.5/30, f. 71.
15 Mr Warren Baker, personal communication, August 1997.

ing into the head of that arm, both named by John Franklin; and Dease Strait between Kent Peninsula and the south coast of Victoria Island, named by Thomas Simpson.

On completion of the expedition, Thomas Simpson spent several weeks at Fort Simpson, polishing his journal for publication, working out the calculations for his surveys, and drawing the map of the expedition to the east.[16] By 22 November the Mackenzie River had frozen solid and on 2 December Simpson set off up the river by dog sledge with a party of ten men. They reached Fort Resolution on the 13th. From here Simpson sent most of the men back to Fort Simpson, while he himself set off for Red River, travelling in a carriole and accompanied by two men, McKay and Sinclair, on 28 December. They reached Ile-à-la-Crosse on 5 January 1840, Carlton on the 15th, Fort Pelly on the 22nd, and Red River on 2 February; this represented a winter trip of over 2800 kilometres in sixty-one days, including stops.

While still at Fort Simpson, Thomas had written to George Simpson (on 25 October)[17] and to the Governor and Committee in London (on 18 October)[18] to present a plan for exploring the remaining "gap" in the arctic coast of North America, from the Castor and Pollux River on the west coast of Boothia Peninsula to Fury and Hecla Strait. He indicated that Dease had waived any claim to the command of such an expedition; this was almost certainly the case, in the light of Dease's problems with his eyesight and other health problems. Simpson was proposing to establish his base at Fort Reliance, George Back's old winter quarters at the east end of Great Slave Lake, to descend the Back River, to explore the coast eastwards from its mouth, and either to return by the same route or to head south to York Factory via the western shores of Hudson Bay. In his letter to George Simpson he even offered to contribute his life savings of £500 and his future income towards the cost of the expedition.

On reaching Fort Garry, Simpson waited impatiently for some reaction from either George Simpson or London to his proposal. When the packet from England arrived on 24 March 1840, with no response from the Governor and Committee, Thomas Simpson resigned himself to not receving any response till spring,[19] but in the meantime set some preparations in motion, alerting Donald Ross (at Norway House) to what might be needed in the way of boats, supplies, and so on. Writing to Ross again on 25 May, in an official letter he made arrangements for flour, grease, and pemmican to be forwarded by the Portage La Loche boats to Lake Athabasca in readiness for the forthcoming expedition.[20] And in a private letter of the same date he indicated that if George Simpson did not come from Lachine to Norway House or Fort Garry for the annual meeting of the Council, he (Thomas) would

16 Simpson, *Narrative*, 399.

17 Letter from Thomas Simpson to George Simpson, 25 October 1839, Alexander Simpson, *Life and Travels*, 338–40.

18 Letter from Thomas Simpson to the Governor and Committee, 18 October 1839, Alexander Simpson, *Life and Travels*, 340–2.

19 Letter from Thomas Simpson to Donald Ross, 25 March 1840, BCA ADD.MS 625.

20 Official letter from Thomas Simpson to Donald Ross, 25 May 1840, BCA ADD.MS 625.

travel to England via Minnesota to put his case to the Governor and Committee in person.[21]

When an express canoe arrived at Red River on 2 June 1840, but still brought no reply from George Simpson or from England, Thomas put his plan into action. But before starting on 3 June he wrote again to Donald Ross making further arrangements for supplies, boats, and men for his proposed expedition, which he now anticipated would start in 1841. He asked Ross to ensure that John Ritch, the boat-builder on the recent expedition, would be made available to join the new expedition in 1841.[22]

Ironically, on that very same date (3 June), the Governor and Committee in London wrote to the governor at Fort Garry, responding positively to Thomas Simpson's proposal, and accepting his offer to lead an expedition to close the final gap in the arctic coast. And on the next day George Simpson wrote with detailed instructions as to how the planned expedition was to be executed.

On that day Thomas Simpson set off southwards from Fort Garry, accompanied by two Métis, John Bird and Antoine Legros.[23] Next day they joined a larger party of Métis, also travelling south. On 10 June, impatient at the slow pace of the larger group, Simpson pushed ahead with four men, his two original companions and two other Métis, James Bruce and Antoine Legros Jr.

The accounts of what happened thereafter are conflicting in detail, and the truth will probably never be known. According to an affidavit made by James Bruce before Henry H. Sibley, JP, Clayton County, Iowa (under whose jurisdiction Minnesota fell for some purposes in 1840), on 13 July 1840 (i.e. about a month after the events), on the evening of 14 June, without provocation Simpson shot John Bird and Antoine Legros Sr. Bruce and Antoine Legros Jr then fled on horseback, back to the camp of the larger party. Next morning Bruce and five others returned to the site of the shooting; they found Simpson lying in his bedroll. As they approached they heard a shot and heard a bullet whistle past them.[24] On cautiously approaching they found Simpson lying dead. This affidavit was largely substantiated by two others, one by Robert Logan Jr, made before A. Ross, JP, at Red River on 14 October 1840, the other by James Flett before John Bunn, magistrate, on 11 October 1840.

Logan testified that:

> Mr. Simpson's body was lying stretched out, with one leg across the other, and the butt end of his double-barreled gun between his legs, the right hand, with the glove off, directed to the trigger, the left hand, with the glove on, holding the gun, near the muzzle, on his breast. All the head above the nose was blown off, and we found a white night-cap lying some ten or fifteen yards off from the body, with a hole in it, as if made by a gun-shot, and much singed by fire, and some of his hair sticking in it.[25]

---

21 Private letter from Thomas Simpson to Donald Ross, 25 May 1840, BCA ADD.MS 625.

22 Letter from Thomas Simpson to Donald Ross, 3 June 1840, BCA ADD.MS 625.

23 Stevenson, "The Unsolved Death of Thomas Simpson Explorer," 65.

24 Stefansson, *Unsolved Mysteries of the Arctic,* 162–4; Alexander Simpson, *Life and Travels*, 362.

25 Alexander Simpson, *Life and Travels,* 363.

This testimony clearly pointed to suicide.

Alexander Simpson, for one, refused to accept this version of events. By his interpretation, John Bird had conceived the notion that among the papers which Simpson was carrying with him was the secret of the Northwest Passage, and that this document was of considerable value. Bird therefore determined to steal Simpson's papers, but the latter, in self-defence, shot Bird and Legros. In Alexander Simpson's view the story of the party hearing a shot as they approached the site of the earlier shooting, and of Simpson's alleged suicide, was pure invention. Either Simpson had died in the fight with Bird and Legros, or he had been dispatched when the party later returned to the site.

The story of Bird thinking that the account of Dease and Simpson having found the Northwest Passage was worth risking his life for appears extremely far-fetched. It is much more likely that, through derogatory remarks and insulting behaviour, Thomas Simpson had finally provoked his Métis companions to violence. This interpretation was certainly supported by a close friend of Simpson's, Chief Factor John Dugald Cameron:

> But – do we know all the truth? I feel convinced that we do not. From my knowledge of Poor Thomas's character – I am sure there must have been a quarrel between him and the others before the work of Blood began. Mr. Simpson was a hardy active walker – anxious to make an expeditious Journey he would have found fault with the slow pace of his fellow travellers – would have made use of harsh remarks which fellows as firey as himself and had no great love for him, would have soon led into quarrels – and from quarrels to the work of Death.[26]

Perhaps the most bizarre suggestion of all (the pros and cons of which were discussed in detail by Stefansson,[27] and which is seriously advanced as a credible explanation by Harper)[28] is that Thomas Simpson was murdered at the instigation of his cousin George Simpson, allegedly jealous of the fame accorded to Thomas for his arctic discoveries, and/or concerned that discovery of the Northwest Passage might have a negative impact on the Company's trade. This is so improbable as to merit no further discussion.

Thomas Simpson's papers (including the manuscript of his journal, which he had prepared for publication prior to leaving Fort Simpson),[29] were recovered from the site of the shooting and returned to Red River. They were turned over to Sir George Simpson at Sault Ste Marie in May 1841:

> While at the Sault [16–17 May 1841], I received through the good offices of Mr. Crooks, the President of the American Fur Company, the letters and papers that were forwarded

26 Letter from J.D. Cameron to James Hargrave, 25 April 1841, in Glazebrook, *The Hargrave Correspondence*, 344–5.

27 Stefansson, *Unsolved Mysteries*, 182–9.

28 Harper, *An Unsolved Mystery*, 9–10.

29 Alexander Simpson, *Life and Travels*, 330.

from Red River last summer by my lamented relative, the late Mr. Thomas Simpson, and the journal, charts and other papers and some articles of clothing etc., etc. belonging to that gentleman. The papers etc. I took with me, and will send them to England by the York ship; the other articles were forwarded to Mr. Christie[30] at Moose, to be disposed of as he, the Executor, may consider expedient, with the exception of a gold time keeper, belonging to Mr. Edward Smith, late Chief Factor, which was lent by that gentleman to the Expedition, and which I requested Mr. Ballenden[31] to forward to him.[32]

Sir George later turned the journal over to Sir Henry Pelly, governor for the Hudson's Bay Company, who on 7 June 1843 signed a memorandum of agreement with the publisher Richard Bentley of London for publication of the narrative.[33] At Sir Henry's request, Sir Edward Sabine agreed to edit the work; he had participated (as astronomer) in Ross's expedition of 1818 on board *Isabella*, and in Parry's expedition of 1819–20 on board *Hecla*.

Initially buried at the site of his death, Thomas Simpson was disinterred soon afterwards and reburied in the churchyard at Kildonan, Red River.[34] His brother Alexander had a commemorative tablet made and installed in the parish church at Dingwall, Scotland:

In Memory of
THOMAS SIMPSON
a Native of this County and Burgh
who departed this life
on the XVth June MDCCCCXL
a few days before he had completed
his Thirty Second Year.
Though removed so early
He had already attained the high
distinction of being the discoverer
of the long sought
NORTH WEST PASSAGE

Subsequently, in light of the suspicions that Simpson might have committed suicide, the parishioners demanded that the tablet be removed from the church, even threatening to remove it by force.[35] Since this incident is also mentioned by Alexander in his biography of his brother (and characterized as an outrage),[36] it must have

30 Chief Factor Alexander Christie.

31 John Ballenden, in charge of the Sault Ste Marie and Lake Huron districts since 1840.

32 Letter from Sir George Simpson to the Governor and Committee, May 1841, HBCA D.4/109, ff. 22–4; Williams, ed. *London Correspondence Inward from Sir George Simpson 1841–42,* 21.

33 MacLaren, "The HBC's Arctic Expedition," 477, n. 7.

34 Stevenson, "The Unsolved Death," 66.

35 Barr, "Richard Cyriax's Note," 114.

36 Alexander Simpson, *Life and Travels*, 386 and 386 n. (original)

occurred before publication of that book in 1845. The tablet was then moved to the county offices in Dingwall, where it was mounted above a door in the entrance hall. It is now on display in the same building, which is now the Dingwall Museum.

## DOCUMENTS

### *1. Letter from Thomas Simpson to the Governor and Committee, 18 October 1839*[37]

Fort Simpson, October 18th 1839

Honourable Sirs,

My much-esteemed friend, Chief Factor Dease, has made over to me his claim to the command of another expedition, which, from the tenor of your Honour's dispatches, you will doubtless determine on setting on foot for exploring the Gulf of Boothia, the only section of the Arctic coast of America now unknown. I am therefore emboldened to lay before your Honours a plan for the accomplishment of this important undertaking, which I humbly think can scarcely fail of success.

The position of the Great Fish River, falling into the sea in the vicinity of the unexplored coast, clearly points it out as the best egress to the scene of operations. True, it possesses no advantages over the Coppermine as a channel of *retreat*; on the contrary, the difficulties would, in such an early winter as the present, be greatly aggravated; but, as a starting point, it offers the immense superiority of bringing the party upon the field one month earlier. As compared with Churchill, its superiority in respect to *time* is still greater; there remains, therefore, not the shadow of a doubt that Fort Reliance[38] ought to be the wintering ground of the proposed expedition.

The party should consist of thirteen men, including the Esquimaux interpreter and a person qualified to act in the double capacity of boat-builder and summer-master; the latter, with two fishermen, to remain at the establishment. The other ten men to accompany myself to the sea; and, with two Indians, form the crews of two small boats, to be built in the course of the winter at Artillery Lake. I have personally bespoke the services "if required" of Mackay and Sinclair, the only steersmen in the country who are acquainted with the long and dangerous navigation of the Great Fish River.[39] The requisite supply of provisions can be procured at Athabasca,[40] and the trading goods brought from the depot[41] next Spring, to be in readiness for starting either in 1840 or 1841, as you may decide. Should it be my good fortune to obtain your Honours' confidence and sanction, there are two points upon which I would require discretionary power. The first, to employ two seasons on the coast, if

37 Alexander Simpson, *Life and Travels,* 340–2.

38 Back's winter quarters at the east end of the East Arm of Great Slave Lake, during the winters of 1833–35.

39 Both had been with Back.

40 Fort Chipewyan.

41 York Factory.

*necessary*; the second, to be permitted either to fall back on the Great Fish River, or to push through the Straits of the Fury and Hecla, and make my way to York Factory. The latter course, if practicable, would be by far the most accordant with my feelings; and I have only to add in conclusion that I am ready and willing to devote not only my life but my means to the achievement of an enterprise that cannot fail to confer new lustre on the Honourable Company's name.

With the utmost respect,
I have the honour to remain, Honourable Sirs,
Your most obedient humble servant,
Thomas Simpson

P.S. A portable barometer and a superior pocket chronometer would be required by the earliest opportunity. The barometer formerly sent us was broken on the way from Canada, and I cannot, with any propriety, retain C.F. Smith's valuable watch longer.

• • •

## *2. Letter from Thomas Simpson to George Simpson, 25 October, 1839*[42]

Fort Simpson, October 25, 1839

So far from wishing to avail myself of the leave of absence which you have so kindly offered unasked, it gives me great uneasiness that a whole year will probably elapse before the final expedition can be set on foot that is destined to accomplish the *North-east*, as my excursion to point Barrow in 1837 achieved the *North-west*, Passage.[43]

Probably you considered the discretion I solicited in my letter of the 31st January too great, at least I infer so from your silence on that head.

As for what remains to be done, I am so far from seeking to convert it to my future advantage, that, with my life, I hereby place at your disposal, towards meeting the expenses of the new expedition, should there be any obstacle, the sum of five hundred pounds, being every shilling I am worth at this moment, besides all the future proceeds of my double commission, till the whole charge of the said expedition shall be redeemed.

Fame I will have, but it must be *alone*. My worthy colleague on the late expedition frankly acknowledges his having been a perfect supernumerary; and to the extravagant and profligate habits of half-breed families I have an insuperable aversion.

The coast, from the Strait of the Fury and Hecla to York Factory, is still more dangerous for boats than that which we have tried so well this season, but my whole soul is set upon it, and I feel an irresistible presentiment that I am destined to bear the Honourable Company's flag fairly through and out of the Polar Sea.

• • •

42 Alexander Simpson, *Life and Travels*, 338–40.

43 This sentence indicates that Thomas Simpson held the peculiar notion that the term "Northwest Passage" applied to the route from Point Barrow to some mid-point on the arctic coast, and that "Northeast Passage" referred to the route from that mid-point to Baffin Bay. See Barr, "Richard Cyriax's Note."

## *3. Letter from Thomas Simpson to Donald Ross, Norway House, 26 October 1839*[44]

*Private*

Fort Simpson, 26th October 1839

Donald Ross Esquire,
My Dear Sir,

I have infinite pleasure in acknowledging your several letters of May and July last, the contents of which were most interesting. I am delighted to hear such favourable accounts of Red River, which I hope ere long to have an opportunity of verifying in person. Of your family and friends I had very particular accounts from your brother-in-law at Fort Norman,[45] with whom and his nice little wife we spent the greater part of the 7th Inst., while the ice was being thawed out of our boats with fire.

Our labours of this season have been rewarded with unexampled success, we having penetrated sixty miles *beyond* Cape Hay – the extreme eastern point seen by Sir George Back. The spring was early, and after the usual perils and hardships amongst the ice we reached Simpson's Pillar of 1838[46] on the 28th July. From thence to the narrow strait separating Boothia from the American Continent,[47] the coast forms one stupendous bay (of which but a very small part was seen by me last year), divided into an endless succession of minor bays, containing the greatest number of islands in the known world. About halfway a river twice the size of the Coppermine[48] finds vent. Our route, as you may suppose, was extremely difficult & intricate; and as we approached the magnetic pole the compass became absolutely useless. My Duties were in consequence most harassing; and we had some very bad weather before we made our triumphant landing on the Montreal Island on the 16th of August. Here McKay soon found King's clandestine cache, containing two bags of pemican *all alive*, a quantity of cocoa – all rotten; besides an old jappd. tin vasculum, three large fish hooks, 2 lbs cannister gunpowder & a box containing about 100 percussion caps. These latter articles are in our possession, irrefragable proofs of our having breakfasted on the identical spot where stood our featherbed precursor's tent, on his return from Point Ogle to the Great fish river, that very day five years before! Not content with this – the *full* performance of the arduous duty we had undertaken in 1836, we crossed the gulf to the farthest visible land, & explored it for the distance already mentioned, till severe northeast gales arrested our progress, and it became perfectly evident that it would be absolute madness to persist furtehr. On "Cape Britannia" – a noble promontory – we raised a gigantic stone pillar; and our shouts & feu de joie resounded along that wild, rocky shore, as we took possession of our splendid discoveries, in the name

44 BCA ADD.MS 625.

45 Adam McBeath. See biography, p. 303.

46 At Cape Alexander, Simpson's farthest east in 1838.

47 Simpson Strait.

48 Ellice River.

of our beloved sovereign. On our return to the Coppermine, we explored a great extent of the southern shores of Boothia[49] and Victoria Land, regaining the continent from the latter by the passage of a magnificent strait, fifty miles wide, with a strong side wind and a rough sea. That's what I call sailing! – in 24 feet bottoms, and ⅜ inch plank, thoroughly worn out to the bargain! Winter commenced on the 29th August, and pursued us with unparallelled rigour, till we forced our way, on the 14th Inst., through the torrent of ice poured out by the Riviere aux Liards, to Mr. McPherson's friendly gate. The day we descended Bear Lake river, after being several times nearly drowned and frozen to death, in crossing that terrible inland sea, the temperature was fourteen degrees below zero; and several of our poor fellows are not yet thoroughly recovered; though happily we have all escaped with life and limb.

The only Esquimaux seen were on the banks of the Richardson (discovered in 1838), a large stream to the westward of the Coppermine, which I explored on foot in the end of June. Native marks & old encampments were, however, numerous to the eastward.

All that now remains unknown of Arctic America is the great gulph of Boothia (otherwise Prince Regent's Inlet) from whence we returned; and "none shall take it out of my hands". Not only my life but the whole proceeds of my commissions shall be devoted to an enterprise, which cannot fail of success and glory. But of this more at another time.

I shall set out for Red River the moment that snowshoe travelling on the Mackenzie becomes practicable.

Meanwhile with kind regards to Mrs. Ross and family, Believe me ever,

My Dear Sir,
Most truly & sincerely Yours
Thomas Simpson

Many thanks for your prompt attention to our own & men's wantages, which are all safe before us at Great Slave Lake.

• • •

## *4. Letter from Thomas Simpson to Donald Ross, Norway House, 25 March 1840*[50]

*Private*

Fort Garry, 25th March 1840

My Dear Sir,

Last night the long looked for packet from England arrived, but as it brought me no word of remembrance from any one, I have of course no news to communicate. Canada seems quiet as yet, under Paullet Thompson, and from negative evidence, the whigs would still appear to rule the destinies of England. The Governor was in

49 King William Island, which Simpson assumed was continuous with the Boothia Peninsula to the east.

50 BCA ADD.MS 625.

the hands of the faculty[51] for his eyes, and your daughter, Mr. Christie says, much improved both in mind & person.

As there is nothing said regarding a future expedition, its initiative in 1840 or 1841 (for I feel as sure as I exist that it will be undertaken), must depend on the spring instructions. In the meantime the boat iron works will be made here and as a sufficiency of goods and provisions are secured from YF and Athabasca, every thing will be in train for a start, if so ordered.

Norway House would, of course, have to equip the people, and furnish one or two boats, besides some odds & ends perhaps; but this you knew before your started. As for Ooligbuck's family, if not up in time for the Portage brigade, they can go in with the Athabasca boats, or with myself,[52] if fortunate enough to get a fair start this year.

Time passes very heavily with me since your departure; indeed I never remember being thoroughly in the blues in my life before —a melancholy reward certainly for all I have done & suffered of late! Mr. Finlayson and the other Gents. are all well, and seemingly in good spirit. I trust you had a pleasant & expeditious journey across the lake, and that my friends proved no impediment to your progress. Upon your arrival I hope you found Mrs. Ross & your fireside (to whom I beg to be warmly remembered) as well as you could wish. Spring at last appears to have commenced; the carioles are being laid "in ordinary" today,[53] and the face of dear Mother earth begins to reappear, after having been, to *me*, seven months submerged in snow.

It is my intention, God willing, to go out to Norway House in spring; till which season of general reunion

I remain, My Dear Sir,
Always Yours Most Sincerely,
Thomas Simpson

Donald Ross Esqr.

• • •

## *5. Letter from John H. Pelly to Lord John Russell, 22 April 1840*[54]

The Rt. Honble Lord John Russell — Hudson's Bay House
22nd April 1840

My Lord,

With reference to two letters under dates 17 April 1838 and 3rd May 1839, which I had the honor of addressing to Lord Glenelg and the Marquis of Normanby as Secretaries of State for the Colonial Department, reporting the discoveries of an expedition, fitted out by the Hudson's Bay Company for the discovery of the North

51 I.e. George Simpson was in the care of the doctor.

52 Ouligbuck had remained in the Mackenzie District, presumably in anticipation of being available for Simpson's planned expedition back to the coast via the Back River. Here Simpson is arranging for his family to travel from Churchill to join him.

53 I.e. sledges were being put in storage for the summer.

54 HBCA A.8/2, f. 232.

West passage, under the command of Messrs Dease & Simpson, I have now the honor of handing Your Lordship copy of a Dispatch bearing date 16/30th October 1839, by which Your Lordship will observe that the exertions of those Gentlemen have been crowned with success.

The operations of the expedition were commenced in the year 1836. In the summer of 1837 they traced the coast from the mouth of McKenzie's River to point Barrow (Mr. Simpson having with great tact performed part of the journey on foot where the navigation was obstructed by ice) thereby connecting the survey and discoveries of Sir John Franklin from the eastward with those of Captn. Beechey from the westward.

In 1838 they descended the Coppermine River, but their boats were obstructed by ice to the westward of Point Turnagain, the extreme point of Sir John Franklin's discoveries to the eastward. Mr. Simpson, however, proceeded on foot for five days at considerable risk at that late season and traced the coast for some distance to the eastward of Point Turnagain, and from thence returned to their winter quarters on the shores of Great Bear Lake. With the breaking up of the ice in 1839, these Gentlemen resumed their harassing labours, and, as Your Lordship will observe by the narrative now enclosed, they accomplished the survey to the mouth of Great Fish River, whereby they connected the surveys and discoveries of Sir John Franklin from the westward with those of Sir George Back from the eastward. They also extended the survey of the Coast for about two degrees of longitude to the eastward of Cape Hay of Sir George Back until they saw open water extending in the direction of the Straits of the Fury and Hecla.

The discovery of a North West passage has been considered an object of sufficient importance to induce Great Britain and Russia to fit out several expeditions at a heavy cost towards its accomplishment. The British Government, moreover, by the 15th Geo. 2 Cap 07 offered a reward of twenty thousand Pounds for the discovery; this offer of a reward was continued by the 58th Geo. 3rd Cap 20, but repealed by the 9th Geo. 4th Cap. 66. Notwithstanding this repeal, however, several of the Officers who commanded expeditions subsequently, fitted out by Govenment at much cost to the Nation, have been rewarded for their partial success. I trust, therefore that your Lordship will consider the services and exertions of Messrs Dease & Simpson of sufficient interest and importance to entitle them likewise to some reward at the hands of Her Majesty's Ministers, more especially so, as the whole expence connected with this expedition, has been defrayed by the Hudson's Bay Company.

The operations of the expedition are to be continued next year under the command of Mr. Simpson, and as it is not probable that we can have any communication with that Gentleman for two or perhaps three years hence after the departure of the Hudson's Bay Company's ships about the latter end of next month, I beg leave to solicit Your Lordship's early consideration of this subject and with much respect,

I have the honor to be
My Lord,
Your Lordship's most obedt. hble. Servt.,
J.H.P. Govr.

• • •

## *6. Letter from the Hudson's Bay Company to Captain Washington, Royal Geographical Society, 22 April 1840*[55]

Capt. Washington, R.N. London, 22nd April 1840
Royal Geographical Society
Sir,

The Governor and Committee have directed me herewith to hand for the information of the Royal Geographical Society copy of a Dispatch from Messrs. Dease & Simpson under date Fort Simpson, McKenzie's River 16th/30th Oct. last, detailing the result of their operations during the last season, by which you will notice that they have accomplished the survey of the Coast of North America Eastward from point Turnagain, crossing the entrance of Capt. Sir George Back's Great Fish River, and proceeding onwards into the gulph of Boothia, from whence they were induced owing to the advanced state of the season to return by the same route leaving but a small portion of the Coast still unexplored, and which I have the satisfaction to say will be resumed in all probability the ensuing summer.

I am, Dear Sir,
Your very obedt.
W.S. Secy.

• • •

## *7. Letter from Thomas Simpson to Donald Ross, 25 May 1840*[56]

Red River, 25 May 1840

Donald Ross Esquire,
Dear Sir,

As a preparation for further discovery service, I beg leave to request, that as many pieces of flour, grease and pemican – say not less than ten bags of each – as the *Portage* boats[57] of this season can conveniently take, in addition to their other ladings, be forwarded to Athabasca, there to remain in deposit till required.

I have the honor to be
dear Sir,
Your obedt, hble. servt.
Thomas Simpson.

• • •

55 HBCA A.5/13, pp. 53–4.
56 BCA ADD.MS 625.
57 The boats bound for Portage La Loche.

## *8. Letter from Thomas Simpson to Donald Ross, Norway House, 25 May 1840* [58]

*Private*

Donald Ross Esquire — Fort Garry, 25 May 1840

My Dear Sir,

Being just starting for Bas de la riviere[59] to meet the canoes, and uncertain whether I shall get back here before the departure of the portage boats, I have written you a few lines of a public letter, requesting you to forward thirty pieces of provisions to Athabasca, by those boats, a precaution of the utmost importance in reference to any further discovery operations. Should the thirty pieces be *absolutely* too much, pray let the reduction be made on the pemican – not on the flour and grease, which keep better. I fear that the contemplated expedition cannot proceed this year, if his Excellency[60] comes not out. In that case it is my intention to proceed direct to England, via St. Peters[61] and the United States, and to urge the matter at home in person. I have little or no doubt of succeeding both there and in the Gulph of Boothia; and the intermediate jaunt, through the *new* to the *old* country, will benefit my health, which I find has suffered more than I suspected during the last four years of toil and anxiety.

In respect to the little fellows at Beren's River,[62] it is my wish that they should be forewarded in the care of Ooligbuck's wife, to Great Slave Lake, but I merely mention this now, that you may know how to dispose of them, in the event of all the parties being at Norway House in time for the portage boats.

By the By, My mother sent me out, I think in 1836 or 1837, a box containing some books, stockings, etc., which I never received. It ought to be still lying at Norway house or York factory. Have the goodness to find it out for me. There are various other matters about which I shall have to write you by the canoes.

The Settlement is dull in the extreme I have broken bread no where since you left;[63] and shall quit the place with much less regret than I ever did before. The spring is bad & tardy and the only topics of interest here are the proposed emigration to Puget Sound and our stone Church, which has been condemnned & abandoned for some time back, now is to be temporarily propped with wood, and finally pulled down & rebuilt next year – after as much discussion, and as great an expenditure of speeches, as attended Catholic Emancipation or the Reform Bill.

With kindest regards to Mrs. Ross,

I ever am, My Dear Sir,
Faithfully Yours
Thomas Simpson

---

58 BCA ADD.MS 625.

59 The mouth of the Winnipeg River at the south end of Lake Winnipeg.

60 George Simpson.

61 St Peter, Minnesota, about a hundred kilometres southwest of Minneapolis.

62 One presumes these two young lads were Thomas Simpson's illegitimate offspring, probably by an Indian woman.

63 I.e. he has not been invited out for a meal.

P.S. 1st June. Finding that the trip to Bas de la rivière would retard my departure either way, I have remained here. The weather for some days back has been oppressively warm, and the plains & woods are now clothed in luxuriant verdure. Every thing promises good crops, and the emigrants, whose main body starts tomorrow, must, I think, quit their green fields with regret. Logan gave us a capital turn out at a *christening* last week. The portage boats will set out on Wednesday – loaded with *fair* passengers, and I shall again drop you a line or two by them. Adieu, T.S.
Your spring favours I duly received. You certainly made a flying march to Norway House. What you mention about repairing the portage boats has been attended to. Please [illegible word] a few oak sledge boards for Fort Reliance.

• • •

### *9. Extract from a letter from Murdoch McPherson to John Bell, Fort Good Hope, 2 June 1840*[64]

...You will afford Oulibuck the Esquimaux Interpreter a passage with the Returns in spring '41 as far as Ft. Good Hope. Thence he will require to proceed without loss of time to Great Slave Lake & you will be so good as to provide him with a Circular to the Gentlemen in charge of posts along the communication requesting them to afford him every facility in their power to enable him to reach his destination in sufficient time to join the A.D. Expedition[65] to which he belongs ...

• • •

### *10. Letter from the Governor and Committee to Governor Christie, Red River, 3 June 1840*[66]

Reverting to the subject of the Arctic Discovery Expedition, the gallantry and excellent management maifested by Messrss. Dease and Simpson in that arduous and interesting service, and the good conduct of the people under their command, entitle them to our warmest commendations. The valuable and important services of Messrs. Dease and Simpson have been brought under the consideration of Her Majesty's Government, who have not, as yet, noticed the subject.

We observe that Mr. Dease avails himself of the leave of absence that has been afforded him, with the intention of visiting Canada this season; and that Mr. Simpson volunteers to conduct another expedition, with the view of continuing the survey from the mouth of the Great Fish River to the Straits of the Fury and Hecla. We have much satisfaction in availing ourselves of that gentleman's proffered services; you will, therefore, be pleased to meet any demands that may be made by Mr. Simpson for men, goods, provisions, craft, etc., etc.; and to take the necessary meas-

---

64 HBCA B.200/b/12, p. 10.

65 Arctic Discovery Expedition: evidently it was intended that Ouligbuck should be a member of Simpson's proposed expedition.

66 Simpson, *Narrative*, xv; Alexander Simpson, *Life and Travels*, 343–4.

ures to give effect throughout the country to that gentleman's views and wishes, in reference to the important and arduous service on which he is about to re-enter.

• • •

## *11. Letter from Thomas Simpson to Donald Ross, Norway House, 3 June 1840*[67]

*Private*

Donald Ross Esquire, Red River, 3rd June 1840

My Dear Sir,

Yesterday evening our anxiety was partly relieved by the arrival of Mr. McDonald with the express canoe, but without the main documents. Our excellent Governor, I am sorry to find, suffers from sore eyes, and his letters are all written by other hands. My own destiny is at length decided, and I must away across the plains and through the States to England. All the arrangements for a new expedition must therefore lie over till next spring, when I shall be up, God willing, to start with, or about the same time as the portage brigade. By McDermot I wrote you requesting that by the boats for the portage *this* season, a quantity of dried provisions, say not less than 10 bags *each*, of pemican, grease and flour, should be forwarded to Athabasca – a most essential precaution for next year. John Ritch has been promised a situation for the intervening year – say either at Norway House, Saskatchewan, or here. York factory would not answer, as the man must be ready to accompany me northward, at the same time as the portage boats. As a friend, I beg of you not to let Ritch go home, as that would most seriously interfere with my plans; place him anywhere he may be most required for boatbuilding, but within reach at the specified time. Ooligbuck's family, as before settled, are to be sent in to Great slave Lake; and my little chaps also, under the woman's care. I suppose that £3 would be a fair remuneration to Harper for his trouble – please to pay him, and give the boys only absolute necessaries, as I cannot afford to lay out much money upon them. The expedition goods will be up from York factory, by the Oxford House boats I suppose. Any portion that you can possibly forward by the Athabasca brigade of this season, would be a great relief & obligation, and I hope to find at Norway House next spring a couple of fine, light inland boats to go to the north in. Pray take the necessary steps to add my name to the testimonial to His Excellency, though I have not yet received my commission. I trust that the canoe brought you *yours*.

The specimens of minerals, plants, birds, etc., that will be out by the McKenzie river boats, you will be pleased to forward to England by the ship. There is a box of books at YF, sent out by the Company to joint address of Mr. Dease & myself – pray retain it at Norway House – and peruse the contents, if worth your notice. Keep a dozen best oak sledge boards, and a few pounds of patent paint, green & other colors, if you have any, for me. George Sinclair will go out in the vessels in charge of

---

67 BCA ADD.MS 625.

some pieces of ironworks, private property, instruments, etc., which you will be kind enough to put in some safe, *dry* corner for me. I am quite ashamed of troubling you with so many commissions, but necessity has no law, and I think I may safely calculate on your friendship, while nothing would afford me greater pleasure than to be able to return your many attentions and kindnesses in *kind*.

With warmest good wishes to Mrs. Ross and your friends, and referring you to Mr. Ballenden for Red River news,

I ever am, My Dear Sir,
Most faithfully & sincerely Yours,
Thomas Simpson

• • •

## *12. Letter from George Simpson, London, to Thomas Simpson, 4 June 1840*[68]

Thomas Simpson Esqr. London 4th June 1840

Dear Sir,

Annexed I beg to hand you an extract from the Governor and Committee's dispatch under date 3d June to C.F. Finlayson, on the subject of the Arctic Discovery Expedition. By that extract you will observe that the Governor and Committee accept the very handsome and spirited tender you have made of your services in conducting a new expedition for the purpose of continuing your late survey from the Mouth of Great Fish River to the Straits of the Fury & Hecla.

As the interesting report of your discoveries and survey from the Coppermine River did not arrive here in sufficient time to be replied to by the spring packet, I am apprehensive that no step has been taken this season towards carrying into effect your projected expedition for continuing the survey; but should the necessary preparations have been made by the council to enable you to enter upon this service next summer I consider the route you propose by Great Fish River as the best to be pursued. But in the absence of any preparatory arrangements for entering upon the survey next summer, I think a year may be gained by starting from Churchill ensuing spring, that is, in March or April 1841, taking two North canoes or very light boats on sleighs as far as ice travelling will serve, forwarding this Autumn, after the arrival of the ship, by one of the small decked craft from York, a quantity of pemican and other provisions as far as she can proceed by open water, there depositing her cargo and returning to Churchill or York, which, from the prevalence of northerly winds at that season, she can do in a very short time. Should that route be pursued, the small quantity of provisions you will be enabled to transport from the deposit to be made this fall, will not admit of your taking all the people who may accompany you from Churchill in the spring for the purpose of hauling your craft and Baggage; you should, therefore, in my opinion take no more than six men along with you beyond

68 HBCA D.4/25, pp. 190–4.

the point where the provisions may be deposited, which I presume would be the limit of your ice travelling). There you ought to leave an abundant supply for your return, in the event of your not being able to proceed to Great Fish River, and from that point send back the remainder of your people in the other craft. Instructions should this fall be forwarded to the Gentleman in charge of Great Slave Lake to have another deposit of provisions made next summer at a given point at the mouth of Great Fish River, by which means, if successful in your survey, you will have the necessary supplies to enable you to get on to Great Slave Lake in the Autumn, or if unsuccessful, you can fall back upon your supplies to be deposited this fall.

In proceeding along the coast to the northward from Churchill, you would in all probability fall in with Esquimaux, who could inform you if there be any nearer channel than the straits of the Fury & Hecla to the large bay you observed eastward from the mouth of Great Fish River. My own belief is that such a channel does exist, probably from Repulse Bay; and if so, it would greatly shorten the voyage.

You have had too much experience in Arctic travelling to render it necessary for me to throw out any suggestions in regard to the supplies and other appointments for the voyage, but I cannot too strongly impress upon you the importance of being well provided with Ammunition, snaring wire and other materials for picking up a scanty subsistence for yourself and party in the event of your being under the necessity of wintering in the dreary and desolate region which is about to become the field of your operations, likewise fishing material, as Salmon Trout are abundant at the mouths of all the rivers falling into the Bay to the northward of Churchill, which would enable you to economize your provisions in the event of your being either ice or wind bound.

The principal object of pointing out the Churchill route is to save a year; but if the necessary preparatory arrangements have been made, to enable you to go by Great Fish River next year, that I consider the preferable route. It is, however, to be understood that you are at liberty to adopt either route you may consider most likely to lead to the speedy and successful accomplishment of the object of your mission.

The party, in either case, I think ought not to exceed six or eight men besides yourself, because, by either route, I contemplate the possibility of your being under the necessity of wintering at some one of the Esquimaux camps you may fall in with, and from the very limited means of subsistence these poor people have, it is scarcely possible that, even of Blubber, they can have a sufficient quantity to maintain during a long Arctic Winter, so large a body of people as the crews of two boats.

By the accompanying extract from the Governor & Committee's Dispatch you will observe that you are at liberty to command any means you may consider necessary for outfitting the expedition in the most complete and ample manner; you will, therefore, make such calls on the Depots and Posts, for servants, Goods, provisions, craft, etc., etc. as you may require. The Chronometer and Mountain Barometer you applied for, are forwarded by the Prince Rupert, likewise some Barometrical Thermometers, with instructions for the use of them; and a narrative of Baron Wrangel's voyage to Siberia & the Polar Sea, which will be interesting to you.

You so well know the value of time during the short season of open water, that it is not necessary to dwell on the importance of not allowing your scientific researches to interfere with the most rapid prosecution of your voyage; it is nevertheless exceedingly desirable that as much of your attention be directed to those pursuits as time & circumstances may admit, and I beg to impress upon you that the completion of the survey to any of the inlets leading into Hudson's Bay, is the great & important object in view; an object which the Hudson's Bay Company are determined sooner or later to accomplish, although it has baffled the exertions of the numerous Government expeditions that have attempted it during the past three hundred years; and such accomplishment I trust is destined to immortalize your name.

Have the goodness to send home by the return of the ship the journals in detail of your two last voyages, with any observations you may have made on the character of the country in reference to its formation and productions, likewise such specimens of Natural History, Minerals etc. as you may have collected, and charts of the coast, Thermometrical and Barometrical registers and observations on the dip of the Needle; in short, let every information connected with science be transmitted to the Company, which is likely to be either useful or interesting.

Self preservation being the first law of nature, you will of course have a due and proper regard to the safety of the party. And that the enterprizing and arduous task you have now in hand may be crowned with the most complete success, is the fervent wish of, Dear Sir,

Your mo. obedt. humble servt.
Geo. Simpson

• • •

### *13. Letter from Thomas Simpson to Alexander Simpson, 5 June 1840*[69]

My dearest Alexander,

I am just on the move for England via the United States; a journey which will, I think, be beneficial to me, as my stomach has been out of order, and my spirits low, for a great part of the Spring ...

Tomorrow morning I take my departure with two companions well mounted and armed, as we expect to fall in with the Sioux, and war is said to be raging on the plains; so that I hope to see something of prairie life.

Farewell, my dearest brother, and while breath remains believe me ever,

Your most affectionate brother
Thomas Simpson

• • •

69 Alexander Simpson, *Life and Travels,* 350–1.

### *14. Letter from Lord John Russell to Sir John H. Pelly, 3 July 1840* [70]

Downing Street,
3rd July 1840

Sir,

I have had the honor to lay before The Queen your letter of the 22nd of April last, submitting the claims of Messieurs Simpson and Dease for some mark of the Royal favor, on account of their successful exertions in furtherance of Arctic Discovery.

I have the pleasure to inform you that the Queen has been pleased to signify Her gracious intention of granting to each of those Gentlemen a Pension, on the Civil List of Her Majesty, of One Hundred Pounds a year, for their exertions towards completing the discovery of the North West Passage.

The Queen has likewise been pleased to declare Her intention of granting to you the Dignity of a Baronet, for your acknowledged worth and long & meritorious services as Governor of the Hudson's Bay Company.

It is also Her Majesty's intention to confer the Honor of knighthood on Mr. Simpson, the Governor of the Settlements of the Hudson's Bay Company.

I have the honor to be, Sir,
Your most obedient Servant
J. Russell

• • •

### *15. Minutes of Governor and Committee, 21 October 1840* [71]

Chief Factors John Charles & Peter Warren Dease with Chief Traders Donald Manson & Murdoch McPherson were introduced to the Board on their return from the Northern Department.

• • •

### *16. Letter from Peter Dease to Dr John Richardson, 3 December 1840* [72]

28 Arundel Street

Decr. 3rd. 1840

My Dear Doctor,

Your last letter came to hand, and I am sorry to hear that Mrs. Richardson is yet unwell. I am just going with Messrs. Charles and McPherson to Liverpool from whence, God Willing, I will either proceed to Scotland or Cross the Channel to Ireland, to be back in about a month if I can so time my movements. You will be kind enough to either send back the books or keep them untill I have the pleasure of seeing you again.

---

70 HBCA A.13/2, p. 165.

71 HBCA A.1/61.

72 MCM M2715.

I had the Honour of meeting Sr. J. Barrow, Captn. Washington and many of your friends among whom was Captn. Beaufort. Indeed I was more gratified than I can Express and more honoured than I deserve. I spoke to Govr. Simpson with respect to your kind offer of Examining the Specimens which he was highly pleased to hear. I am sorry his health is not good but hope it will mend.

Please offer my Kindest regards to Mrs. Richardson and all your family and

Believe me
Yours truly
Peter W. Dease

Dr. J. Richardson M.D.
Haslar Hospital, Near Gosport

• • •

# Chapter Twelve
# Assessment

By any objective assessment, the expedition mounted by Dease and Simpson was one of the most successful ever attempted in the Arctic. In terms of efficiency and low cost (£1000), it ranks with the achievements of Dr John Rae's explorations of Melville Peninsula, Pelly Bay and area, and Victoria Island,[1] Otto Sverdrup's methodical exploration of much of the Queen Elizabeth Islands in 1898–1902,[2] or Georgiy Alekseyevich Ushakov's remarkably cost-efficient survey and exploration of the entire archipelago of Severnaya Zemlya (with a team of only four men) in 1930–32.[3]

In financial terms, at only £1000[4] the cost was minimal, only a fraction of that of any of the Royal Navy's numerous attempts at searching for the Northwest Passage, or at finding the Franklin expedition when it disappeared during its attempt at the search. The contrast with Captain George Back's concurrent expedition on board HMS *Terror*, which achieved absolutely nothing in terms of exploration, at the cost of the ship being severely damaged by the ice, can hardly have been more striking.

The cost in men's health and lives was also minimal; the only fatality was that of Peter Taylor, who died of a long-standing complaint while conveying mail to the expedition's base at Fort Confidence, rather than on any of the exploring journeys *sensu stricto*. Here again the contrast with Franklin's expeditions, for example, is stag-

1 Rae, *Narrative of an Expedition to the Shores of the Arctic Sea in 1846 and 1847*; Rich, ed., *John Rae's Correspondence with the Hudson's Bay Company on Arctic exploration 1844–1855*.

2 Sverdrup, *New Land*.

3 Georgiy Alekseyevich Ushakov, *Po nekhozhenoy zemle* [The untrodden land] (Moscow-Leningrad, 1951).

4 Sampson, "Dease, Peter Warren."

gering; on his first overland expedition 11 men died out of a total of 20, while on his final expedition 129 men died.

There can be no doubt that much of the credit for this success must go to Thomas Simpson. His drive, ambition, and sheer toughness and endurance contributed enormously to the expedition's achievements. But one should not underestimate the contribution made by Dease's steadiness, tact, and experience at running remote posts and base-camps, and especially his experience at getting the most out of the voyageurs.

In the light of this contribution, Simpson's egotistical and defamatory remarks about Dease appear particularly strident, especially when one notes that at no point in his journal does Dease criticize Simpson. Thus after the first season's exploring trip (west to Point Barrow), writing to George Simpson on 8 September 1837, Thomas Simpson says: "Mr Dease is a good, honourable man. I believe I have acquired his friendship, for in everything, even to the plan of our little Fort Confidence, he has adopted my advice, and has left the direction of the march entirely to me."[5]

There may be some truth in this latter claim of Simpson's. One must remember that in his entry about Dease in his *Character Book*, George Simpson had described him as "wanting in ambition to distinguish himself in any measure out of the usual course." But this does not excuse Thomas Simpson's subsequent outrageous claims. Part way through the following winter at Fort Confidence, on 29 January 1838, in another letter to George Simpson, he wrote: "Mr. Dease is a very worthy man but he is dull and indolent. We live together, old wife and all, on the happiest terms; and I now understand that mysterious science – the management of an inland post – as intimately as if I had spent the last long nine years on the north side of Portage-la-loche."[6] By his closing remark, Simpson is implying that in the course of some five months he has gained all the knowledge and skills that Dease had acquired in the course of a lifetime in the fur trade!

Writing to his brother Alexander, at the same time, Thomas Simpson noted: "Mr. Dease and I live together on the happiest footing; his old wife, a little grandchild, and a strapping wench, a daughter of his brother Charles, joining our mess. Dease is a worthy, indolent, illiterate soul, and moves just as I give the impulse."[7]

He had made remarks in a similar vein in a letter to his friend Donald Ross only ten days earlier:

Mr. Dease is a very worthy man, well acquainted with the management of Indians and of an inland establishment, but it is no vanity to say that everything which requires either planning or execution devolves upon me. He never would have dreamed of planning a

5 Letter from Thomas Simpson to George Simpson, 8 September 1837, Alexander Simpson, *Life and Travels,* 256.

6 Letter from Thomas Simpson to George Simpson, 29 January 1838, Alexander Simpson, *Life and Travels,* 272–3.

7 Letter from Thomas Simpson to Alexander Simpson, 29 January 1838, Alexander Simpson, *Life and Travels*, 275–6.

third campaign, and though we live together on the happiest terms, he evidently feels that he is not in his right place.[8]

Again writing to George Simpson after completing the first trip eastwards from the Coppermine, when Dease and Simpson had been forced by ice to turn back at Kent Peninsula, not much farther east than Franklin's farthest at Point Turnagain, Thomas Simpson remarked:

My worthy senior, like Franklin and Back, was alarmed by the storms, the snow, and frost, in August, circumstances which always attended the disruption of any great body of ice ... Had I been alone in the command, the latter circumstance, and the rapid disruption of the ice as we returned towards the boats, would have determined me to make a bold attempt to reach Ross' Pillar, or the Great Fish *this* season; but being somewhat in the situation of Sinbad in the tale, I was not altogether free to follow my own designs. That they would have proved, by this date, successful, I have little doubt from the remarkable beauty of the weather which we still enjoy. But my excellent senior is so much engrossed with family affairs that he is disposed to risk nothing; and is, therefore, the last man in the world for a discoverer. I write not in anger but in sorrow; I esteem Mr. Dease for his upright private character, while I cannot help regarding him and his followers as a dead weight upon the expedition.[9]

Later, on his return from the final season's exploration, Simpson is even more viciously critical of Dease: "Had I not been, like Sinbad the sailor, hampered with an old man on my back I should immediately have turned eastward with both boats, but the apprehensions of my useless senior and of the crews overpowered my single voice."[10]

And finally, writing to George Simpson from Fort Simpson to propose a new expedition, Thomas Simpson wrote of Dease: "My worthy colleague on the late expedition frankly acknowledges his having been a perfect supernumerary; and to the extravagant and profligate habits of half-breed families I have an insuperable aversion."[11] The latter remark is an oblique reference to Dease's family, who joined him for the two winterings at Fort Confidence on Great Bear Lake.

There is some evidence that Dease was prepared to take a back seat to Simpson when it came to vigorously pursuing the exploration of the coast, and to taking the credit for the expedition's achievements. Writing to a friend, John Stuart, of Forres Scotland, on 18 December 1838 (after the first trip eastwards from the Coppermine), and commenting on the publication of their report on the first season's explorations

8 Letter from Thomas Simpson to Donald Ross, 18 January 1838, BCA ADD.MS 625.

9 Letter from Thomas Simpson to George Simpson, 18 September 1838, Alexander Simpson, *Life and Travels*, 299–300.

10 Letter from Thomas Simpson to Alexander Simpson, 18 September 1838, Alexander Simpson, *Life and Travels*, 303–4.

11 Alexander Simpson, *Life and Travels,* 339.

(west to Point Barrow), he noted: "I see our report has been published and highly applauded; the merit is due to Mr. T.S. not me. As you know scientific pursuits are out of my reach therefore do not aspire to undeserved distinction."[12] But it was one thing for Dease modestly to admit his own limitations, and quite another for Simpson to slander a competent, reliable veteran, without whose practised handling of the voyageurs and careful attention to details of provisioning and equipment the expedition would probably have foundered.

Stone has expressed this point even more emphatically. In his biographical sketch of Thomas Simpson, Stone accuses him of conceit and bigotry,[13] of seeming "to have had an inadequate view of his own potential,"[14] of "fundamental unfitness for the position he occupied" in the light of his virulent antipathy to Métis;[15] and of becoming "increasingly self-centred and neurotic" and even partially mentally unhinged.[16] Most tellingly, Stone suggests that "it is possible to conjecture that were it not for the presence of Dease, the tragedy [Simpson's puzzling death] that took place once the expedition was over might have occurred during it and, if this had happened, an even greater mystery would have remained for polar historians to ponder over."[17]

It is particularly in the light of Thomas Simpson's remarks about Peter Dease that it is important that Dease's journal should belatedly be presented to the public. It reveals him to be a competent organizer, a perceptive observer, and a careful diarist, and in general gives the lie to Simpson's jaundiced and self-promoting comments.

12 Letter from Peter Dease to John Stuart, 18 December 1838, Warren Baker collection.

13 Stone, "Thomas Simpson," 1.

14 Stone, "Thomas Simpson," 3.

15 Stone, "Thomas Simpson," 4.

16 Stone, "Thomas Simpson," 11.

17 Stone, "Thomas Simpson," 16.

# Appendix: Biographical Sketches

JOHN BALLENDEN was born at Stromness, Orkney, around 1812. He joined the Company as an apprentice clerk under James Hargrave at York Factory in 1829. He next served at Red River and in 1836 was promoted to accountant at Upper Fort Garry, replacing Thomas Simpson (see p. 26). In December of that year he married Sarah McLeod, daughter of Chief Trader Alexander McLeod. In 1840 he was made clerk-in-charge at Sault Ste Marie and in 1844 chief trader in charge of the Lake Huron District. Promoted chief factor in charge of the Lower Red River District, based at Upper Fort Garry in 1848, he was on his way west to take up this new position when he suffered a stroke with partial paralysis. In 1850 a court case against an individual who had made slanderous charges against his wife placed additional stress on Ballenden and he took a year's furlough in the fall. His next posting was to Fort Vancouver, in charge of the Columbia District, but his poor health continued and after another stroke he took a further furlough in 1853–54. He returned to Red River in 1854 but ill health forced him to return to Scotland in 1855 and he retired officially on 1 June 1856 and died in Edinburgh on 7 December of that year. (Fleming, ed., *Minutes of Council, Northern Department of Rupert Land, 1821–31,* 426–7; Van Kirk, "Ballenden, John")

FRANÇOIS BEAULIEU, the patriarch of the Beaulieu "clan" of Salt River, is recorded by different writers as having been born variously in 1771 and 1793, son of François Beaulieu and Ethiba, a woman of Cree and Chipewyan descent. Around the turn of the century he was employed by the North West Company as post-hunter at their Great Bear Lake post. Thereafter he gained a reputation as a man to be feared and, from 1816 until the merger of the North West Company and the Hudson's Bay Company in 1821, he was employed by the former as an enforcer and "hit-man." At Peter Dease's suggestion he had participated as interpreter and post-hunter in Sir John Franklin's second overland expedition of 1825–27. Thereafter he developed a reputation as a trading chief and operated as a free trader, although from 1848 he was in charge of the Hudson's Bay Company's post at Fort Resolution. At one point in his life he had seven wives, but discontinued six of these relationships when he was baptized in 1848 and his marriage to Catherine St Germain was officially confirmed by the Catholic church. For the later part of his life he held a monopoly from the Hudson's Bay Company for providing salt (naturally produced at the Salt Plains) for use at the various posts in the Athabasca and Mackenzie River districts. (Hanks, "François Beaulieu II"; Keith, "Journals of an Athabascan Fur Trader," 243; Neatby, "Beaulieu, François")

JOHN BELL was born in Mull, Argyll (now Strathclyde), c. 1799. He joined the North West Company as an apprentice in 1818 and was at Fort Alexander in 1820–21. After the coalition he served in the Winnipeg River area until 1824, when he was posted to the Mackenzie River District. In 1826 he took charge of Fort Good Hope and remained there until 1840. During part of this period he served under Peter Dease and married his daughter Nancy; one is a little surprised that Dease did not see fit to mention this relationship.

Governor George Simpson and the Board in London were keen to expand trade westward from the Mackenzie. Specifically they thought that the Colville, whose mouth was discovered by Dease and Simpson in 1837, was identical with the Yukon, and that it would offer a profitable area for expansion. Bell was instructed to explore westwards from the lower Mackenzie and in 1839 he explored the Peel River as far as he could. In 1840 he established Peel River Post (now Fort McPherson) a short distance above the Peel's confluence with the Mackenzie, and was in charge there until 1845. In 1841 he was appointed chief trader. In 1842 he made another exploratory trip westwards and discovered the relatively easy Rat Portage across to the headwaters of the Rat River (now the Bell), a tributary of the Porcupine. In 1845 he explored the Porcupine to its confluence with the Yukon. Thus Bell's explorations paved the way for the Company's expansion into the Yukon basin, with posts at Fort Yukon and later Rampart House.

In 1846–47 Bell took a furlough, largely due to ill health; then in 1847–49 he was attached to the expedition led by Drs John Rae and John Richardson, in search of the missing Franklin expedition. His important contribution was to refurbish and man Fort Confidence as the expedition's wintering base.

In 1851, at his own request, he left the Mackenzie District and took charge of Oxford House, and then the Cumberland House District. From 1853 until 1857 he ran the Athabasca District from Fort Chipewyan, and thereafter served at Sept Iles and Weymontachingue in the Saint Maurice District. He resigned because of ill health in 1860, and retired to Saugeen in Bruce County, where he died on 24 June 1868. (HBCA Search File; Craig, "Bell, John"; Fleming, *Minutes of Council*, 427–8)

FRANCIS BUTCHER was born in 1810 and joined the Company as an apprentice clerk at Lachine in 1832. After a year at York Factory and another at Fort Chipewyan, he was promoted clerk in 1835. He remained in the Athabasca District until 1854, serving variously at Fort Chipewyan, Great Slave Lake, and Dunvegan. He was made chief trader in charge of Athabasca District in 1852. After a furlough in Upper Canada in 1853–54 he served as chief trader at Fort William in 1854–55 and at New Brunswick in Moose District in 1855–56. He retired on 1 June 1854. (HBCA Biography)

ROBERT CAMPBELL was born on 21 February 1808 in Glen Lyon, Perthshire (now Tayside). Having gained a knowledge of sheep from working on his father's farm, in 1830 he was hired by the Hudson's Bay Company to assist Chief Factor James MacMillan to establish an experimental farm at Red River. Disillusioned with this operation, he requested to be transferred to the fur trade, and was posted to the Mackenzie River District in 1834. In 1837 he was promoted to postmaster and instructed to open a post at Dease Lake, now in northern British Columbia. There he spent two frustrating and (for the Company) rather fruitless years because of uncooperative Indians and Russian competition.

In 1840 he was dispatched to explore the northern reaches of the Liard system. He discovered and named Frances Lake (after Governor Simpson's young wife), then pushed north overland to the Pelly River; he was thus the first white man to reach the Yukon basin from the east. In 1842 he returned to Frances Lake, where he opened a post. The next year he descended the Pelly to a large river (the Yukon), which he named the Lewes. Five years later (in 1848), at his direction, James Stewart opened a post at the Pelly/Yukon confluence. Then in 1851 Campbell travelled down the Yukon to the Hudson's Bay Company post at Fort Yukon, thus confirming that both posts were on the same river, i.e. the Yukon.

In 1852 Fort Selkirk was looted by Chilkat Indians, resentful of the Company's intrusion into an area where they had previously had a trading monopoly. In light of the continued poor returns from this post and the extreme difficulties of supplying it via the Liard route, and despite Campbell's objections, Governor Simpson refused to reopen Fort Selkirk.

After a furlough in Britain, Campbell was posted to Fort Liard in 1854 and in 1855 took charge of the Swan River District and was promoted to chief factor. During the Red River rebellion of 1869–70, afraid that the Métis would attack his post, he evacuated his furs and his family via the United States. Governor William Mactavish strongly disapproved of this action, and, while Campbell was on furlough in Scotland, he was informed that he had been dismissed. For the next decade he divided his time between Scotland and Manitoba and finally settled at Merchiston Ranch near Riding Mountain, now in Manitoba. He died there on 9 May 1894. He had married Ellenora Stirling at Norway House on 5 August 1859, and they had three children. (Coates, "Campbell, Robert")

ALEXANDER CHRISTIE was born in 1792, possibly in Glasgow, Scotland. He joined the Hudson's Bay Company in 1809 and was posted to Moose Factory to investigate the possibilities for a lumber industry. After returning to Britain, he was sent back to Moose Factory again, this time to establish a sawmill. By 1817 he was stationed at Rupert's House and was in charge of the Eastmain trade, and especially the whale (i.e. beluga) fishery. By the time of the merger with the North West Company in 1821, he had risen to the rank of chief factor. He was granted a furlough in 1824 and from 1826 until 1830 was in charge of Moose Factory; after a year at York Factory, in 1831 he was given charge of the Red River District and was appointed governor of Assiniboia. He supervised the construction of Lower Fort Garry (still surviving) from 1831 onwards and began construction of Upper Fort Garry at the Forks of the Red and Assiniboine rivers.

On 19 February 1835 his marriage to his wife, Anne, *à la façon du pays*, was confirmed at Red River. During his second term as governor from 1844 until his retirement in 1848, much of his energies were devoted to combatting free traders such as Andrew McDermot and James Sinclair who were challenging the Company's monopoly; since he had no troops to counter their activities, he was largely unsuccessful. He retired to Scotland and died in Edinburgh on 9 December 1872. (Bowsfield, "Christie, Alexander")

CHARLES DEASE, the youngest son of Dr John Dease and thus younger brother of Peter Dease, was born at Côte des Neiges, Montreal, on 3 December 1797. He joined the North West Company as a clerk around 1814. At the time of the coalition of the North West Company and the Hudson's Bay Company he was a clerk at Fort Good Hope, and remained there until 1824. He spent the winter of 1824–25, like his brother Peter, at Captain John Franklin's "base-camp" at Fort Franklin during the latter's Second Land Arctic Expedition. He returned to Fort Good Hope for a further winter (1825–26), but was then transferred to Norway House. There he suffered an accident and was dispatched south to Red River for medical treatment. He returned to Norway House in February 1827, but, afflicted with deafness, he retired to Montreal that summer. He was still alive in 1863. (Fleming, *Minutes of Council Northern Department*, 433)

GEORGE DESCHAMBEAULT was born in Quebec in 1806, the son of Colonel Louis Deschambeault; the latter was a friend of Lord Selkirk and through the latter's influence George joined the Hudson's Bay Company as an apprentice clerk in 1819. He was stationed at Oxford House in 1819–20, and from 1820 until 1822 at Ile-à-la-Crosse. During the period 1822–26 he served in the Saskatchewan District, and from 1826 until 1841 he was in charge of various posts in the English River District, including, as we see from Dease's journal, Ile-à-la-Crosse in 1836. He was in charge of the Great Slave Lake post (Fort Resolution) from 1841 until 1843, and of various posts in the Mackenzie District from 1843 until 1847. Promoted to chief trader in 1847, he next had charge of Cumberland House until 1849, and of Ile-à-la-Crosse in 1849–50. After a year's furlough, he continued to serve in the Athabasca, English River, and Cumberland House districts until another furlough in 1868. Thereafter he served in the Red River District and died at St Boniface on 27 December 1870. (Fleming, *Minutes of Council*, 436)

GEORGE FLETT, an Orkneyman from Birsay, was born around 1803. He joined the Hudson's Bay Company as a sailor in 1833 and sailed for York Factory in the *Prince Rupert*. Changing his vocation within the Company, he moved inland to Red River and in 1836 was hired as a middleman for Dease and Simpson's Arctic Discovery Expedition. On his return south he stayed within the Northern Department but in 1841 returned home to Orkney. In 1843 he signed on with the Company again as a slooper, serving in the York Factory district until 1846. In that year he joined Dr John Rae's expedition to Repulse Bay as a steersman. On his return south in 1847 he signed on at York Factory as a schooner's mate. He was drowned in a boating accident on the lower Hayes River on 15 October 1849. (Rich and Johnson, *John Rae's Correspondence with the Hudson's Bay Company on Arctic Exploration 1844–1855*, 356–7)

JAMES HARGRAVE was born on 19 November 1798 in Hawick, Roxburgh (now Borders), Scotland. He began his working life as a schoolteacher near Galashiels, but persuaded his father that the entire family should emigrate to North America. James joined the North West Company in 1820, serving initially at Sault Ste Marie. After the merger with the Hudson's Bay Company in 1821 he was posted as a clerk to York Factory under Chief Factor John MacTavish. From 1823 until 1827 he served in the Lower Red River District. But then from 1827 until 1851 he was back at York Factory. He was made chief trader and was given charge of York Factory in 1833. He married Letitia MacTavish in

Britain on 8 June 1840 and brought her back out to York Factory. He was made chief factor in 1844. Sir George Simpson had a very high opinion of Hargrave's managerial skills and kept him at York Factory despite repeated requests from Hargrave to be posted to a less remote location. Finally in 1851 he was posted to Sault Ste Marie.

Letitia's death in September 1854, followed by that of their youngest child a few months later, had a devastating effect on Hargrave. His final posting was back to York Factory from 1856 until his retirement in 1859. He died at Brockville, Ontario, on 16 May 1865, and is buried at St James Church, Toronto. James Hargrave maintained an enormous correspondence, only a fraction of which has been published. (Van Kirk, "Hargrave, James")

JOHN HUTCHISON, a nephew of Governor Robert Semple, was born in Glasgow around 1798. He joined the Company as a clerk in 1816 and served in that capacity at Fort Edmonton (1818–21), Severn (1821–23), and the Mackenzie River District (Fort Norman, Fort Liard, etc.) (1821–31). In 1829 under Peter Dease he took part in the West Branch Expedition from Fort Liard; he established Fort Halkett and served there under Dease in 1829–30, taking charge of the post in 1830 for two years. In 1832 he sailed to Britain for a year's furlough for the benefit of his health. After a year as a "clerk disposable" in the Northern Department, he took charge of Great Slave Lake (Fort Resolution) in 1834–35 and Fort Halkett (1835–37). In 1837 he retired to Scotland. (HBCA Search File; Rich, ed., *Colin Robertson's Correspondence Book, September 1817 to September 1822*, 223–4)

ADAM McBEATH served the Company first as postmaster and latterly as clerk from 1829 until 1885. He was in charge of posts in the Mackenzie District (Fort Simpson, Fort Norman, Rivière aux Liards, and Fort Good Hope) from 1836 until 1855. Later (from 1857 until 1874) he was in charge of Shoal River Post in the Swan River District. He married Mary McKenzie at Fort Simpson on 25 May 1838, and died in July 1890 at Fort Qu'Appelle. (HBCA Biography and Search File)

ANDREW McDERMOT was a wealthy Red River merchant. Born at Belanagare, Castlerea, County Roscommon, Ireland, in 1790, he emigrated to Rupert's Land in 1812. He worked for the Company as writer, clerk, and trader at Red River, Pigeon River, Berens River, Fort Dauphin, Big Point, Thieving River, Netley Creek, and Pembina. On leaving the Company he settled at Red River, where he soon became one of its richest and most influential merchants. He became a councillor of Assiniboia in 1839. He married an Indian woman, Mary, at Norway House in 1821, and later, Sarah McNab, at York Factory. Sixteen children were born of this latter marriage, thirteen of whom lived to adulthood. The McDermots lived in a large house, Emerald Grove, on the banks of the Red River; with its outbuildings it became the nucleus of Winnipeg. McDermot Avenue, which marked the northern limit of his property, commemorates his name. He donated the land for Winnipeg's first post office and for Winnipeg General Hospital. He died in 1881 at the age of ninety-one. (HBCA Biography and Search File; Hyman, "McDermot, Andrew")

BENJAMIN MACKENZIE was born about 1807, son of Chief Factor Roderick Mackenzie Sr and an Indian woman from the Lake Nipigon area. Educated in Montreal, he assisted his father unofficially in the Lake Superior District from 1823 onwards. Recommended to become an apprentice by Governor William Williams in September 1825, he in fact did not join the Company until 1827, when he was appointed postmaster at Fort William, then at the outpost of Lac la Fleche in 1827–28 and 1828–29 respectively. From 1829 until 1831 he served as postmaster at La Cloche in the Lake Huron District; then in 1831 he was posted to Fort St James in New Caledonia. In 1834 he was promoted clerk in the Columbia District and in 1835 was appointed clerk at Fort Vancouver. He contracted consumption (tuberculosis) and in June 1837 sailed for Honolulu en route to England. He died in Honolulu in 1838 or 1839. Shortly before sailing for Hawaii (on 28 February 1837) he married Catherine Campbell, daughter of Chief Trader Colin Campbell, at Fort Vancouver. (HBCA Search File; Fleming, *Minutes of Council*, 447–8)

ALEXANDER McLEOD was born around 1782 and joined the North West Company in 1802. For almost two decades he served in the Peace River and Athabasca Departments. With the amalgamation of the North West Company and the Hudson's Bay Company in 1821, he was made chief trader, in charge of the Athabasca District. From 1822 until 1825 he was in charge of the Mackenzie River District, serving at Fort Simpson and at Fort Good Hope. Appointed to the Columbia District in 1825, over the next six years he mounted a number of exploring and trapping expeditions southwards to the Umpqua, Klamath, and Sacramento valleys. He was strongly criticized by George Simpson for the management of these expeditions, however. In the summer of 1831 he was transferred to the Mackenzie River District, and from 1833 until 1835 he was attached to Captain George Back's expedition to explore the Back River. Thus it was McLeod who was responsible for building Fort Reliance, the expedition's winter quarters. Then from 1835 until 1837 he was in charge of Fort Resolution, and was promoted chief factor in 1836. Given charge of the Athabasca District in 1837, he made his headquarters at Dunvegan, remaining there until a furlough in 1839. His furlough was extended for a further year and he died on 11 June 1840, survived by his wife, two sons, and five daughters. (Fleming, *Minutes of Council*, 448–50)

MURDOCH McPHERSON, a native of Gairloch, Ross and Cromarty (now Highland) in the West Highlands, joined the North West Company in 1816, serving in the Montreal and Athabasca districts. After the coalition he continued to serve (as a clerk) in the Athabasca District, and was transferred to the Mackenzie River District in 1823. From 1826 until 1834 he was in charge of Rivière des Liards; he then took charge of Fort Simpson. He was promoted chief trader in March 1834, and in 1835 took charge of the Mackenzie River District. He held this position until 1840 and was thus closely involved with supporting Dease and Simpson's expedition.

After a furlough in 1840–41 he was transferred to Tadoussac, in charge of the King's Posts and the post at Mingan, until 1844, when he was moved back to Fort Simpson to run the Mackenzie District for the remainder of his career. He was on furlough from 1849 until 1851, in which year he retired. His wife, Jean, was a daughter of Edward Smith; they had two sons and a daughter. (HBCA Search File and Biography)

WILLIAM MOWAT had joined the Company in 1817. Until 1833 he served as a labourer at Lac la Pluie, Island Lake, and in the English River District. He was in charge of Fort Norman from 1833 until 1839, and thus was closely involved in supporting Dease and Simpson's expedition. In 1839 he moved north to Fort Good Hope, then in 1840 he took charge of the new post at Peel's River. He died in 1841. (HBCA Search File)

SIR JOHN PELLY, governor of the Hudson's Bay Company, was born on 3 March 1777; his father was a captain with the East India Company. He became a director of the Hudson's Bay Company in 1806, then deputy-governor and finally governor. He became a director of the Bank of England in 1840, and its governor in 1841. He was made a baronet by Queen Victoria, in large part as a result of Dease and Simpson's Arctic Discovery Expedition on 6 July 1840. (*Dictionary of National Biography*, 15:720)

JOHN RITCH was born around 1806 in Graemsay on the island of Hoy in the Orkney Islands. He joined the Company as a boat builder in 1828, and was stationed at Norway House in that capacity until he joined the expedition. On his return from the expedition he continued to serve as a boat builder at Red River and appears to have retired to the Orkneys in 1843. (HBCA Biography)

DONALD ROSS was born in Stornoway on the Island of Lewis around 1797 and joined the Hudson's Bay Company in 1816 in the capacity of a writer. For his first season he was stationed at York Factory but then transferred to Cumberland House as writer and accountant from 1817 until 1824. After a further two years at York Factory, Governor Simpson appointed him his confidential secretary, at Lachine (Montreal). In this capacity he accompanied Simpson on his trip to and from the Columbia District, in 1828–29, wintering at Fort Vancouver.

On his return east, Ross was promoted to chief trader and was granted a year's furlough in Scotland. In 1830 he was placed in charge of Norway House, a charge which he retained for twenty-one years. In 1839 he was made chief factor, and was also appointed sheriff of Rupert's Land. He was granted another furlough in 1851; this was extended because of ill health, and he died at Lower Fort Garry on 19 November 1852. He is buried at the Upper Church, Red River. George Simpson's assessment of him in his "Character Book" is one of the most positive in that remarkable document; indeed, for George Simpson, it has to be described as glowing. It begins: "A very steady regular well behaved man who understands the whole routine of the interior business better than any of his class from the circumstance of his having been my confidential Clerk or Secty. for seven years and a most confidential Man I always found Him." (Fleming, *Minutes of Council*, 453; Williams, ed., *Hudson's Bay Miscellany, 1670–1870*, 198)

JOHN ROWAND was born in Montreal in 1787 and joined the North West Company in 1803. His first posting was to Fort Augustus (Fort Edmonton), and apart from brief absences he would reside there for the rest of his life. Around 1810 he married Lisette Humphraville, probably a daughter of Edward Umfreville, *à la façon du pays*. In 1820 he was made a partner in the North West Company, and after the merger became a chief trader in the Hudson's Bay Company. In 1822, along with Chief Factor Donald McKenzie, he led the rather abortive Bow River Expedition, aimed at capturing the trade of the plains of southern Alberta. In 1823 he took charge of the Saskatchewan District, with his headquarters at Fort Edmonton. He regularly attended the annual meetings of the Council of the Northern Department, from which he was returning to Fort Edmonton when Dease encountered him at Cumberland House. Rowand accompanied Sir George Simpson for part of his round-the-world trip in 1841; he travelled with him out to the coast, then north to Fort Stikine and south to California and on to Hawaii. On his retirement he was heading east to Montreal when he died of a stroke in May 1854 at Fort Pitt, while breaking up a fight among the voyageurs. Initially buried at Fort Pitt, in 1856 he was disinterred and reburied in Mount Royal Cemetery in Montreal. (Van Kirk, "Rowand, John")

ALEXANDER SIMPSON, Thomas Simpson's younger brother and biographer, was born on 14 June 1811 in Dingwall, son of Alexander Simpson Sr and his wife, Mary. Inspired by a visit by his cousin George (by then governor of the Hudson's Bay Company in North America) in 1825, he joined the Company in 1827 and was posted to Lachine in 1828. He clashed frequently with his superior, James Keith. In 1834 he was transferred to Moose Factory, where he kept the books. His next posting, in 1839, was to Fort Vancouver under Dr John McLoughlin; here he was entrusted with organizing the Company's activities in the Sandwich Islands (Hawaii), and California, and the running of the Puget Sound Agricultural Company. He travelled to Hawaii and California in 1839. McLoughlin sent him to Hawaii on a permanent basis in 1840; while there, in June 1841 he was made chief trader. With George Simpson's permission he travelled home to Scotland following Thomas Simpson's mysterious death (see p. 276), returning to Hawaii in 1842. When George Simpson proposed posting him to the Northern Department, Alexander threatened to resign because of the painful associations that department held for him in connection with his brother's death.

Still in Hawaii, Alexander Simpson actively promoted British annexation of the Hawaiian kingdom, even when he was made acting consul in September 1842. He left Hawaii after Britain annexed the islands in March 1843. On 1 June 1843 he resigned from the Company and returned to Scotland. The details of his later life are very vague: he probably married, and he died possibly in the early 1870s. (Abel, "Simpson, Alexander")

GEORGE SIMPSON, the Company's governor-in-chief in North America, based at Lachine near Montreal, was born in Lochbroom Parish, i.e. the Ullapool area of the Scottish Highlands, around 1786 or 1787. Born out of wedlock, he was brought up by his aunt, Mary Simpson (Thomas Simpson's mother), whose husband was the schoolmaster in Dingwall. Around 1800 he travelled south to London where he was employed by the sugar brokerage firm of Graham and Simpson, in which his uncle Geddes Mackenzie Simpson was a partner. In 1812 this company merged with Wedderburn and Company. Andrew Wedderburn (later Colvile) was a member of the Committee of the Hudson's Bay Company, and by this chain of coincidence George Simpson became involved with that company.

In 1820, at Colvile's urging, George Simpson was appointed governor-in-chief, *locum tenens*, for the Company in North America; it was thought that his demonstrated business acumen and energies would equip him well for the organization of the combined operations of the Hudson's Bay Company and the North West Company following their merger in 1821. Simpson travelled to North America in 1820 and, after successfully holding his own against the officers of the North West Company during a turbulent winter at Athabasca, took up residence first at York Factory, then at Lachine, from where he quickly demonstrated his amazing talents in the running of the Company's far-flung commercial empire.

For almost forty years he made long annual canoe trips to Norway House for the annual meeting of the Council of the Northern Department, or to more remote Company posts. Throughout his life he prided himself on the speed with which he made these trips. In 1841–2 he also made a trip around the world, promoting Company interests and handling Company business at Sitka, Honolulu, and Okhotsk, from where he travelled across Russia to St Petersburg in connection with negotiations with the Russian government and with the Russian American Company.

Throughout his life Simpson maintained an amazing volume of correspondence, both official and private; in both cases his mastery of detail was astounding, as his correspondence associated with the Arctic Discovery Expedition will reveal. Where Company business was concerned he was extremely hard-nosed, but he could be kind and generous to friends.

Simpson married his cousin Frances Simpson in London in 1830 and by her had five children. Prior to this marriage, however, he had had liaisons with several Métis women, à *la façon du pays*, resulting in several children. Simpson supported these children, but they (and their mothers) were kept strictly separate from his "other" family.

Simpson was deeply interested in, and supportive of, arctic exploration. Apart from Dease and Simpson's Arctic Discovery Expedition (for which he was knighted in 1841), he had supported Sir John Franklin's first expedition to some degree and had played a critical role in the organization and support of Franklin's second expedition in 1825–27. After the disappearance of Franklin's last expedition in 1845 he was responsible for several expeditions mounted to search for it, including those led by Rae and Richardson in 1847–49 and by Rae in 1850–51 (both of which operated from Fort Confidence). It was on yet another expedition authorized by Simpson, in 1853–54, that Rae found the first clues to the fate of the Franklin expedition. And finally, Simpson was entirely responsible for James Anderson and James Stewart's Land Arctic Searching Expedition of 1855, aimed at pursuing the clues discovered by Rae.

Simpson died on 7 September 1860 and is buried in Mount Royal Cemetery in Montreal. The enduring success of the Hudson's Bay Company as a commercial empire was due in large part to George Simpson's remarkable drive and ability. He would have become a business leader of world-wide reputation in any age. For further details see Galbraith, *The Little Emperor,* and "Simpson, Sir George."

GEORGE SINCLAIR, a Métis, was born in 1790 and entered the service of the Hudson's Bay Company in 1810. For the next forty years he worked as a steersman, living at

Swan River when not travelling. He had been one of the crew of Captain George Back's boat which descended the Back River in 1834. In 1849 he was appointed interpreter at Swan River, and returned as a freeman to Red River in 1856. (Back, *Narrative of the Arctic Land Expedition*, 56; HBCA Search File)

JAMES SINCLAIR was an influential merchant from Red River. Born in 1811, the son of William Sinclair and his Cree wife, he worked for the Company for one year (1826–27) at Fort Albany. But in 1827 he set himself up as a private trader at Red River, where he became Andrew McDermot's partner. In 1841 Sinclair led a party of colonists recruited by the Hudson's Bay Company from Red River across the mountains to Fort Vancouver in the Columbia District. Having returned to Red River, he formally joined the Company again and led another party of colonists (including his wife and family) west to the Columbia in 1854. He was killed during an Indian attack at his settlement near Walla Walla on 26 March 1856. Sinclair was married twice: to Elizabeth Bird on 3 December 1829 and to Mary Campbell on 20 April 1848. He had thirteen children. (Spry, "Sinclair, James")

EDWARD SMITH first joined the XY Company, then after its amalgamation with the North West Company in 1804–5 worked for that company first as a clerk, then as a partner, until its merger with the Hudson's Bay Company in 1821. He had managed the Athabasca District from Fort Chipewyan from 1821 until 1823, and the Mackenzie District from Fort Simpson from 1823 until 1832. After a two-year furlough he resumed his position at Fort Chipewyan from where, clearly, he was visiting Norway House when Dease encountered him. He went on furlough in 1837 and retired on 1 June 1839. (HBCA Biography)

JOHN STUART was born in Strathspey, Scotland, in 1779. He joined the North West Company in 1799 and was posted to the Peace River District. In 1806 he accompanied Simon Fraser west to New Caledonia, where he served at Fort St James on Stuart Lake (which Fraser named after him). In 1808 he was with Fraser on his pioneer canoe trip down the Fraser River. In 1809 he took charge of the New Caledonia District and remained in charge for the next fifteen years. With the merger with the Hudson's Bay Company in 1821 he was made chief factor and in 1822 was made a councillor of the Northern Department. From 1824 until 1827 he was in charge of the Saskatchewan District, initially based at Fort Carlton and then at Lesser Slave Lake. His next posting, in 1827, was to Fort Alexander (Bas de la Rivière), from where he ran the Winnipeg River District. Finally, from 1832 until 1835 he was in charge of the Mackenzie River District, based at Fort Simpson. In 1835 he was granted a furlough and travelled home to Scotland, where he settled in Forres. The furlough was extended until 1839 when he retired. He died on 14 January 1847. (Rich, *Journal of Occurrences in the Athabasca Department by George Simpson, 1820 and 1821, and Report*, 469)

PETER TAYLOR joined the Company as an apprentice in 1826, and was first stationed in the Columbia District. For the first five years with the Company he served on board a coasting vessel but in 1831 transferred, as middleman, to the Mackenzie District. From 1833 until 1835 he participated in Captain George Back's expedition down the Great Fish (Back) River. Having next served as middleman in the Lac La Pluie (Rainy Lake) District, he joined the Dease and Simpson expedition in 1836. He died of exhaustion and starvation on 12 December 1837 on Great Bear Lake while carrying dispatches from Fort Simpson to the expedition base at Fort Confidence. See Dease's journal entry for 29 December 1837. (HBCA Biography)

DR WILLIAM TODD was born in Ireland in the mid-1780s; he joined the Hudson's Bay Company in 1816 and served for two years as clerk and surgeon at Cumberland House. In 1819 he volunteered for service at Fort Wedderburn (Lake Athabasca), embroiled in intense competition with the North West Company. After a year's furlough he spent 1821–22 in the Lower Red River District, 1822–27 at York Factory, and 1827–29 at Fort Vancouver in the Columbia District. He returned to Red River in 1829, serving as a clerk at Brandon House. In 1836 George Simpson dispatched him back to York Factory, mainly to try to solve the mystery of a deadly disease which had broken out there. He contracted the disease himself soon after his arrival and never fully recovered; his health was undermined for the rest of his life.

In 1837 he was placed in charge of the Swan River District, based at Fort Pelly with the rank of chief trader. While there he heard of the start of the severe smallpox epidemic which had broken out at Fort Union on the Missouri River. Anticipating (correctly) that it would spread widely and rapidly, he began an extensive campaign of vaccination in his own district and further afield. This was the first time that the Jenner type of vaccination was used in the West. Severe as the epidemic was, it would have been much worse without Todd's efforts. One suspects, too, that it was through Todd's influence that Peter Dease was able to carry out his localized campaign of vaccination at Fort Confidence in the winter of 1837–38. Todd retired from the Company in 1851 and settled at Red River, where he died in December of that year.

Todd's career as a trader was not in any way outstanding, but he was probably the most prominent medical practitioner in the West prior to 1850. He served as personal physician to both George Simpson and his wife, Frances. (Ray, "Todd, William")

# Published Sources


Abel, Kerry. "Simpson, Alexander." *Dictionary of Canadian Biography*. Vol. 7: *1846 to 1850*. Toronto: University of Toronto Press, 1988, 808–9.

Armour, D.A. "Dease, John." *Dictionary of Canadian Biography*. Vol. 5: *1801 to 1820*. Toronto: University of Toronto Press, 1983, 228–30.

Back, George. *Narrative of the Arctic Land Expedition to the Mouth of the Great Fish River, and along the Shores of the Arctic Ocean, in the Years 1833, 1834 and 1835*. London: John Murray, 1836.

— *Narrative of an Expedition in H.M.S. Terror, undertaken with a View to Geographical Discovery on the Arctic Shores, in the Years 1836–37*. London: John Murray, 1838.

Banfield, A.W.F. *The Mammals of Canada*. Toronto: University of Toronto Press, 1974.

Barr, William. *Back from the Brink: The Road to Muskox Conservation in the Northwest Territories*. Calgary: Arctic Institute of North America, 1991.

— "Richard Cyriax's Note concerning Thomas Simpson's Claim of Having Discovered the Northwest Passage." *Polar Record*, 36, no. 197 (2000), 113–16.

Barr, William. ed. *Searching for Franklin, the Land Arctic Searching Expedition 1855. James Anderson's and James Stewart's Expedition via the Back River, 1855*. London: Hakluyt Society, 1999.

Barr, William, and Glyndwr Williams, eds. *The Voyage of Christopher Middleton 1741–1742*. Vol. 1 of *Voyages in Search of a Northwest Passage, 1741–1747*. London: Hakluyt Society, 1994.

— *The Voyage of William Moor and Francis Smith, 1746–1747*. Vol. 2 of *Voyages in Search of a Northwest Passage, 1741–1747*. London: Hakluyt Society, 1995.

Beaglehole, J.C., ed. *The Journals of Captain James Cook on His Voyages of Discovery*. Vol. 3: *The Voyage of the Resolution and Discovery 1776–1780*. 2 vols. Cambridge: Cambridge University Press for the Hakluyt Society, 1967.

Beechey, Frederick W. *Narrative of a Voyage to the Pacific and Beering's Strait, to co-operate with the Polar Expeditions: performed in His Majesty's Ship Blossom, under the Command of Captain F.W. Beechey … in the Years 1825, 26, 27, 28*. London: Henry Colburn and Richard Bentley, 1831.

Bockstoce, John, ed. *The Journal of Rochfort Maguire 1852–1854. Two Years at Point Barrow, Alaska, aboard HMS* Plover *in search of Sir John Franklin*. 2 vols. London: Hakluyt Society, 1988.

Bowsfield, Hartwell. "Christie, Alexander." *Dictionary of Canadian Biography*. Vol. 10: *1871 to 1880*. Toronto: University of Toronto Press, 1972, 167–8.

Campbell, F.H.A. "Geology of the Helikian Rocks of the Bathurst Inlet Area, Coronation Gulf, Northwest Territories." *Geological Survey of Canada Paper* 78-1A (1978), 97–106.

Christie, S. Hunter. "Discussion of the Magnetical Observations made by Captain Back, R.N. during his Late Arctic Expedition." *Philosophical Transactions of the Royal Society of London* 126 (1836), 377–415.

Coates, Kenneth. "Campbell, Robert." *Dictionary of Canadian Biography*. Vol. 12: *1891 to 1900* (Toronto: University of Toronto Press, 1990, 155–6.

— *The Commerce of Discovery: The Hudson's Bay Company and the Simpson and Dease Expeditions.* Paper presented at the Symposium on Early Investigations of the Western Arctic, Twentieth Annual Conference of the Canadian Archeological Association. Calgary, 1987.

Cooke, Alan, and Clive Holland. *The Exploration of Northern Canada 500 to 1920: A Chronology.* Toronto: Arctic History Press, 1978.

Craig, J. "Bell, John." *Dictionary of Canadian Biography*. Vol. 9: *1861 to 1870*. Toronto: University of Toronto Press, 1976, 42–3.

Cundy, R. *Beacon Six*. London: Eyre and Spottiswoode, 1970.

Davies, K.G., ed. *Northern Quebec and Labrador Journals and Correspondence 1819–35*. London: Hudson's Bay Record Society, 1963.

Davis, Richard C., ed. *Sir John Franklin's Journals and Correspondence: The First Arctic Land Expedition 1819–1822*. Toronto: Champlain Society, 1995.

— *Sir John Franklin's Journals and Correspondence: The Second Arctic Land Expedition, 1825–1827*. Toronto: Champlain Society, 1998.

Dease, Peter Warren, and Thomas Simpson. "An Account of the Recent Arctic Discoveries." *Journal of the Royal Geographical Society* 8 (1838), 213–25.

— "An Account of Arctic Discovery on the Northern Shore of America in the Summer of 1838." *Journal of the Royal Geographical Society* 9 (1839), 325–30.

— "Narrative of the Progress of Arctic Discovery on the Northern Shore of America, in the Summer of 1839." *Journal of the Royal Geographical Society* 10 (1840), 268–74.

*Dictionary of National Biography*. "Sir John Pelly." *Dictionary of National Biography*. Vol. 15. London: Oxford University Press 1949–50, 720.

Ens, Gerhard J. *Homeland to Hinterland: The Changing Worlds of the Red River Métis in the Nineteenth Century*. Toronto: University of Toronto Press, 1996.

Fleming, R.H., ed. *Minutes of Council, Northern Department of Rupert Land, 1821–31*. Toronto: Champlain Society for the Hudson's Bay Record Society, 1940.

Franklin, John. *Narrative of a Journey to the Shores of the Polar Sea in the Years 1819, 20, 21 and 22*. London: John Murray, 1823.

— *Narrative of a Second Expedition to the Shores of the Polar Sea in the Years 1825, 1826, and 1827 ... Including an Account of the Progress of a Detachment to the Eastward by John Richardson*. London: John Murray, 1828.

Fraser, J.A. "Geological Notes on Northeastern District of Mackenzie, Northwest Territories" *Geological Survey of Canada Paper* 63–40 (1964).

Galbraith, John S. *The Little Emperor: Governor Simpson of the Hudson's Bay Company.* Toronto: University of Toronto Press, 1976.

— "Simpson, Sir George." *Dictionary of Canadian Biography*. Vol. 8: *1851 to 1860*. Toronto: University of Toronto Press, 1985, 812–18.

Gibson, James R. *The Lifeline of the Oregon Country: The Fraser-Columbia Brigade System 1811–47*. Vancouver: University of British Columbia Press, 1998.

Glazebrook, G.P. de T. *The Hargrave Correspondence, 1821–1843*. Toronto: Champlain Society, 1938.

Hanbury, David. *Sport and Travel in the Northland of Canada.* London: Edward Arnold, 1904.

Hanks, Chris. "François Beaulieu II: The Origins of the Métis in the Far Northwest." In *Selected Papers of Rupert's Land Colloquium 2000, May 24–28, 2000, Vancouver, Washington*. Winnipeg: Centre for Rupert's Land Studies, 2000, 111–26.

Harper, Marjory. *An Unsolved Mystery: The Story of Thomas Simpson, Arctic Explorer.* Dingwall: Dingwall Museum, n.d.

Hearne, Samuel. *A Journey from Prince of Wales's Fort in Hudson's Bay, to the Northern Ocean Undertaken by Order of the Hudson's Bay Company, for the Discovery of Copper Mines, a North West Passage, etc., in the Years 1769, 1770, 1771 & 1772.* London: A. Strahan and T. Cadell, 1795.

Hyman, B.E. "McDermot, Andrew." *Dictionary of Canadian Biography.* Vol. 11: *1881 to 1890.* Toronto: University of Toronto Press, 1982, 545–6.

Keith, H. Lloyd. "Journals of an Athabascan Fur Trader. James Keith at Fort Chipewyan, 1823–1826." In *Papers of the 1994 Rupert's Land Colloquium*, ed. Ian MacLaren et al. Winnipeg: Centre for Rupert's Land Studies, University of Winnipeg, 1997.

— *North of Athabasca: Slave Lake and Mackenzie River Documents of the North West Company, 1800–1821.* Montreal and Kingston: McGill-Queen's University Press, 2001.

King, Richard. *Narrative of a Journey to the Shores of the Arctic Ocean in 1833, 1834 and 1835, under the Command of Capt. Back, R.N.* London: Richard Bentley, 1836.

Kupsch, Walter. "A Valley View in Verdant Prose: The Clearwater River from Portage la Loche." *Musk-Ox* 20 (1977), 28–49.

Mackenzie, Alexander. *Voyages from Montreal, on the River St. Lawrence, through the Continent of North America, to the Frozen and Pacific Oceans; in the years 1789 and 1793.* London: T. Cadell Jun., 1801.

Ian S. MacLaren, "The HBC's Arctic Expedition 1836–1839: Dease's Field Notes as Compared to Simpson's Narrative." In *The Fur Trade Revisited: Selected Papers of the Sixth North American Fur Trade Conference, Mackinoc Island, Michigan, 1991*, ed. J.S.H. Brown, W.J. Eccles, and D.P. Heldman. East Lansing: Michigan State University Press for Mackinac State Historic Parks, 1994, 465–79.

McClellan, Catharine, and Glenda Denniston. "Environment and Culture in the Cordillera." In *Handbook of North American Indians.* Vol. 6: *Subarctic*, ed. June Helm. Washington: Smithsonian Institution, 1981, 372–86.

Macleod, Margaret A., ed. *The Letters of Letitia Hargrave.* Toronto: Champlain Society, 1947.

Morton, A.S. *A History of the Canadian West to 1870–71.* 2nd edition. Toronto: University of Toronto Press, 1973.

Neatby, Leslie H. "Beaulieu, François." *Dictionary of Canadian Biography.* Vol. 10: *1871 to 1880.* Toronto: University of Toronto Press, 1972, 38.

"North-West Passage." *The Times.* 16 January 1836, p.6.

Pannekoek, Frits. *A Snug Little Flock: The Social Origins of the Riel Resistance of 1869–70.* Winnipeg: Watson and Dwyer, 1991.

Parry, William Edward. *Journal of a Voyage for the Discovery of a North-West Passage from the Atlantic to the Pacific; performed in the Years 1819–20, in His Majesty's Ships Hecla and Griper.* London: John Murray, 1821.

— *Journal of a Second Voyage for the Discovery of a North-West Passage from the Atlantic to the Pacific, performed in the Years 1821–22–23, in His Majesty's Ships Fury and Hecla, under the Orders of Captain William Edward Parry.* London: John Murray, 1824.

Pullen, H.F., ed. *The Pullen Expedition in Search of Sir John Franklin.* Toronto, Arctic History Press, 1979.

Rae, John. *Narrative of an Expedition to the Shores of the Arctic Sea in 1846 and 1847.* London: T. and W. Boone, 1850.

Ray, Arthur. "Todd, William." *Dictionary of Canadian Biography.* Vol. 8: *1851 to 1860.* Toronto: University of Toronto Press, 1951, 888–90.

Rich, Edwin Ernest. *The History of the Hudson's Bay Company 1670–1870.* Vol. 2: *1763–1870.* London: Hudson's Bay Record Society, 1959.

Rich, Edwin Ernest, ed. *Journal of Occurrences in the Athabasca Department by George Simpson, 1820 and 1821, and Report.* London: Hudson's Bay Record Society, 1938.

— *Colin Robertson's Correspondence Book, September 1817 to September 1822.* Appendix A, Biographical. London: Champlain Society/Hudson's Bay Record Society, 1939, 223–4.

— *A Journal of a Voyage from Rocky Mountain Portage in Peace River to the Sources of Finlays Branch and North West Ward in Summer 1824* [by Samuel Black]. London: Hudson's Bay Record Society, 1955.

Rich, Edwin Ernest, and A.M. Johnson, eds. *John Rae's Correspondence with the Hudson's Bay Company on Arctic Exploration 1844–1855.* London: Hudson's Bay Record Society, 1953.

Richardson, John. "Dr. Richardson's Narrative of the Proceedings of the Eastern Detachment of the Expedition." In John Franklin, *Narrative of a Second Expedition.* London: John Murray, 1828.

— *Arctic Searching Expedition: a Journal of a Boat-voyage through Rupert's Land and the Arctic Sea, in Search of the Discovery Ships under Command of Sir John Franklin.* 2 vols. London: Longman, Brown, Green and Longmans, 1851.

Ross, John. *Narrative of a Second Voyage in Search of a North-West Passage, and of a Residence in the Arctic Regions during the Years 1829, 1830, 1831, 1832, 1833.* London: A.W. Webster, 1835.

Royal Geographical Society. "Communications on a North-West Passage." *Journal of the Royal Geographical Society* 6 (1836), 34–50.

St-Onge, Denis. "Glacial Lake Coppermine, North-Central District of Mackenzie, Northwest Territories." *Canadian Journal of Earth Sciences* 17, no. 9 (1980), 1310–15.

Sampson, W.R. "Dease, Peter Warren." *Dictionary of Canadian Biography.* Vol. 9: *1861 to 1870.* Toronto: University of Toronto Press, 1976, 196–9.

Scott, W.B., and E.J. Crossman. *Freshwater Fishes of Canada.* Ottawa: Fisheries Research Board of Canada, 1973.

Simpson, Alexander. *The Life and Travels of Thomas Simpson, the Arctic Discoverer.* London: Richard Bentley, 1845.

Simpson, Thomas. *Narrative of the Discoveries on the North Coast of America: effected by the Officers of the Hudson's Bay Company during the Years 1836–39.* London: Richard Bentley, 1843.

Spry, Irene M. "Sinclair, James." *Dictionary of Canadian Biography.* Vol. 8: *1851 to 1860.* Toronto: University of Toronto Press, 1985, 819–20.

Stefansson, Vilhjalmur. *Unsolved Mysteries of the Arctic.* New York: Macmillan, 1951.

Stevenson, John A. "The Unsolved Death of Thomas Simpson Explorer." *Beaver* 266, no. 1 (1935), 17–20; 64–6.

Stone, Ian. "Thomas Simpson, Arctic Explorer." *Aberdeen University Review* 53, no. 181 (1989), 1–17.

Sverdrup, Otto. *New Land: Four Years in the Arctic Regions.* London and New York: Longmans Green, 1904.

Tyrrell, James W. *Across the Sub-Arctics of Canada. A Journey of 3,200 Miles by Canoe and Snowshoe through the Barren Lands.* London: T. Fisher Unwin, 1898.

Van Kirk, Sylvia. "Ballenden, John." *Dictionary of Canadian Biography.* Vol. 8: *1851 to 1860.* Toronto: University of Toronto Press, 1985, 59–60.

— "Hargrave, James." *Dictionary of Canadian Biography.* Vol. 9: *1861 to 1870.* Toronto: University of Toronto Press, 1976, 364–6.

— "Rowand, John." *Dictionary of Canadian Biography.* Vol. 8: *1851 to 1860.* Toronto: University of Toronto Press, 1985, 777–9.

VanStone, James W., ed. *A.F. Kashevarov's Coastal Explorations in Northwest Alaska, 1838.* Fieldiana Anthropology, vol. 69. Chicago: Field Museum, 1977.

Ushakov, Georgiy Alekseyevich. *Po nekhozhenoy zemle* [The Untrodden Land]. Moscow-Leningrad, 1951.

Wallace, Hugh G. *The Navy, the Company and Richard King: British Exploration in the Canadian Arctic, 1829–1860.* Montreal and Kingston: McGill-Queen's University Press, 1980.

Wallace, W.S., ed. *John McLean's Notes of a Twenty-five Years' Service in the Hudson's Bay Territory.* Toronto: Champlain Society, 1932.

Williams, Glyndwr, ed. *London Correspondence Inward from Sir George Simpson 1841–42.* London: Hudson's Bay Record Society, 1973.

— *Hudson's Bay Miscellany, 1670–1870.* Winnipeg: Hudson's Bay Record Society, 1975.

# Index

RUPERT'S LAND RECORD SOCIETY SERIES
Jennifer S.H. Brown, Editor

1 The English River Book
A North West Company Journal and Account Book of 1786
*Edited by Harry W. Duckworth*

2 A Country So Interesting
The Hudson's Bay Company and Two Centuries of Mapping, 1670–1870
*Richard I. Ruggles*

3 Arctic Artist
The Journal and Paintings of George Back, Midshipman with Franklin, 1819–1822
*Edited by C. Stuart Houston*
*Commentary by I.S. MacLaren*

4 Ellen Smallboy
Glimpses of a Cree Woman's Life
*Regina Flannery*

5 Voices from Hudson Bay
Cree Stories from York Factory
*Compiled and edited by Flora Beardy and Robert Coutts*

6 North of Athabasca
Slave Lake and Mackenzie River Documents of the North West Company, 1800–1821
*Edited with an Introduction by Lloyd Keith*

7 From Barrow to Boothia
The Arctic Journal of Chief Factor Peter Warren Dease, 1836–1839
*Edited and annotated by William Barr*